Linear and Non-linear Mechanical Behavior of Solid Materials

Christian Lexcellent

Linear and Non-linear Mechanical Behavior of Solid Materials

Springer

Christian Lexcellent
Mécanique Appliquée
FEMTO-ST
Besançon
France

ISBN 978-3-319-85707-7 ISBN 978-3-319-55609-3 (eBook)
DOI 10.1007/978-3-319-55609-3

Translation from the French language edition: *Comportements linéaires ou non linéaires des matériaux solides. cours et exercises* by Christian Lexcellent, © Cepadues 2016. All rights reserved.

Softcover reprint of the hardcover 1st edition 2017

Printed on acid-free paper

This Springer imprint is published by Springer Nature
The registered company is Springer International Publishing AG
The registered company address is: Gewerbestrasse 11, 6330 Cham, Switzerland

Preface

I have taught for at least 40 years, the engineering mechanics under various aspects such as the continuum mechanics of deformable bodies, the plasticity, the viscoplasticity, the mechanic of damage and fracture, the magnetic and classical shape memory alloys and also the fluid mechanic.

The people interested by these courses has changed over the years from "the maitrise de technologie de construction" to "the maitrise de mécanique", and also "the DEA de sciences des matériaux".

Including the students of the "l'école nationale supérieure de mécanique et des microtechniques de Besançon (ENSMM Besançon)".

Concerning the mechanic of solid materials, the book of J. Lemaitre and J.L. Chaboche (1985) became the reference. It also expanded with two supplementary authors named "A. Benallal et R. Desmorat", to reach 577 pages in its third edition in 2009.

Following the drafting of about 500 pages of handouts, I decided to write a textbook, incorporating "in situ" exercises to facilitate a better understanding of the students.

My sources were mainly the following:

- J. Mandel. Introduction à la mécanique des milieux continus déformables. Académie Polonaise des Sciences, 1974.
- G. Cailletaud, M. Tijani, S. Cantournet, L. Corte, S. El Arem, S. Forest, E. Herve-Luanco, M. Mazier, H. Proudhon, and D. Ryckelink. Mécanique des Matériaux solides (Notes de cours). Mines-Paris-Tech, 2011.
- Q.S. Nguyen. Problémes de plasticité et de rupture. Publications Mathématiques d'Orsay (Université de Paris-Sud), 1982. Cours de DEA
- P. Suquet. Rupture et Plasticité. Ecole Polytechnique, 2003. Cours de Master I.
- J.B. Leblond. Mécanique de la rupture fragile et ductile. (the two first chapters) Hermés- Lavoisier, 2003.

So if a teaching book should be a consistent and synthetic compilation, it addresses generally "known things" even if sometimes it overflows on research.

Thus, I wish to warmly thank my colleagues who have inspired me in writing this book.

Thank you to my colleagues of the ENSMM: Frédérique Trivaudey, Jacques Dufaud, Violaine Retel-Guicheret and Sylvaine Mallet who accompanied me in teaching the mechanics of deformable continuous media.

Finally, Joel Abadie drew quarantine figures.

Without him again and Scott Cogan, I would not have been able to extricate myself from meandering LYX (word processor used in the writing of this book).

English language was seriously amended by Benoit Vieille from INSA Rouen (France).

To address the linear and nonlinear behaviors of solid materials, the concepts of deformation, displacement, and stress will be introduced.

Most of the time, the developments herein will be restricted to small perturbations, with the exception of shape memory alloys, which may have substantial deformations, whose magnitude can reach 8% and shape memory polymers or materials with superplastic deformations, and whose deformations can reach several hundred percent.

Chapter 1 provides a brief overview of solid mechanics, experimental methods, and classes of material behaviors.

Chapter 2 lays the foundations for the thermodynamic modeling framework based on the aforementioned concepts of deformation, displacement, and stress.

Linear elastic behavior and thermoelasticity will be presented in Chap. 3 along with exercises in continuum mechanics.

Chapter 4 introduces the criteria to the elastic yield domain with special attention to the asymmetry in the tension and compression behaviors.

Chapter 5 deals with the plastic behaviors in the framework of generalized standard materials.

Chapter 6 presents viscoelasticity with a description of the advantages and drawbacks of conventional models such as Kevin-Voigt, Maxwell, and Zener.

Chapter 7 is devoted to the study of viscoplasticity in a classical range. At the end of the chapter, a parallel is made between plastic and viscoplastic behaviors.

Chapters 8 and 9 are devoted to conventional and magnetic shape memory alloys, so-called "smart materials" which may be used as actuators or sensors in adaptive structures.

Finally, Chap. 10 examines fracture mechanics and damage behavior.

I leave the task of developing numerical models in structural mechanics to my colleagues.

Besançon, France Christian Lexcellent

Contents

Notations

$\overrightarrow{x^0}$	Initial position vector of material point M_0 at $t = 0$, $\overrightarrow{O_0M_0} = \overrightarrow{x^0}$
x_i^0	Cartesian coordinates in reference landmark R linked to the non "deformed" solid state
$\overrightarrow{x}$	Position vector of material point M at time t, $\overrightarrow{O_0M} = \overrightarrow{x}\left(\overrightarrow{O_0M_0}, t\right)$
$x_i = x_i(t)$	Cartesian coordinates in reference landmark R linked "during deformation" at time t of solid
$\overrightarrow{u} = \overrightarrow{M_0M} = \overrightarrow{x} - \overrightarrow{x^0}$	Displacement vector
$\overrightarrow{V} = \overrightarrow{V}(\overrightarrow{\mathrm{x}}, \mathrm{t})$	Rate vector
u_i	Displacement components
$\underline{F} = \underline{\nabla}_{x^0}\overrightarrow{x} = \frac{\partial \overrightarrow{x}}{\partial \overrightarrow{x^0}}$	Second order tensor (represented by a matrix (3×3)) defining the transformation gradient tensor
$F_{ij} = \frac{\partial x_i}{\partial x_j^0}$	Component on line i and column j of the $\underline{F}$ matrix
$\det(\underline{F})$	$\underline{F}$ discriminant
$\underline{H} = \underline{\nabla}_{x^0}\overrightarrow{u} = \frac{\partial \overrightarrow{u}}{\partial \overrightarrow{x^0}}$	Gradient tensor of vectorial displacement
$u_{i,j} = \frac{\partial u_i}{\partial x_j^0}$	Component on line i and column j of the $\underline{H}$ matrix
$\underline{\varepsilon}$	"Small deformation" tensor
$\underline{\varepsilon^{el}}$	Elastic strain tensor
$\underline{\varepsilon^{\mathrm{pl}}}$	Plastic strain tensor
p	Cumulated plastic strain
p^{vp}	Cumulated viscoplastic strain
$\underline{\varepsilon^{vp}}$	Viscoplastic strain tensor
$\overline{\varepsilon^{pl}}$	Equivalent plastic strain
$\underline{\varepsilon^{tr}}$	Phase transformation strain tensor
$\underline{\varepsilon^{th}}$	Thermal strain tensor

ε_{ij}	Strain components on line i and column j of the $\underline{\varepsilon}$ matrix
$\underline{\dot{\varepsilon}}$	Strain rate tensor
$\underline{\omega}$	Rotation tensor
ε	Unit extension
$\varepsilon_1,\ \varepsilon_2,\ \varepsilon_3$	Main strains
$\overrightarrow{X_1},\ \overrightarrow{X_2},\ \overrightarrow{X_3}$	Main directions
$\theta_1,\ \theta_2,\ \theta_3$	$\underline{\varepsilon}$ main invariants
$\overrightarrow{D}_{n^0}$	Strain vector
$\varepsilon_{n^0n^0}$	Axial strain
$\varepsilon_{n^0t^0}$	Transversal strain
$\underline{E}$	Lagrangian strain tensor
$\underline{A}$	Eulerian strain tensor
$\overrightarrow{f^{\star}(M)}$	Volume density dynamic force
$\overrightarrow{f(M)}$	Volume density static force
J	Jacobian of the geometric transformation
$\overrightarrow{\gamma}(M)$	Acceleration of the point M
$\rho(M)$	Mass density in M
$\overrightarrow{F(P)}$	Surface density force
$\overset{v}{\overrightarrow{T}}(M)$	Stress vector
$\underline{\sigma}$	Cauchy stress tensor
$\underline{\sigma_{rev}}$	Reversible stress tensor
$\underline{\sigma_{irr}}$	Irreversible stress tensor
$\underline{X}$	Internal kinematic stress tensor
σ_i	Internal stress (1D)
$R(p)$	Isotropic hardening
$N = \sigma_{\nu\nu}$	Normal stress
$T = \sigma_{\nu t}$	Tangential stress
τ	Shearing stress
σ_{ij}	Stress component on line i and column j of the matrix $\underline{\sigma}$
$\Theta_1, \Theta_2, \Theta_3$	$\underline{\sigma}$ main invariants
$J_1,\ J_2,\ J_3$	Deviator of $\underline{\sigma}$ main invariants
$\overline{\sigma}$	Huber–Von Mises equivalent stress
$\overline{X}$	Huber–Von Mises internal kinematic equivalent stress
σ_y	Initial elastic yield
W	Heat internal source
T	Temperature (°K)
e	Specific internal energy
s	Specific entropy
ψ	Specific free energy
Φ or D	Dissipation
$\overrightarrow{q}$	Vector heat flux

z	Martensite volume fraction
G	Specific Gibbs free enthalpy
$\underline{\underline{C}}$	Elastic stiffness fourth-order tensor
$\lambda,\ \mu$	Lamé constants
λ_p	Plastic multiplier
E	Young modulus
ν	Poisson coefficient
K	Compressibility modulus
$A(x_1, x_2)$ or $A(r, \theta)$	Airy function
α	Linear expansion coefficient of the material
$\underline{\underline{S}}$	Compliance fourth-order tensor
E_r	Relaxed Young modulus
η	Viscosity
E	Complex modulus
E'	Storage modulus
E''	Loss modulus
$\Omega(f)$	Viscoplastic potential
z	Martensite volume fraction
z_σ	Martensite volume fraction stress induced
z_T	Thermal martensite volume fraction
z_k	k variant volume fraction
$\underline{U_i}$	Variant i Bain matrix
$\underline{\underline{E_{kl}^{re}}}$	Strain tensor associated to the reorientation of variant k in variant l
$\underline{\underline{Q}}, \underline{\underline{R}}$	Rotation matrix
$\overrightarrow{h}$	Magnetic field
α	Proportion of Weiss domain inside austenite
θ	Rotation angle of magnetisation vector $\overrightarrow{m}$ under magnetic field $\overrightarrow{h}$ inside the REV
C_p	Specific heat
$\overrightarrow{m}$	Magnetisation vector
m_s	Magnetisation at saturation
K_I et K_{II} K_{III}	Stress intensity factor of modes I et II; III
G	Energy release rate
J	Rice integral
D	Damage isotropic variable
$\underline{\underline{D}}$	Damage fourth order tensor
Y	Damage dual variable

Chapter 1
Introduction: Elementary Concepts

Abstract In reference to its desired use, the materials' properties are described. A particular attention is paid to the description of experimental methods for the material characterization. A short list of possible behavior models is given.

1.1 Materials Properties

The selection and choice of material's type is classically done according to the desired use. For example, for the development of aircraft engines, mechanical properties will be decisive.

One can make a brief inventory of the material properties:

- Mechanical properties:
  - elastic modulus, elasticity yield, hardening, ductility,
  - viscosity, creep rate,damping,
  - fracture strength, fatigue resistance, wear.
- Physical properties:
  - electrical conductivity, aimantation, magnetization,
  - thermal conductivity, specific heat, phase transformation latent heat,
  - surface energy, connecting energy.
- Chemical properties:
  - resistance to corrosion, oxidation, irradiation.

The choice of a material is completely dependent upon the function to be performed and is often subject to a compromise.

As Z.Q. Feng (Feng, 1991) said, Aluminum is sometimes used in automotive cylinder heads despite its low melting temperature, but due to its low density and its good thermal conductivity. Other considerations concern the material performance with particular attention to cost and aesthetics (building facades, car body),

C. Lexcellent, *Linear and Non-linear Mechanical Behavior of Solid Materials*, DOI 10.1007/978-3-319-55609-3_1

– disponibility, fiability, machinability, formability layout, weldability,
– no harm, recyclability, cost and appearance.

1.2 Knowledge and Use of Materials

1. Knowledge of the microstructure and its evolution following the thermomechanical conditions imposed.
2. Knowledge of the physical mechanisms associated with the elementary deformations and their impact on the macroscopic scale. In short, scale transitions "micro-meso-macro".
3.Implementation of models in structural analysis codes.

Point 1 is the area of metallurgists and chemists. Point 2 is one of the mechanicians of solid materials. Point 3 is the field of structural mechanics or mechanics called "numerical" ones.

1.3 The Main Class of Materials

In general, numerical finite element modeling is not limited to a single material! but rather to a class of materials. Thus, the developed models may concern:

metals, classical or magnetic shape-memory alloys, ceramics, polymers, composites, woods, concretes, soil (sand and rocks), biomaterials (bone-tissue).

1.4 Experimental Methods: Types of Tests

Many tests are used to characterize the mechanical properties of materials. Some of them are standard:
AFNOR: Association Française de NORmalisation
ISO: International Standardisation Organisation
ASTM: American Society for Testing and Material.
There are also indirect characterization methods such as measurements of "electrical resistance" during a tensile test, to determine the fraction of the phase produced in a mother phase (the shape memory alloys, for example).

Differential thermal analysis (DTA) or "differential scanning calorimetry" (DSC) to measure the phase transition temperatures characteristics and the latent heat of phase transformation.

For nuclear materials, measurements of radiation levels by Geiger counter, can be quite essential.

Measurements of strain fields, temperature through infrared cameras, quite successful now, may prove very useful.

1.4.1 The Tensile Test

This is the basic test for mechanical characterization of material. It is generally carried out at constant strain rate. The results are obtained in term of strength F versus elongation $\Delta l = l - l_0$. In the case of the assumption of "small perturbations" (which will be explained in the next chapter), one converts the results in a stress–strain curve $(\sigma - \varepsilon)$ (Fig. 1.1). For metals and composites, the test samples are threaded cylinder heads or rectangular plates.

In the case of simple compression, if one does not want to "test" barrels, we must pay particular attention to the boundary conditions.

The tensile is characterized by a linear portion (elasticity), followed by a non-linear increasing portion (corresponding to the hardening of the material) and then decreases. One has to note that this change to negative slopes is generally related to the fact that the deformations of the field are not uniform (e.g., the phenomenon of necking).

If the stress–strain curve is the same regardless of the imposed strain rate; we speak of elastic–plastic behavior.

Be called:

- R_e: elasticity limit,
- $R_{0.2}$: conventional elasticity limit,
- R_m: resistance to tension,
- A_h: elongation for maximal stress,
- A_r: fracture elongation.

The response of a material whose behavior is sensitive to the strain rate, is shown schematically in Fig. 1.2.

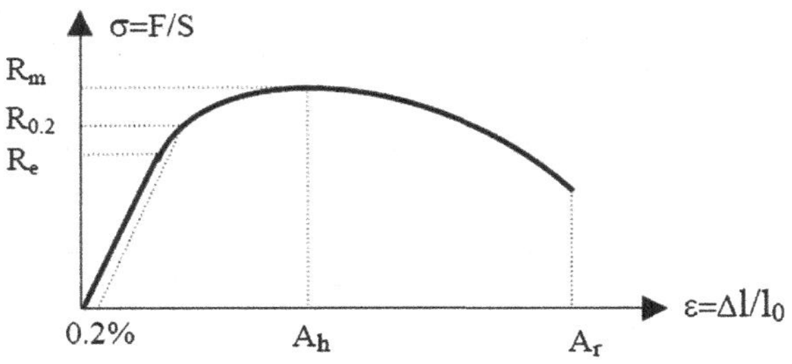

Fig. 1.1 Schematic representation of a tensile test

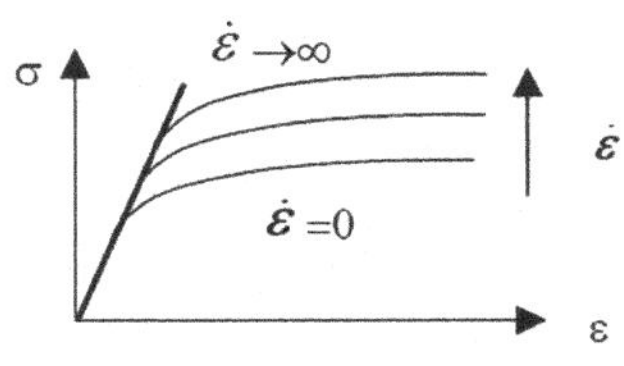

Fig. 1.2 Schematic representation of the tensile behavior of a viscoplastic material

The experimental curves will be between two theoretical curves corresponding limits, one at infinite speed (in practice: high) and the other at zero speed (in practice: very low). The very low speed curve corresponds to a series of thermodynamic equilibrium states virtually reversible.

Note that the change in temperature "in situ" during the test plays a great role. A test at very low speed can be considered isothermal if one excludes the phase transformations. A test at high speed can be considered adiabatic. It is the case of a multiphysics coupling and the heat equation will separate the aspect of "structure" of the specimen from its viscosity (Chrysochoos et al., 2009).

1.4.2 Creep Test

It consists in maintaining a specimen under a constant external mechanical loading. In general, a continuous deformation occurs during the hold time. If the deformation remains the same over time, we say that the material has no viscosity. In fact, over very long time, deferred deformations can be observed on real materials.

However, it was decided that the viscoelastic materials and viscoplastic can have creep deformation and not the elastoplastic materials. This choice depends greatly on the loading time duration.

1.4.3 Relaxation Test

It consists in maintaining a specimen at a fixed strain state. Let note that a material which exhibits creep, also exhibits relaxation and « vice-versa ». Sometimes, we refer to the relaxation test as "dual" to the creep one.

1.4.4 Different Possible Mechanical Tests

1D: tension, compression, shearing; torsion on thin tubes,

3 points and 4 points bending (this allows to have a central region where the bending momentum is constant) (Fig. 1.3),

bulging,

indentation test,

Charpy test (testing flexural notched impact effectued by a pendulum; measuring resilience).

Let note that the word "resilience" which is the ability of materials to withstand shock, spread to the ability of an individual to psychically withstand the trials of life.

2D: plates biaxial traction, traction–torsion of tubes.

3D: traction–torsion-+internal pressure on tubes.

Bending 3 points   Bending 4 points

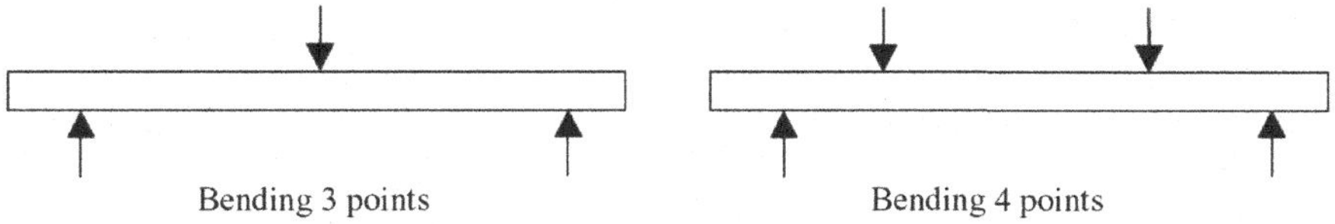

Fig. 1.3 Schematic bending test

1.5 Behavior Models

It is what represents the answer to some basic mechanical loading sufficient to identify the type of behavior.

In one approach to the kind of rheological models are assembled basic elements:

1. The spring that symbolizes the linear elasticity.

2. The shock absorber which represents the linear viscosity (Fig. 1.4b) or the nonlinear viscosity (Fig. 1.4c).

3. The pad corresponding to the onset of permanent deformation threshold.

These elements can be combined together to form rheological models. They are a schematic view of the behavior and imperfectly represent physical phenomena causing deformations.

The answers of the models are the following:

(a): linear elastic solid $\sigma = E\varepsilon$,

(b): viscoelastic solid (Kelvin–Voigt model) which includes a spring and a damper in parallel $\sigma = \eta\dot{\varepsilon} + H\varepsilon$,

(c): perfect elastic–plastic solid (Saint Venant model) consists of a spring and a serial pad,

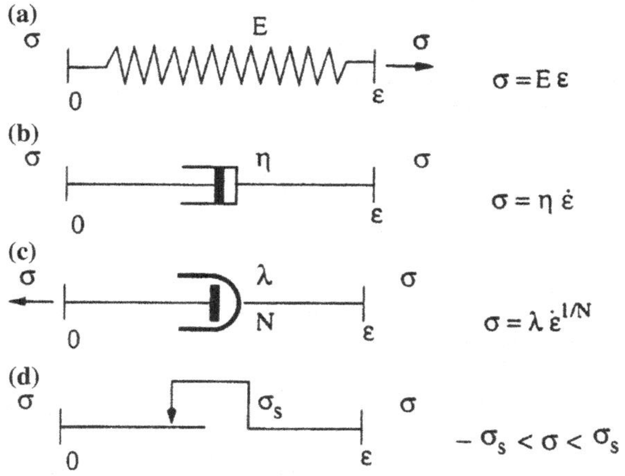

Fig. 1.4 The basic elements for the representation of material behavior

(a) $\sigma = E\varepsilon$

(b) $\sigma = \eta\dot{\varepsilon}$

(c) $\sigma = \lambda\dot{\varepsilon}^{1/N}$

(d) $-\sigma_s < \sigma < \sigma_s$

(d): elastoplastic hardenable solid (generalized Saint Venant model) which gives a linear stress–strain curve in two pieces,

(e): perfect elastic–viscoplastic solid (Bingham–Norton model) which includes a linear spring in series with a damper and a pad in parallel,

(f): elastic–viscoplastic solid with hardening, requiring a more complicated treatment.

1.6 Formulation and Choice of Behavior Models

1.6.1 Formulation of Behavior Models

Except in the case of elasticity, materials have the memory of their anterior thermomechanical loading; their behavior at time t depends on their history.

This can be addressed by the choice of functional $\sigma(t) = \Gamma_{\tau<t}(\varepsilon(\tau))$ or to assume that the effect of the history can be addressed by "internal variables" chosen to represent the state of the material at the time t.

The writing of behavior models will meet three principles according to Feng (Feng, 1991):

1. the principle of "local state" which states that the behavior in a point only depends on variables defined at this point and not from the neighborhood;
2. the principle of "material simplicity" that assumes that only the first gradient of the transformation occurs in laws;
3. the notion of invariance under the change and configuration repository.

1.6.2 Choice of Behavior Models

Clearly, for each type of material, its behavior models depend first on the range of operating temperature and the intensity of the mechanical loading.

Let us try to classify!

1.6.2.1 Viscoelastic Behavior

- thermoplastic polymers in the vicinity of the melting temperature,
- glasses in the vicinity of the glass transition temperature,
- fresh concrete.

1.6.2.2 Perfect Elastoplastic Behavior

- soil survey,
- Limit analysis (boundary effort or movement),
- Shaping of metals.

1.6.2.3 Elastoplastic Behavior with Hardening

- metals at temperatures below a quarter of their melting temperature,
- Soils and rocks.

1.6.2.4 Viscoplastic Behavior

- Metals at intermediate and high temperature,
- Wood, soils (and also salt),
- Ceramics at very high temperatures.

1.6.2.5 Models and Reality

Note that any model is an approximation of the real and the choice of a particular model of behavior depends on the intended application.

As Feng (1991) said: « Thus, a steel at room temperature can be considered as linear elastic for calculating deflections of a mechanical structure; viscoelastic for a vibration damping problem; perfectly plastic for limit load calculation; elastic-viscoplastic for the study of residual stress... »

"A polymer can be considered as a solid for a shock problem and as a fluid to study its stability over long periods."

« The engineer responsible for the design of nuclear waste storage caverns in a salt massif cannot neglect the delayed behavior of this rock as the durability of the structure is required for decades or even centuries. While a mining tunnel which is only need for a few days, can be modeled as part of the elastoplasticity (neglecting viscosity). In the case of metallic materials operating above one third of their melting temperatures, the incorporation of viscoplasticity is also necessary for long periods (e.g., to certify the holding nuclear power plant components on a quarantine years), even for relatively short operations, so for example, in aircraft turbine blades ».

Chapter 2
Thermodynamics Framework for Modeling Solid Materials

Abstract This chapter is aimed at defining a constitutive framework for modeling solid materials. The purpose of continuum mechanic of deformable bodies is to deliver databases of solid behavior investigation. At first, the concepts of « vectorial displacement » strain and stress are introduced. In the framework of thermodynamic of irreversible process, the modeling scheme is explicited.

2.1 Introduction

This chapter is aimed at defining a constitutive framework for modeling solid materials. The elastic, elastoplastic, or elastoviscoplastic material and also classical shape memory alloys (SMA) and magnetic ones (SMAs) behavior will be investigated in this framework.

The purpose of continuum mechanic of deformable bodies is to deliver databases of mechanic of solid. At first, the concepts of « vectorial displacement », strain, and stress will be introduced.

In the framework of thermodynamic of irreversible process, the modeling scheme will be explicited.

Modeling is the art of producing an abstract representation (a mathematical model) of a natural phenomenon (a physical problem) (Duvaut 1984). Physics is classically ruled by two types of laws:

– conservation laws: they are universal and exact laws, which are valid irrespective of the medium's behavior;

– constitutive laws: they aimed at describing the response of the medium to given thermomechanical loading (or, for instance, thermo-magneto-mechanical loading for magnetic SMAs). They are approximated laws, determined by the identification of coefficients (e.g., the measurements of the Young's modulus associated with a uniaxial tensile test for a linear elastic solid).

These laws only partially express reality and lead to uncertainties.

C. Lexcellent, *Linear and Non-linear Mechanical Behavior of Solid Materials*, DOI 10.1007/978-3-319-55609-3_2

In the case of complex materials and loads, we often need to adapt the model in order to make it corresponding as close as possible to the physical reality. This is the model-adjustment phase. However, it has a cost: in terms of readability, the more sophisticated a model becomes.

2.2 Concept of Deformation (or Strain), Kinematical Study of Continuum Media

2.2.1 Vectorial Displacement Notion $\vec{u}\left(\overrightarrow{O_0M_0}, t\right)$

When one wants to study, at the macroscopic point of view, the deformable bodies, one must consider the material as an assembly of distinct particles with an invariable mass. Around a material point of the body, it exists a matter distribution whose physical properties are continuum, and differentiable functions of the considered point coordinates, and at first the volumetric mass.

One admits the continuity of geometrical transformations, e.g., two points infinitely closed at time $t_0 = 0$ stay infinitely closed at time t and reciprocally.

Fig. 2.1 Transport of an elementary vector $\overrightarrow{M_0M_0'} \Longrightarrow \overrightarrow{MM'}$

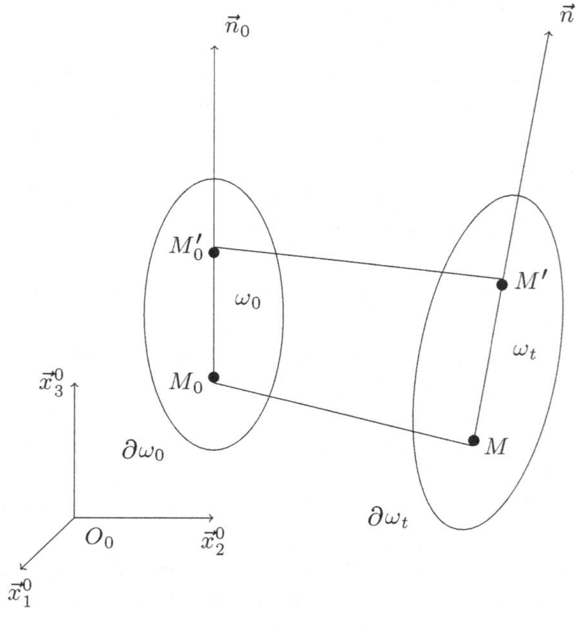

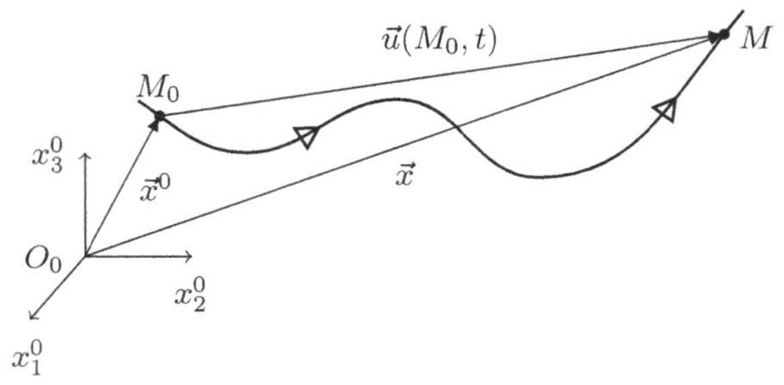

Fig. 2.2 Trajectory of a point: notion of « vectorial displacement » $\vec{u}$

Let a body of volume be ω_0 (subset of the global body V_0) and border be $\partial\omega_0$ at the initial time $t_0 = 0$. Under the effect of external mechanical loading, this body occupies a volume ω_t (subset of the global body V_t) and border $\partial\omega_t$ at time t (Fig. 2.1).

Let us consider the length elements $\overrightarrow{dx^0}$ and $d\vec{x}$:

$$\overrightarrow{dx^0} = \overrightarrow{M_0M_0'} = \vec{n_0}dx^0 \quad |\ \vec{n^0}\ |= 1$$

and

$$d\vec{x} = \overrightarrow{MM'} = \vec{n}\,dx \quad |\ \vec{n}\ |= 1.$$

Note that the $x_i^0 (i = 1, 2, 3)$, the coordinates of $M_0 : x_1^0, x_2^0, x_3^0$ will be designated as the Lagrangian coordinates (and t). The curve in Fig. 2.2 refers to the path taken by a current point which is in M_0 at $t_0 = 0$ and in M at the time t. Note that the $x_i (i = 1, 2, 3)$ the coordinates of (x_1, x_2, x_3, t) will be designated as the Eulerian coordinates (which change with time t).

Let $\overrightarrow{O_0M} = \overrightarrow{O_0M_0} + \overrightarrow{M_0M}$ or $\vec{x} = \vec{x^0} + \vec{u}\left(\vec{x^0}, t\right)$ where $\vec{u} = \overrightarrow{M_0M}$ is called « displacement vector » at the point M_0 at the time t.

2.2.2 Notion of Body "Deformability"

A body is said to be « non-deformable » if whatever the couples of points (M_0, M_0') at $t_0 = 0$ and (M, M') at t are such as $|\ \overrightarrow{dx^0}\ |=|\ d\vec{x}\ |$.

This notion mainly concerns the « general mechanic theory » where a body is often represented by its center of gravity G by its total mass M.

This also means that the body geometrical shape does not change. It may only undergo a rotation and a translation.

A body is known as "deformable" if there are couples (M_0, M_0') à $t_0 = 0$ (M, M') at t such that $|\ \overrightarrow{dx^0}\ |\neq|\ d\vec{x}\ |$. Clearly, the body changes of geometry over time. It is this change that introduces the mechanics of deformable media.

Let $d\overrightarrow{x^0} \Longrightarrow d\overrightarrow{x}$ by $\underline{F}$, e.g., $d\overrightarrow{x} = \underline{F}.d\overrightarrow{x^0}$ for given t.

By definition $\underline{F}$ is called "transformation gradient tensor". This is a second-order tensor which can be represented by a 3 x 3 matrix:

$$\underline{F} = \begin{bmatrix} \frac{\partial x_1}{\partial x_1^0} & \frac{\partial x_1}{\partial x_2^0} & \frac{\partial x_1}{\partial x_3^0} \\ \frac{\partial x_2}{\partial x_1^0} & \frac{\partial x_2}{\partial x_2^0} & \frac{\partial x_2}{\partial x_3^0} \\ \frac{\partial x_3}{\partial x_1^0} & \frac{\partial x_3}{\partial x_2^0} & \frac{\partial x_3}{\partial x_3^0} \end{bmatrix}.$$

We introduce the concept of continuity on the assumption that the functions $x_i = x_i\left(x_i^0, t\right)$ are continuous and continuously differentiable.

One can speak about continuum mechanic (deformable).

Note: we surreptitiously introduced the second-order tensor (for example $\underline{F}$) or others, in order to describe the phenomena in a form independent of the chosen coordinate system. The vector is a first-order tensor and the scalar is a zero-order tensor.

2.2.3 Definition of Finite Transformations

In any case

$$d\overrightarrow{x} = \underline{F}.d\overrightarrow{x^0}$$

at given t.

2.2.3.1 Lagrangian Approach

It constitutes the view of deformable body as reference to non-deformable body at initial time t_0.

Let us evaluate $\left(dx^2\right) - \left(dx^0\right)^2$:

$$\begin{aligned} \left(dx^2\right) - \left(dx^0\right)^2 &= \left(d\overrightarrow{x}\right)^t . \left(d\overrightarrow{x}\right) - \left(d\overrightarrow{x^0}\right)^t . \left(d\overrightarrow{x^0}\right) \\ &= \left(d\overrightarrow{x^0}\right)^t .\underline{F}^t.\underline{F}. \left(d\overrightarrow{x^0}\right) - \left(d\overrightarrow{x^0}\right)^t .\underline{1}. \left(d\overrightarrow{x^0}\right). \end{aligned}$$

Then

$$E = \frac{(dx)^2 - \left(dx^0\right)^2}{2\left(dx^0\right)^2} = \left(\overrightarrow{n_0}\right)^t . \frac{1}{2}\left(\underline{F}^t.\underline{F} - \underline{1}\right) . \left(\overrightarrow{n_0}\right).$$

One calls

$$\underline{E} = \frac{1}{2}\left(\underline{F}^t.\underline{F} - \underline{1}\right),$$

the lagrangian strain tensor whose components are

$$E_{ij} = \frac{1}{2}\left(\frac{\partial u_i}{\partial x_j^0} + \frac{\partial u_j}{\partial x_i^0} + \sum_{k=1}^{k=3} \frac{\partial u_k}{\partial x_i^0}\frac{\partial u_k}{\partial x_j^0}\right).$$

2.2.3.2 Eulerian Approach

Its constitutes the view of deformable body as reference to the body during its deformation at a time t.

$$\overrightarrow{dx^0} = \underline{F^{-1}}.d\vec{x}.$$

Let us evaluate once again $(dx)^2 - \left(dx^0\right)^2$:

$$A = \frac{(dx)^2 - \left(dx^0\right)^2}{2\,(dx)^2} = (\vec{n})^t.\frac{1}{2}\left(\underline{1} - \left(\underline{F}^{-1}\right)^t.\underline{F}^{-1}\right).(\vec{n}).$$

One calls

$$\underline{\mathrm{A}} = \frac{1}{2}\left(\underline{1} - \left(\underline{F}^{-1}\right)^t.\underline{F}^{-1}\right),$$

the Eulerian strain tensor whose components are

$$A_{ij} = \frac{1}{2}\left(\frac{\partial u_i}{\partial x_j} + \frac{\partial u_j}{\partial x_i} - \sum_{k=1}^{k=3} \frac{\partial u_k}{\partial x_i}\frac{\partial u_k}{\partial x_j}\right).$$

2.2.4 Additive Decomposition: Strain Tensors and Rotation Tensors

Knowing that for each second-order tensor $\underline{T}$, it is entitled to write

$$\underline{T} = \frac{1}{2}\left(\underline{T} + \underline{T}^t\right) + \frac{1}{2}\left(\underline{T} - \underline{T}^t\right) = \underline{T}^s + \underline{T}^{as}.$$

$\underline{T}^t$ is defined as the transposed of $\underline{T}$ $\left(T_{ij}^t = T_{ji}\right)$.

$\underline{F}$ can also be decomposed in a symmetric and an antisymmetric tensor:

$$\underline{F} = \frac{1}{2}\left(\underline{F} + \underline{F}^t\right) + \frac{1}{2}\left(\underline{F} - \underline{F}^t\right).$$

Let $\underline{F}^s = \frac{1}{2}\left(\underline{F} + \underline{F}^t\right)$ be the symmetric part of $\underline{F}$: $\left(F^s_{ij} = F^s_{ji}\right)$ and $\underline{F}^{as} = \frac{1}{2}\left(\underline{F} - \underline{F}^t\right)$ be the antisymmetric part of $\underline{F}$ $\left(F^{as}_{ij} = -F^{as}_{ji}\right)$.

Similarly, we can write $\overrightarrow{dx} = d\overrightarrow{x^0} + \overrightarrow{du}\left(\overrightarrow{x^0}, t\right)$ where $\underline{F} = \underline{1} + \underline{\nabla}_{x^0}\overrightarrow{u}$ with

$$\underline{\nabla}_{x^0}\overrightarrow{u} = \begin{bmatrix} \frac{\partial u_1}{\partial x_1^0} & \frac{\partial u_1}{\partial x_2^0} & \frac{\partial u_1}{\partial x_3^0} \\ \frac{\partial u_2}{\partial x_1^0} & \frac{\partial u_2}{\partial x_2^0} & \frac{\partial u_2}{\partial x_3^0} \\ \frac{\partial u_3}{\partial x_1^0} & \frac{\partial u_3}{\partial x_2^0} & \frac{\partial u_3}{\partial x_3^0} \end{bmatrix}$$

$\underline{F} = \underline{1} + \frac{1}{2}\left(\underline{\nabla}_{x^0}\overrightarrow{u} + \left(\underline{\nabla}_{x^0}\overrightarrow{u}\right)^t\right) + \frac{1}{2}\left(\underline{\nabla}_{x^0}\overrightarrow{u} - \left(\underline{\nabla}_{x^0}\overrightarrow{u}\right)^t\right)$,
where

$$d\overrightarrow{x} = \underline{1}.d\overrightarrow{x^0} + \underline{\varepsilon}.d\overrightarrow{x^0} + \underline{\omega}.d\overrightarrow{x^0}.$$

And the partition is made of the gradient of the transformation $\underline{F}$ in a symmetric part $\underline{\boldsymbol{\varepsilon}} = \frac{1}{2}\left(\underline{\nabla}_{x^0}\overrightarrow{u} + \left(\underline{\nabla}_{x^0}\overrightarrow{u}\right)^t\right)$ called deformation tensor in the hypothesis of small perturbations and an antisymmetric part $\underline{\omega} = \frac{1}{2}\left(\underline{\nabla}_{x^0}\overrightarrow{u} - \left(\underline{\nabla}_{x^0}\overrightarrow{u}\right)^t\right)$ called rotation tensor plus the tensor identity $\underline{1}$.

The component ε_{ij} of the symmetric matrix representative of $\underline{\varepsilon}$ will be defined as (line i and column j):

$$\varepsilon_{ij} = \frac{1}{2}\left(\frac{\partial u_i}{\partial x_j^0} + \frac{\partial u_j}{\partial x_i^0}\right) = \varepsilon_{ji}.$$

Note that the components of the deformation tensor are "dimensionless" while the displacement is measured in meters.

The component ω_{ij} of the antisymmetric matrix $\underline{\boldsymbol{\omega}}$ also known as rotation matrix is written as

$$\omega_{ij} = \frac{1}{2}\left(\frac{\partial u_i}{\partial x_j^0} - \frac{\partial u_j}{\partial x_i^0}\right) = -\omega_{ji}.$$

We can state a necessary and sufficient condition such that $\underline{\boldsymbol{\omega}}$ will be antisymmetric and written as (see Exercise 2.1)

$$\forall \overrightarrow{dx^0} \ni \vec{\alpha} \text{ such that } : \underline{\omega}.\overrightarrow{dx^0} = \vec{\alpha} \wedge \overrightarrow{dx^0}.$$

2.2.4.1 Exercise 2.1

Establish the necessary and sufficient condition for a tensor $\underline{\omega}$ to be antisymmetric, e.g.,

$$\forall \overrightarrow{dx^0} \ni \vec{\alpha} \text{ such that } : \underline{\omega}.\overrightarrow{dx^0} = \vec{\alpha} \wedge \overrightarrow{dx^0}.$$

Correction:

$$\vec{\alpha} \begin{bmatrix} \alpha_1 = \omega_{32} = -\omega_{23} = \frac{1}{2}\left(\frac{\partial u_3}{\partial x_2^0} - \frac{\partial u_2}{\partial x_3^0}\right) \\ \alpha_2 = \omega_{13} = -\omega_{31} = \frac{1}{2}\left(\frac{\partial u_1}{\partial x_3^0} - \frac{\partial u_3}{\partial x_1^0}\right) \\ \alpha_2 = \omega_{21} = -\omega_{12} = \frac{1}{2}\left(\frac{\partial u_2}{\partial x_1^0} - \frac{\partial u_1}{\partial x_2^0}\right) \end{bmatrix}$$

or

$$\vec{\alpha} = \tfrac{1}{2}\text{rot}\,\vec{u}.$$

It may be noted that the « antisymmetric action » $\underline{\omega}$ operates a rotation $\vec{\alpha}$ on the elementary vector $\overrightarrow{dx^0}$.

It is still necessary to define the action $\underline{\varepsilon}.\overrightarrow{dx^0}$ which will be identified as a deformation action. Indeed, by evaluating ε we will call the unitary extension:

$$\varepsilon = \frac{dx - dx^0}{dx^0}.$$

Note that it is easier to assess $(dx)^2$ compared to $(dx^0)^2$.

In fact $(dx^0)^2 = (dx_1^0)^2 + (dx_2^0)^2 + (dx_3^0)^2 = dx_i^0 dx_i^0$ and $(dx)^2 = (dx_1)^2 + (dx_2)^2 + (dx_3)^2 = dx_i dx_i$ and also $dx_i = dx_i^0 + du_i = dx_i^0 + \frac{\partial u_i}{\partial x_j^0} dx_j^0$.

Conventions: i present one in the monomial: Free Index (the previous term is a component in i) j repeated 2 times: silent index (we sum j from 1 to 3).

For the sake of simplicity, the term $\frac{\partial u_i}{\partial x_j^0}$ will be replaced by $u_{i,j}$ and the same approach applies to higher order derivations $\frac{\partial u_i^3}{\partial x_j^0 \partial x_k^0 \partial x_l^0} = u_{i,jkl}$.

Moreover, $\varepsilon = \frac{(dx - dx^0)(dx + dx^0)}{dx^0(dx + dx^0)} = \frac{dx^2 - dx^{02}}{2dx^{02}(1 + \varepsilon/2)}$ with $dx = (1 + \varepsilon)\,dx^0$.

These two possible situations can be considered:

(i) Either, the unitary extension ε is an infinitesimal of order 1 (e.g., ε very small toward 1).

In this case, it is the « small deformations » elastic behavior domain of solid media. In this situation, the geometrical shape changes little over time. Practically when ε is of the order of a few 10^{-3}, hypothesis of « small perturbations » is made (sometimes it is referred to the term "small strain"). (See Exercise 2.2.).
(ii) Either, the unitary extension ε is not an infinitesimal of order 1. This is the case of the elastoplastic behavior, for example, of stamped sheet metal (car racks) where parts can undergo deformation greater than 1.
In fluid mechanics, the unitary extension can be very large (example, for air, inflating a balloon).
In this case, the initial geometric shape can change significantly over time.

In the situation (i), hypothesis of « small perturbations », one can write

$$\varepsilon = \frac{(dx)^2 - \left(dx^0\right)^2}{2\left(dx^0\right)^2(1+{}^{\varepsilon}/2)} \simeq \frac{(dx)^2 - \left(dx^0\right)^2}{2\left(dx^0\right)^2}.$$

That is to also write that $\frac{\partial u_i}{\partial x_j^0}$ is an infinitesimal of order 1 and $\frac{\partial u_k}{\partial x_i^0}\frac{\partial u_k}{\partial x_j^0}$ infinitesimal of order 2 is negligible compared to the previous term.
Introducing

$$n_i^0 = \frac{dx_i^0}{dx^0},$$

the director cosines of $\overrightarrow{n^0}$ carrier of $d\overrightarrow{x^0}$ and $du_i = \frac{\partial u_i}{\partial x_j^0} dx_j^0$.
One obtains

$$\varepsilon = \sum_{i,j=1}^{3} \frac{1}{2}\left(\frac{\partial u_i}{\partial x_j^0} + \frac{\partial u_j}{\partial x_i^0}\right) n_i^0 n_j^0 = \varepsilon_{ij} n_i^0 n_j^0.$$

Thus, the unitary extension ε of a vector element $d\overrightarrow{x^0} = \overrightarrow{n^0} dx^0$ is written as

$$\varepsilon = \left(\overrightarrow{n^0}\right)^t . \underline{\varepsilon} . \overrightarrow{n^0} = \left(\overrightarrow{n^0}\right)^t . \overset{\overrightarrow{n^0}}{D} = \varepsilon_{n^0 n^0}.$$

Thus, the vector deformation is introduced as $\overset{\overrightarrow{n^0}}{D} = \underline{\varepsilon}.\overrightarrow{n^0}$:

$$\overset{\overrightarrow{n^0}}{D} = \varepsilon_{n^0 n^0} \overrightarrow{n^0} + \varepsilon_{n^0 t^0} \overrightarrow{t^0},$$

with $\varepsilon_{n^0 n^0}$ axial deformation and $\varepsilon_{n^0 t^0}$ transverse deformation (Fig. 2.3).
So

$$\varepsilon_{ij} = \overrightarrow{x_i^0}.\underline{\varepsilon}.\overrightarrow{x_j^0}.$$

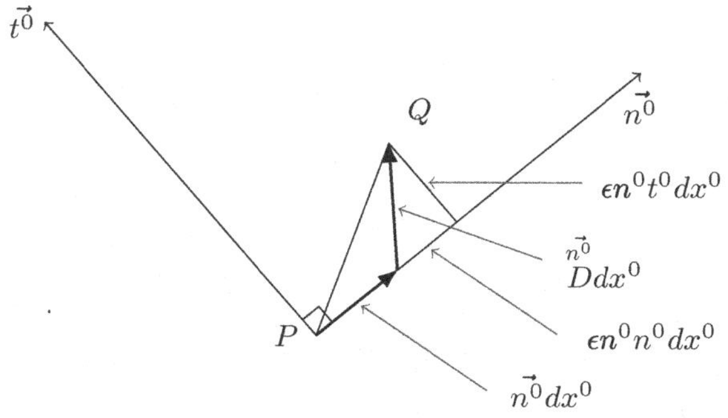

Fig. 2.3 Geometrical illustration of "deformation vector"

For example,

$$\varepsilon_{11} = \overrightarrow{x_1^0}.\underline{\varepsilon}.\overrightarrow{x_1^0} \; and \; \varepsilon_{12} = \overrightarrow{x_1^0}.\underline{\varepsilon}..\overrightarrow{x_2^0}.$$

With a simple geometric calculation, one can give that

$$cos\theta = cos\theta_0 + \varepsilon_{ij}\left(n_i^0 n_j^{'0} + n_i^{'0} n_j^0\right),$$

with (Fig. 2.4)

$$\overrightarrow{n^0} = \frac{\overrightarrow{dx^0}}{dx^0}, \; \overrightarrow{n^{'0}} = \frac{\overrightarrow{dx^{'0}}}{dx^{'0}}.$$

Let us consider that

$$\theta_0 = \left(\overrightarrow{x_1^0}, \overrightarrow{x_2^0}\right) = \frac{\pi}{2} \Rightarrow cos\theta_0 = 0 \Rightarrow cos\theta = cos\left(\frac{\pi}{2} - \alpha\right) \simeq \alpha = 2\varepsilon_{12}.$$

ε_{12} can be interpreted as the half-distortion of the initial right angle $\left(\overrightarrow{x_1^0}, \overrightarrow{x_2^0}\right)$.

Exercise 2.2 allows us to compare the tensors $\underline{E}$, $\underline{A}$ and $\underline{\varepsilon}$ and shows us that their measurements are in agreement in the case of small perturbations.

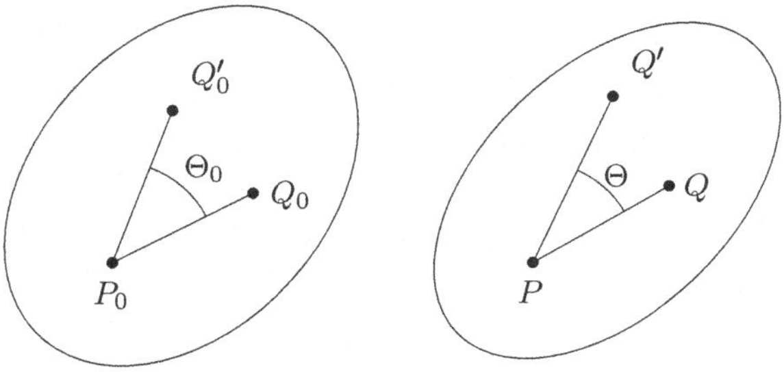

Fig. 2.4 Variation of the diedre angle θ_0 at t_0 and θ at t

2.2.4.2 Exercise 2.2

Considering the planar transformation defined by

$$\vec{x}\begin{cases} x_1 = (1+\lambda)\cos\theta x_1^0 - (1+\mu)\sin\theta x_2^0 \\ x_2 = (1+\lambda)\sin\theta x_1^0 + (1+\mu)\cos\theta x_2^0 \\ x_3 = x_3^0 \end{cases}$$

where λ, μ, and θ are positive dimensionless constants.

1. Determine $\underline{F}$, $\underline{E}$ and $\underline{A}$.
2. Decompose $\underline{H} = \underline{\nabla}_{x^0}\vec{u}$ in its symmetric part $\underline{\varepsilon}$ and antisymmetric one $\underline{\omega}$.
3. In the case where λ, μ, *and* θ are infinitely small toward 1, compare tensors $\underline{E}$ $\underline{A}$ and $\underline{\varepsilon}$ *and* $\underline{\omega}$.

Correction: *in 2D* 1. $\underline{F}=\begin{bmatrix} (1+\lambda)\cos\theta & -(1+\mu)\sin\theta \\ (1+\lambda)\sin\theta & (1+\mu)\cos\theta \end{bmatrix}$ $\underline{E} = \begin{bmatrix} \lambda + \frac{\lambda^2}{2} & 0 \\ 0 & \mu + \frac{\mu^2}{2} \end{bmatrix}$.

$$\underline{A} = \begin{bmatrix} a_{11} & a_{12} \\ a_{21} & a_{22} \end{bmatrix} \quad with \quad \left\{ \begin{array}{c} a_{11} = \frac{1}{2}\left(1 - \frac{\cos^2\theta}{(1+\lambda)^2} - \frac{\sin^2\theta}{(1+\mu)^2}\right) \\ a_{12} = a_{21} = -\frac{1}{2}\cos\theta\sin\theta\left(\frac{1}{(1+\lambda)^2} - \frac{1}{(1+\mu)^2}\right) \\ a_{22} = \frac{1}{2}\left(1 - \frac{\sin^2\theta}{(1+\lambda)^2} - \frac{\cos^2\theta}{(1+\mu)^2}\right) \end{array} \right\}.$$

2. $\underline{\varepsilon} = \begin{bmatrix} (1+\lambda)\cos\theta - 1 & \frac{\lambda-\mu}{2}\sin\theta \\ \frac{\lambda-\mu}{2}\sin\theta & (1+\mu)\cos\theta - 1 \end{bmatrix}$.

$$\underline{\omega} = \begin{bmatrix} 0 & -\left(1+\frac{\lambda+\mu}{2}\right)\sin\theta \\ \left(1+\frac{\lambda+\mu}{2}\right)\sin\theta & 0 \end{bmatrix}.$$

3. $E = \underline{A} = \underline{\varepsilon} = \begin{bmatrix} \lambda & 0 \\ 0 & \mu \end{bmatrix}$ $\underline{\omega} = \begin{bmatrix} 0 & -\theta \\ \theta & 0 \end{bmatrix}$.

2.2.5 *Intrinsic Properties of Symmetrical Second-Order Tensor*

We concentrate our attention on $\underline{\boldsymbol{\varepsilon}}$, knowing that the other tensors $\underline{E}$ and $\underline{A}$ have strictly the some properties.

2.2.5.1 Principal Deformations, Principal Directions

For each symmetrical matrix, it exists as orthonormous landmark $M_0\left(\vec{X_1}, \vec{X_2}, \vec{X_3}\right)$ called principal such that

$$\underline{\varepsilon}\,(M_0) = \begin{bmatrix} \varepsilon_{11}\,(M_0) & \varepsilon_{12} & \varepsilon_{13} \\ \varepsilon_{21} & \varepsilon_{22} & \varepsilon_{23} \\ \varepsilon_{31} & \varepsilon_{32} & \varepsilon_{33} \end{bmatrix}_{\overrightarrow{x_1^0}, \overrightarrow{x_2^0}, \overrightarrow{x_3^0}}$$

$\Longrightarrow$ *by rotation* $\underline{\varepsilon}\,(M_0) = \begin{bmatrix} \varepsilon_1\,(M_0) & 0 & 0 \\ 0 & \varepsilon_2 & 0 \\ 0 & 0 & \varepsilon_3 \end{bmatrix}_{\overrightarrow{X_1}, \overrightarrow{X_2}, \overrightarrow{X_3}}$. Classically, one researches the roots of the characteristic equation:

$$\det\left(\underline{\varepsilon}\,(M_0) - \lambda\underline{1}\right) = 0.$$

And for a symmetric matrix, it can be proved that the three roots $\varepsilon_1, \varepsilon_2, and \varepsilon_3$ are real, distinct, or not.

The principal directions $\left(\overrightarrow{X_1}, \overrightarrow{X_2}, \overrightarrow{X_3}\right)$ are obtained by the vectorial equation:

$$\overset{\overrightarrow{X_i}}{D} = \underline{\varepsilon}.\overrightarrow{X_i} = \varepsilon_i \overrightarrow{X_i} \ with \ |\ \overrightarrow{X_i}\ | = 1,\ \overrightarrow{X_i} = \begin{bmatrix} \alpha_i \\ \beta_i \\ \gamma_i \end{bmatrix}_{\overrightarrow{x_1^0}, \overrightarrow{x_2^0}, \overrightarrow{x_3^0}}.$$

2.2.5.2 Principal Invariants of the Strain Tensor

They are the coefficients of the characteristic equation: $-\lambda^3 + \theta_1\lambda^2 - \theta_2\lambda + \theta_3 = 0$

$$\theta_1 = \varepsilon_1 + \varepsilon_2 + \varepsilon_3 = \varepsilon_{11} + \varepsilon_{22} + \varepsilon_{33} = \mathrm{tr}\left(\underline{\varepsilon}\right)$$
$$\theta_2 = \varepsilon_1\varepsilon_2 + \varepsilon_2\varepsilon_3 + \varepsilon_3\varepsilon_1 = \varepsilon_{11}\varepsilon_{22} - \varepsilon_{12}\varepsilon_{21} + \varepsilon_{22}\varepsilon_{33} - \varepsilon_{23}\varepsilon_{32} + \varepsilon_{33}\varepsilon_{11} - \varepsilon_{31}\varepsilon_{13}$$
$$\theta_3 = \varepsilon_1\varepsilon_2\varepsilon_3 = \det\left(\underline{\varepsilon}\right).$$

These invariants will be mainly used in order to define criterions of elastic yield limits.

2.2.5.3 Mohr Circle and Tri-Circle

Let us consider the strain vector (Fig. 2.5):

$$\overset{\overrightarrow{n^0}}{D} = \varepsilon_{n^0 n^0}\overrightarrow{n^0} + \varepsilon_{n^0 t^0}\overrightarrow{t^0}.$$

If we represent it in the plane $(\varepsilon_{n^0n^0}, \varepsilon_{n^0t^0})$, N extremity of the strain vector $\overset{\overrightarrow{n^0}}{D}$. N is inside the « mixtiligne » triangle between the three half-circles (or on one of these three half-circles) (Fig. 2.6).

This intrinsic property of the symmetrical second-order tensor leads to the definition of a « maximum shearing criterion ».

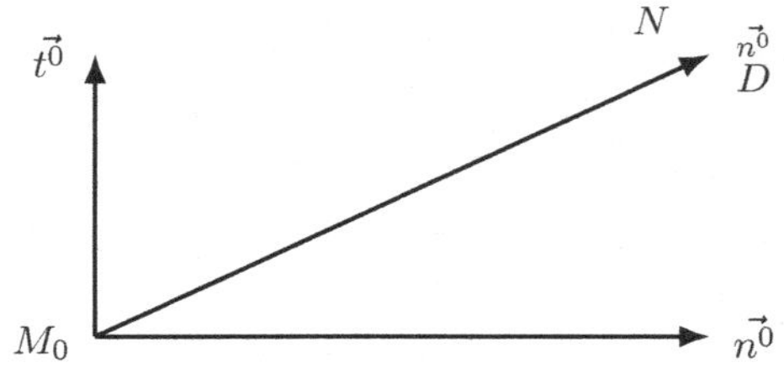

Fig. 2.5 Strain vector with axis $\overrightarrow{n^0}$ *and* $\overrightarrow{t^0}$

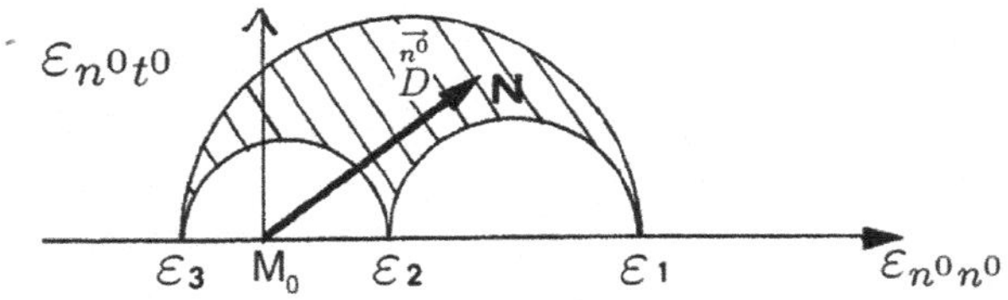

Fig. 2.6 Mohr tri-circle

$$(\varepsilon_{n^0t^0})_{max} = \left(\frac{\varepsilon_1 - \varepsilon_3}{2}\right)$$

acceptable or not for the material.

2.2.5.4 Compatibility Equations

When a vectorial displacement $\vec{u}$ is given sufficiently continuously derivable, $\underline{\varepsilon}$ and $\underline{\omega}$ are obtained by partial derivations of the components u_i of $\vec{u}$.

However, the reverse path may be more tortuous!

Let us consider that $\underline{\varepsilon}\left(x_1^0, x_2^0, x_3^0\right)$ known: does $\vec{u}\left(x_1^0, x_2^0, x_3^0\right)$ can be obtained ?

This constitutes a system with 6 partial derivatives equations with three unknown functions: $u_1\left(x_1^0, x_2^0, x_3^0\right)$, $u_2\left(x_1^0, x_2^0, x_3^0\right)$, $u_3\left(x_1^0, x_2^0, x_3^0\right)$

$$\frac{1}{2}\left(\frac{\partial u_i}{\partial x_j^0} + \frac{\partial u_j}{\partial x_i^0}\right) = \varepsilon_{ij}\left(x_1^0, x_2^0, x_3^0\right) \ knowns\ i, j = 1, 2, 3.$$

There are some integratibility conditions (that the mechanician calls « compatibility conditions »).

The first two can be written as

$$\varepsilon_{11,23} = \left(\varepsilon_{13,2} + \varepsilon_{21,3} - \varepsilon_{32,1}\right)_{,1}$$
$$2\varepsilon_{12,12} = \varepsilon_{11,22} + \varepsilon_{22,11}.$$

The four others are obtained by circular permutation of 1, 2, 3.

From the physical point of view, the knowledge of the ε_{ij} at point M_0 provides the geometry of the deformed element around M_0 (Fig. 2.7). We need all the assembled elements to form a continuum medium.

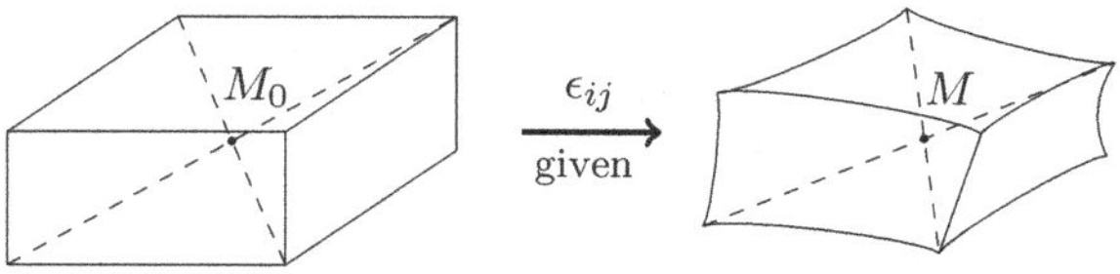

Fig. 2.7 Geometry of the deformed element associated to the data of $\underline{\varepsilon}\,(M_0)$

Exercise 2.3 will give the calculation of the vectorial displacement $\overrightarrow{u}\,(M_0)$ from the knowledge of $\underline{\varepsilon}\,(M_0)$ by first checking compatibility equations. A first integration will give $\underline{\omega}\,(M_0)$ and a second $\overrightarrow{u}\,(M_0)$.

2.2.6 *Exercise 2.3*

Let us give the most general form of displacement $\overrightarrow{u}$ generated by a tensor $\underline{\varepsilon}$ such that

$$\underline{\varepsilon} = \begin{bmatrix} 0 & 0 & \varphi\left(x_1^0, x_2^0\right) \\ 0 & 0 & \psi\left(x_1^0, x_2^0\right) \\ \varphi\left(x_1^0, x_2^0\right) & \psi\left(x_1^0, x_2^0\right) & 0 \end{bmatrix}.$$

Applications:

$$\begin{aligned} \varphi\left(x_1^0, x_2^0\right) &= 2Bx_2^0\left(x_1^0 + a\right) \\ \psi\left(x_1^0, x_2^0\right) &= B\left(\left(x_1^0\right)^2 - \left(x_2^0\right)^2 - 2ax_1^0\right) \end{aligned}$$

B and a are constants.
Correction:
Compatibility equations: they deliver the following condition:

$$\varphi_{,2} - \psi_{,1} = K = Cte.$$

We will use the conventional method of double integration.

Knowing that $u_{i,j} = \varepsilon_{ij} + \omega_{ij}$, we will first calculate the ω_{ij} through the medium of the formula:

$$\begin{aligned} \omega_{ij,l} &= \varepsilon_{il,j} - \varepsilon_{jl,i} \\ \omega_{ii} &= 0, \ \omega_{ji} = -\omega_{ji} \end{aligned}$$

with $i \neq j$.

which gives

$$\left\{ \begin{aligned} \omega_{12} &= -\omega_{21} = Kx_3 + \alpha_3 \\ \omega_{23} &= -\omega_{32} = -\psi - Kx_1 + \alpha_1 \\ \omega_{31} &= -\omega_{13} = \varphi - Kx_2 + \alpha_2 \end{aligned} \right\},$$

then $u_{i,j} = \varepsilon_{ij} + \omega_{ij}$ can be obtained and then by a simple integration components of vector displacement $\overrightarrow{u}$:

$$\left\{ \begin{array}{c} u_1 = Kx_2x_3 + \alpha_3x_2 - \alpha_2x_3 + \beta_1 \\ u_2 = -Kx_1x_3 - \alpha_3x_1 + \alpha_1x_3 + \beta_2 \\ u_3 = \int (2\varphi - Kx_2)\,dx_1 + \int (2\psi + Kx_1)\,dx_2 + \alpha_2x_1 - \alpha_1x_2 + \beta_3 \end{array} \right\}.$$

The compatibility equations deliver $K = 4Ba$ and

$$u_3 = 2B\left(x_1^2x_2 - \frac{x_2^3}{3}\right) + \alpha_2x_1 - \alpha_1x_2 + \beta_3.$$

One can recognize the torsion of a beam with section an "equilateral triangle".

Important Remark

As far the calculation of « small deformations » is concerned, such as distorted body geometry is slightly different from the non-deformed natural state, the reference system chosen will be R (related to the non-deformed state of the natural solid). Thus, in this case, the stress calculation (to be continued) will be also performed in R.

To lighten the scriptures, we will replace the notation x_i^0 by x_i.

This important approximation cannot be made for "finite strains" and in fluid mechanics.

2.3 Stress Concept

This is a generalization of the concept of pressure (which is expressed in Newton per square meter).

2.3.1 Inventory of Possible External Mechanical Loadings

Before introducing the stress concept, we will make an inventory of possible external mechanical loadings. There are external forces acting inside the body called volume forces and forces "also" exterior, acting on the border of the body called surfaces forces.

2.3.1.1 Volume Force Density

External forces acting on a volume element are "distance actions" (gravity, electrostatic action, ...)

On a volume element $d\omega_t$ (independently of the surrounding volume elements) is exerted $\overrightarrow{f^\star\,(M)d\omega_t}$ with $\overrightarrow{f^\star\,(M)}$ defined as the volume density of dynamic force (of dimension N/m^3) (Fig. 2.8). This density is expressed as follows:

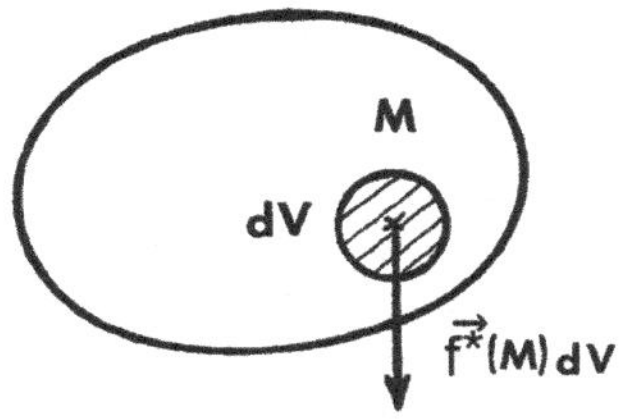

Fig. 2.8 Volume density of dynamic force $\overrightarrow{f^{\star}}(M)$

$$\overrightarrow{f^{\star}(M)} = \overrightarrow{f(M)} - \rho(M)\,\overrightarrow{\gamma(M)},$$

where $\overrightarrow{f(M)}$ constitutes the volume density of static force, $\vec{\gamma}(M)$ the point M acceleration and $\rho(M)$ the density:

$$\vec{\gamma}(M) = \frac{d\vec{v}(M)}{dt} = \frac{\partial^2 \vec{u}\left(\vec{x}, t\right)}{\partial t^2}.$$

A typical static volumetric density is gravity :$\overrightarrow{f(M)} = \rho\overrightarrow{g(M)}$.

2.3.1.2 Surface Force Density

External forces acting on one surface element are "contact forces" generated at the surface of the element by the adjoining elements.

On $\partial\omega_t$, surface element of the boundary of the body is exerted $\overrightarrow{F(P)}\partial\omega_t$ where $\overrightarrow{F(P)}$ is the surface density of the external forces (whose dimension is N/m^2) (Fig. 2.9).

One classical example is the external pressure (Fig. 2.10):

$$\overrightarrow{F(P)} = -p(P)\,\vec{v},$$

where $\vec{v}$ unitary outward normal at P to the body surface.

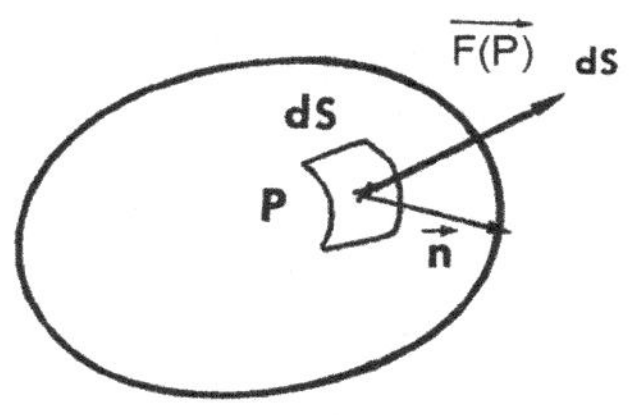

Fig. 2.9 Surface force density $\vec{F}(P)$

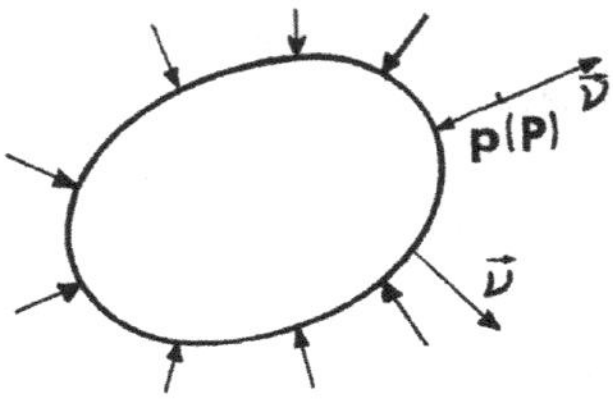

Fig. 2.10 Pressure $p\,(P)$ applied at the surface body

2.3.2 Introduction of the Stress Concept

Let us consider S subjected to external mechanical loading and splitted into two parts S_1 and S_2 (Fig. 2.11). So that S_1 REST IN EQUILIBRIUM, we must replace the action of S_2 on S_1 by a surface force density called $\overset{\nu}{\overrightarrow{T}}(M)$ applied to the cutting surface Σ. $\overrightarrow{\nu}$ denotes on exterior unitary normal to S_1. Note that the action of S_1 on S_2 is equal and of opposite sign to that of S_2 on S_1, therefore meaning that $\overset{\nu}{\overrightarrow{T}}(M) = -\overset{-\nu}{\overrightarrow{T}}(M)$.

For three particular $\overrightarrow{\nu}$ directions, $\overrightarrow{x_1}, \overrightarrow{x_2}, \overrightarrow{x_3}$, one defines three vectors:

$$\left\{\begin{aligned} \overset{\overrightarrow{x_1}}{T} &= \sigma_{11}\overrightarrow{x_1} + \sigma_{21}\overrightarrow{x_2} + \sigma_{31}\overrightarrow{x_3} \\ \overset{\overrightarrow{x_2}}{T} &= \sigma_{12}\overrightarrow{x_1} + \sigma_{22}\overrightarrow{x_2} + \sigma_{32}\overrightarrow{x_3} \\ \overset{\overrightarrow{x_3}}{T} &= \sigma_{13}\overrightarrow{x_1} + \sigma_{23}\overrightarrow{x_2} + \sigma_{33}\overrightarrow{x_3} \end{aligned}\right\}.$$

There are nine defined components forming a matrix:

$$\underline{\sigma}\,(M) = \begin{bmatrix} \sigma_{11}\,(M) & \sigma_{12} & \sigma_{13} \\ \sigma_{21} & \sigma_{22} & \sigma_{23} \\ \sigma_{31} & \sigma_{32} & \sigma_{33} \end{bmatrix},$$

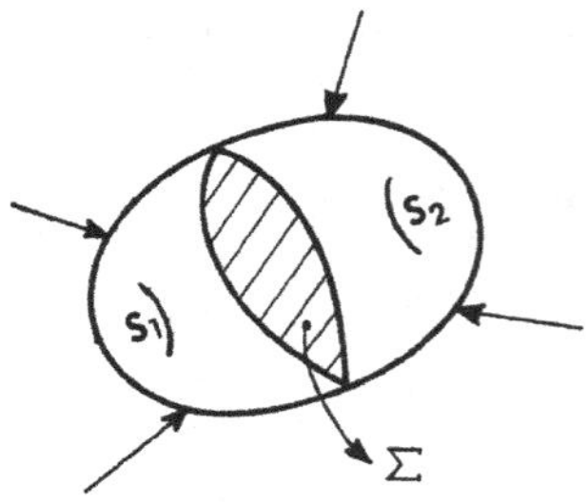

Fig. 2.11 Partition of the body into two parts in order to introduce the stress concept

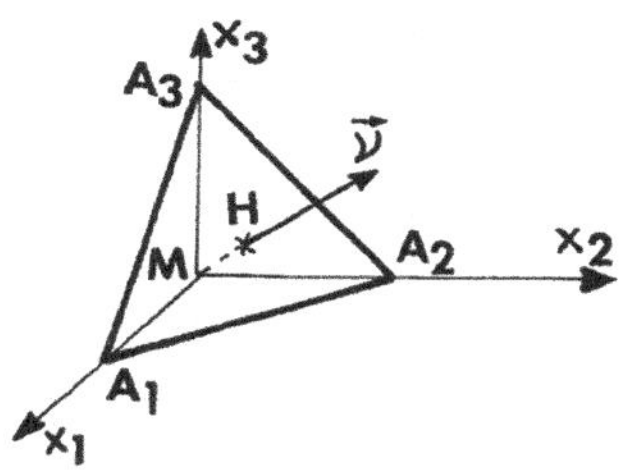

Fig. 2.12 Tetrahedron equilibrium

showing that the knowledge of these nine components permits to calculate $\overset{\nu}{\vec{T}}(M)\ \forall \vec{\nu}$.

Let us write the balance of a tetrahedron in the neighborhood of any point M (Fig. 2.12).

Data: $MH = h,\quad MA_i = \frac{h}{\nu_i},\quad area\ A_1A_2A_3 = dS,\quad area\ MA_1A_2 = \nu_3 dS,\quad area\ MA_2A_3 = \nu_1 dS,\ area\ MA_3A_1 = \nu_2 dS,\ vol.\ tetrahedron = \frac{1}{3}hdS,$

Leading to,

$$\overset{\nu}{\vec{T}} + \overset{-x_1}{\vec{T}}\nu_1 + \overset{-x_2}{\vec{T}}\nu_2 + \overset{-x_3}{\vec{T}}\nu_3 + \frac{1}{3}h\vec{f^\star} = \vec{0}.$$

Let h tends to 0, e.g., H to M.

It comes

$$\overset{\nu}{\vec{T}}(M) = \begin{bmatrix} \sigma_{11}(M)\,\nu_1 & \sigma_{12}\nu_{12} & \sigma_{13}\nu_3 \\ \sigma_{21}\nu_1 & \sigma_{22}\nu_2 & \sigma_{23}\nu_3 \\ \sigma_{31}\nu_1 & \sigma_{32}\nu_2 & \sigma_{33}\nu_3 \end{bmatrix} = \begin{bmatrix} \sigma_{11}(M) & \sigma_{12} & \sigma_{13} \\ \sigma_{21} & \sigma_{22} & \sigma_{23} \\ \sigma_{31} & \sigma_{32} & \sigma_{33} \end{bmatrix} \begin{bmatrix} \nu_1 \\ \nu_2 \\ \nu_3 \end{bmatrix} = \underline{\sigma}(M)\,.\,\vec{\nu}.$$

Normal stress is N and shearing stress is T (Fig. 2.13).

One can note that the stress dimension σ_{ij} is the same as a pressure, e.g., in N/m^2. One often expresses stress in MPa (1 MPa $= 10^6$ N/m^2) (Fig. 2.14).

We can write the dynamic equilibrium equation of a volume element, with inertia center $M\,(x_1, x_2, x_3)$; but we will integrate its expression in the conservation laws.

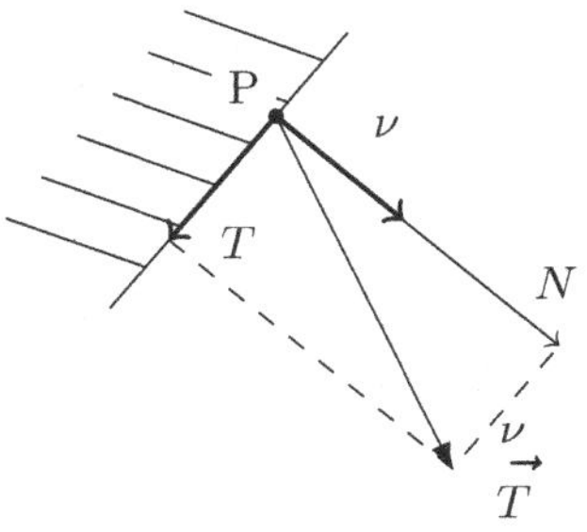

Fig. 2.13 Stress vector $\overset{\nu}{\vec{T}}(M)$ with normal component N and shear one T

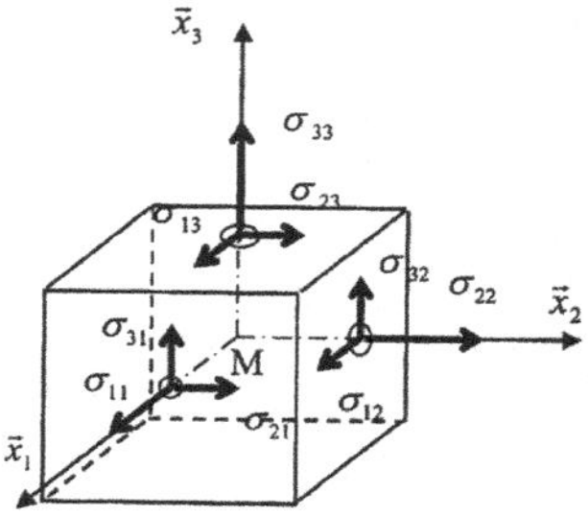

Fig. 2.14 Cubic volume element around M. Visualization of stress components σ_{ij}

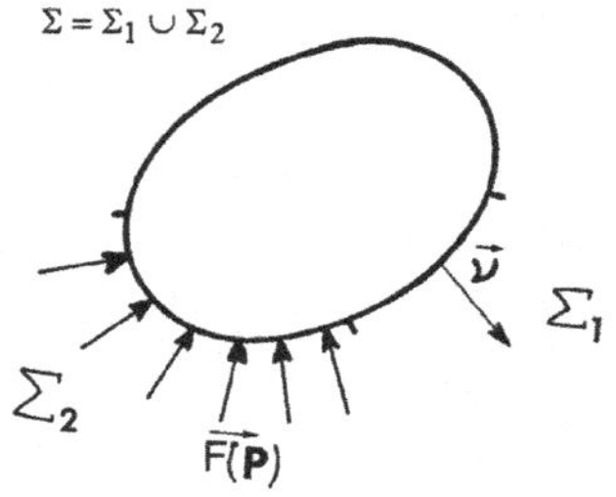

Fig. 2.15 Σ_1 free surface; on mechanical loading; Σ_2 under $\overrightarrow{F\,(P)}$

2.3.2.1 Yield Conditions in Terms of Stress (Fig. 2.15)

On the boundary surface $\Sigma_2 : \overset{\nu}{\vec{T}}\,(P) = \overrightarrow{F\,(P)}$. On a free surface Σ_1 of mechanical loading: $\overset{\nu}{\vec{T}}\,(P) = \vec{0}$.

2.3.2.2 Saint Venant' Principle

The case of concentrated (or punctual) forces applied to the boundary is a serious problem. The Saint Venant' principle can help us to solve it.
Saint Venant' principle formulation:

If on a part Σ_1 of the boundary Σ, we replace a surface density force $\overrightarrow{F\,(P)}$ by an another called $\overrightarrow{F'\,(P)}$, the two distributions having the same resulting force $\overrightarrow{R_{\Sigma_1}}$ and the same momentum $\overrightarrow{M_{\Sigma_1}}$ (that is to say the same torsor), the other loading conditions in the volume and on Σ_2 remaining unchanged; while away from Σ_1 (practically close enough), the solution in terms of vector displacement, deformations, and constraints remains unchanged.

Exercise 2.4 will consist in studying the stress tensor at a point M and using the boundary conditions to determine the stress tensor $\underline{\sigma}\,(M)$.

2.3.2.3 Exercise 2.4

Let us study the stress tensor at a point M (Fig. 2.16).

Let yield conditions are in terms of stress at a point M of the deformable body.

At a point M of the upstream face of a dam, the pressure water p is known and the normal component $(-q)$ of the stress vector is acting on a horizontal facet.

The angle of the facing with the vertical is ϕ.

Let us determine the stress tensor at point M in reference to horizontal axis $\overrightarrow{Mx_1}$ and vertical $\overrightarrow{Mx_2}$.

Let us determine first the principal stresses and directions.

Correction:

Precept: In M, for each unitary face of normal $\vec{\nu}$, let its stress vector be $\overset{\nu}{\vec{T}}(M)$.

Pressure effect of p: $\overset{\nu}{\vec{T}}(M) = -p\,(M)\,\vec{\nu}$

$$\begin{bmatrix} \sigma_{11} & \sigma_{12} \\ \sigma_{21} & \sigma_{22} \end{bmatrix} \begin{bmatrix} \cos\phi \\ \sin\phi \end{bmatrix} = -p \begin{bmatrix} \cos\phi \\ \sin\phi \end{bmatrix}.$$

Knowledge of $(-q)$: $\overset{x_2}{\vec{T}}(M) = \sigma_{12}\vec{x_1} + (-q)\,\vec{x_2} \Longrightarrow \sigma_{22} = (-q)$

The system resolution gives us

$$\sigma_{11} = -p + (p-q)\,\mathrm{tg}^2\phi$$
$$\sigma_{12} = -\,(p-q)\,\mathrm{tg}\phi.$$

Principal stress and directions:

The equation $\overset{\nu}{\vec{T}}(M) = -p\,(M)\,\vec{\nu}$ délivers $\sigma_1 = -p$ and $\vec{X_1} = \vec{\nu}$, *and* the invariance of tr $(\underline{\sigma})$ gives $\sigma_2 = \sigma_{11} + \sigma_{22} - \sigma_1 = -q + (p-q)\,tg^2\phi$ and $\vec{X_2} = \vec{t}$ (tangent to the facing).

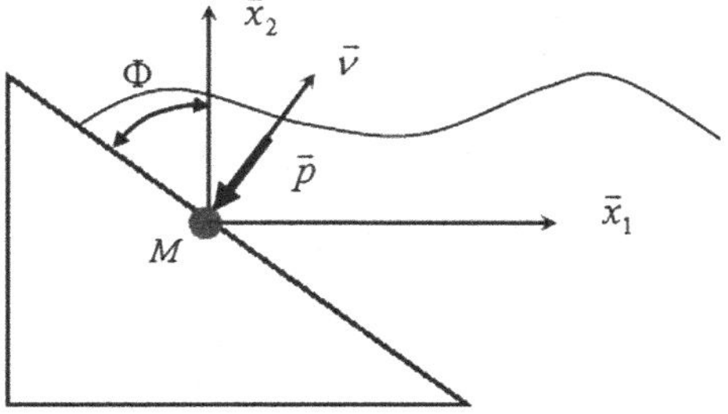

Fig. 2.16 Dam under pressure

2.4 Conservation Laws

These are as follows:
– mass balance;
– momentum balance;
– first and second principles of thermodynamics;
– diffusion;
– etc.

2.4.1 Concept of a Material System (Fig. 2.17*)*

Consider ω_0 to be any "fairly small" subdomain of the domain Ω_0 (Lagrange domain).

At all times, we can match the subdomain ω_t *of* Ω_t (Euler domain) such that

$$\forall \begin{cases} \overrightarrow{x^0} \epsilon \omega_0 & \overrightarrow{x} = \overrightarrow{f}\,(\overrightarrow{x^0}, t) \epsilon \omega_t \\ t > 0 \end{cases} \tag{2.1}$$

$$\overrightarrow{x^0} \begin{pmatrix} x_1^0 \\ x_2^0 \\ x_3^0 \end{pmatrix} \tag{2.2}$$

$x_i^0{}_i$ i = 1, 2, 3, t Lagrange variables:

$$\overrightarrow{x} \begin{pmatrix} x_1 \\ x_2 \\ x_3 \end{pmatrix} \tag{2.3}$$

x_i i = 1, 2, 3, t Euler variables.

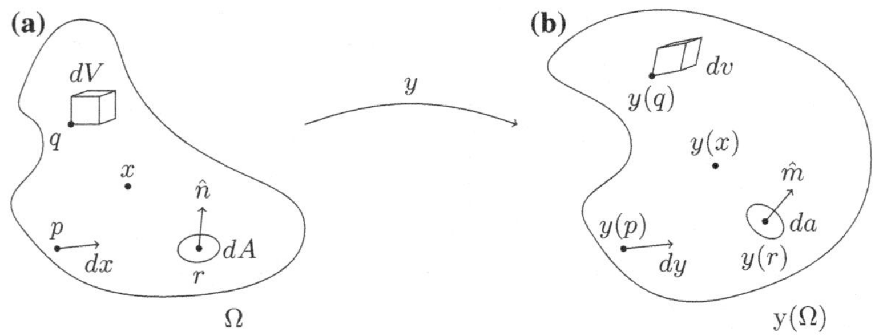

Fig. 2.17 Transport of a point, of an elementary vector, a surface element, a volume element

On the one hand, the Lagrange variables describe the trajectory over time of a particle located at a point $\overrightarrow{x_0}$ at time $t_0 = 0$ the conventional origin in time graphs. On the other hand the Euler variables hold the velocity field to be known at every point in space time:

$$\overrightarrow{V} = \overrightarrow{V}\left(\overrightarrow{x}, t\right). \tag{2.4}$$

2.4.2 Concept of a Particulate Derivative

Let us consider G an application matches $G\left(\overrightarrow{x}, t\right)$, for the coordinate pair $\left(\overrightarrow{x}, t\right)$. It can also be written as $G(\overrightarrow{f}\left(\overrightarrow{x^0}, t\right), t)$ *or* $G\left(\overrightarrow{x}(t), t\right)$.

The time derivative is written as

$$\frac{dG}{dt} = \frac{\partial G}{\partial t} + \sum_{i=1}^{i=3} \frac{\partial G}{\partial x_i}\frac{dx_i}{dt} = \frac{\partial G}{\partial t} + \overrightarrow{\nabla_x G}.\overrightarrow{V}. \tag{2.5}$$

This derivative, known as a particle derivative, represents the unitary variation of G in relation to time as we follow the trajectory of the particle animated with velocity $\overrightarrow{V}$: we denote $\frac{DG}{Dt}$.

The partial derivative $\partial/\partial t$ measures a change rate in a point of space.

The result extends naturally to vector functions:

$$\frac{\mathrm{d}\overrightarrow{U}}{\mathrm{dt}} = \frac{\partial \overrightarrow{U}}{\partial \mathrm{t}} + \underline{\nabla}_{\mathrm{x}} \overrightarrow{U}.\overrightarrow{V} \tag{2.6}$$

with

$$\underline{\nabla}_{\mathrm{x}} \overrightarrow{U} = \begin{pmatrix} U_{1,1} & U_{1,2} & U_{1,3} \\ U_{2,1} & U_{2,2} & U_{2,3} \\ U_{3,1} & U_{3,2} & U_{3,3} \end{pmatrix}. \tag{2.7}$$

The derivative of the integral of a scalar function on a time-dependent domain ω_t is therefore

$$\frac{d}{dt}\int_{\omega_t} G(\overrightarrow{x}(t), t)d\omega_t = \frac{d}{dt}\int_{\omega_0} G(\overrightarrow{x}(t), t)Jd\omega_0 = \int_{\omega_0}\left(\frac{dG}{dt}J + G\frac{dJ}{dt}\right)d\omega_0 \tag{2.8}$$

with

$$\frac{dJ}{dt} = (\mathrm{div} V)J. \tag{2.9}$$

J constitutes the jacobian of the transformation.

This gives

$$\frac{d}{dt}\int_{\omega_t} G(\overrightarrow{x}(t),t)d\omega_t = \int_{\omega_t}\left(\frac{dG}{dt} + G\mathrm{div}\overrightarrow{V}\right)d\omega_t. \tag{2.10}$$

By introducing the relation

$$\mathrm{div}(G\overrightarrow{V}) = \overrightarrow{V}.\overrightarrow{\mathrm{grad}}G + G\mathrm{div}\overrightarrow{V}, \tag{2.11}$$

one obtains

$$\begin{aligned}\tfrac{d}{dt}\textstyle\int_{\omega_t} G(\overrightarrow{x}(t),t)d\omega_t = \int_{\omega_t}\left(\frac{\partial G}{\partial t} + \mathrm{div}(G\overrightarrow{V})\right)d\omega_t = \\ \textstyle\int_{\omega_t}\frac{\partial G}{\partial t}d\omega_t + \int\int_{\partial\omega_t} G(\overrightarrow{x}(t),t)\overrightarrow{V}.\overrightarrow{\nu}dS_t.\end{aligned} \tag{2.12}$$

This derivation can be interpreted as being the sum of a term representing the eigen variation of G in relation to time and a flux on the boundary $\partial\omega_t$ ($\overrightarrow{\nu}$ unitary normal to this contour).

2.4.3 *Mass Balance*

2.4.3.1 Lagrange Variables

Let us consider $d\omega_0 \rightarrow d\omega_t$ so $\rho_0 d\omega_0 = \rho d\omega_t$ with $d\omega_t = Jd\omega_0$.

J is the jacobian of the transformation that gives

$$\rho J = \rho_0. \tag{2.13}$$

2.4.3.2 Euler Variables

Let us consider $\rho(\overrightarrow{x},t)$ the density at point $\overrightarrow{x}$ *for any time t*: $m(t) = \int_{\omega_t}\rho(\overrightarrow{x},t)d\omega_t$.

The principle of mass conservation leads to $\frac{dm(t)}{dt} = 0$, e.g.,

$$\frac{\partial\rho}{\partial t} + \mathrm{div}(\rho\overrightarrow{V}) = 0 \tag{2.14}$$

or

$$\frac{d\rho}{dt} + \rho \operatorname{div} \overrightarrow{V} = 0. \tag{2.15}$$

This means that an incompressible solid (ρ constant) is governed by the equation

$$\operatorname{div} \overrightarrow{V} = 0. \tag{2.16}$$

2.4.4 Equation of Motion

The fundamental law of the dynamics indicates that for every system ω_t, if one calls $\overrightarrow{V}\left(\overrightarrow{x}, t\right)$ the velocity field, the time derivative of the torsor of linear momentum is equal to the torsor of applied external forces, which means $\frac{d}{dt}\int_{\omega_T} \rho \overrightarrow{V} d\omega_t = \int \int_{\partial\omega_t} \underline{\sigma}.\overrightarrow{v} dS_t + \int_{\omega_T} \overrightarrow{f} d\omega_t$ that gives $\int_{\omega_t} \rho \frac{d\overrightarrow{V}}{dt} d\omega_t = \int_{\omega_T} (\operatorname{div}\underline{\sigma} + \overrightarrow{f}) d\omega_t$ and finally

$$\operatorname{div}\underline{\sigma} + \overrightarrow{f^{\star}} = \overrightarrow{0} \tag{2.17}$$

$$\overrightarrow{f^{\star}} = \overrightarrow{f} - \rho \overrightarrow{\gamma}, \tag{2.18}$$

with $\overrightarrow{f^{\star}}$ defined as the volume density of dynamic force and $\overrightarrow{f}$ defined as the volume density of static force with acceleration $\overrightarrow{\gamma} = \frac{\mathrm{d}\overrightarrow{V}}{\mathrm{dt}}$.

It represents the « Cauchy equation » (1822) or dynamic equilibrium equation. The second equation

$$\frac{d}{dt}\int_{\omega_t} \overrightarrow{OM} \wedge \rho \overrightarrow{V} dV_t = \int\int_{\partial\omega_t} \overrightarrow{OP} \wedge \underline{\sigma}.\overrightarrow{v} dS_t + \int_{\omega_T} \overrightarrow{OM} \wedge \overrightarrow{f} d\omega_t \tag{2.19}$$

allows to obtain the symmetry of the stress tensor ($\sigma_{ij} = \sigma_{ji}$).

2.4.5 Energy Balance: First Thermodynamics Law

For any material system, the time derivative of the total energy is equal to the power of the external forces increased by the heat input.

To begin with, there is a scalar e, e.g., the internal energy input per unit mass, such that the total energy is the sum of this internal energy and the mass kinetic energy.

So $\frac{d}{dt}\left[\int_{\omega_t} \rho e d\omega_t + \int_{\omega_t} \rho \frac{V^2}{2} d\omega_t\right] = \int_{\omega_t} W d\omega_t - \int\int_{\partial\omega_t} \overrightarrow{q}.\overrightarrow{v}\, dS_t + \int_{\omega_T} \overrightarrow{f}.\overrightarrow{V}\, d\omega_t + \int\int_{\partial\omega_t} \underline{\sigma}\,\overrightarrow{v}.\overrightarrow{V}\, dS_t$ with W the heat internal source and $\overrightarrow{q}$ the heat-flow vector; classically, with the divergence theorem:

$$\int\int_{\partial\omega_t} \overrightarrow{q}.\overrightarrow{v}\, dS_t = \int_{\omega_t} \mathrm{div}\,\overrightarrow{q}\, d\omega_t \tag{2.20}$$

and

$$\int\int_{\partial\omega_t} \underline{\sigma}\,\overrightarrow{v}.\overrightarrow{V}\, dS_t = \int_{\omega_t} \mathrm{div}(\underline{\sigma}\,\overrightarrow{V})d\omega_t \tag{2.21}$$

with

$$\mathrm{div}(\underline{\sigma}\,\overrightarrow{V}) = \underline{\sigma} : \underline{\dot{\varepsilon}} + .\mathrm{div}\underline{\sigma}.\overrightarrow{V} \tag{2.22}$$

and $\underline{\dot{\varepsilon}}$ is the Eulerian tensor of strain rates such that

$$\dot{\varepsilon}_{ij} = \frac{1}{2}\left(\frac{\partial V_i}{\partial x_j} + \frac{\partial V_j}{\partial x_i}\right). \tag{2.23}$$

Finally one obtains

$$\rho\frac{de}{dt} = \underline{\sigma} : \underline{\dot{\varepsilon}} + W - \mathrm{div}\,\overrightarrow{q}. \tag{2.24}$$

2.4.6 Entropy Variation: Second Law of Thermodynamic

There is a function s, per unit mass, called entropy, such that, for any material system, the entropy variation is greater than the integral of the ratio of the heat input, divided by the absolute temperature T:

$$\frac{d}{dt}\left(\int_{\omega_t} \rho s d\omega_t\right) \geqq \int_{\omega_t} \frac{W}{T} d\omega_t - \int\int_{\partial\omega_t} \frac{\overrightarrow{q}}{T}.\boldsymbol{v} dS_t. \tag{2.25}$$

Alternatively, using the divergence theorem,

$$\int_{\omega_t} \rho\frac{ds}{dt} d\omega_t \geqq \int_{\omega_t} \left(\frac{W}{T} - \mathrm{div}\left(\frac{\overrightarrow{q}}{T}\right)\right) d\omega_t; \tag{2.26}$$

this being true regardless ω_t, one has

$$\rho \frac{ds}{dt} \geqq \frac{W}{T} - \text{div}\left(\frac{\vec{q}}{T}\right). \tag{2.27}$$

Synthesis

$$\frac{d\rho}{dt} + \rho \text{div}\, \vec{V} = 0 = 0\ (I) \tag{2.28}$$

$$\text{div}\underline{\sigma} + \vec{f}^{\star} = \vec{0}\ (II) \tag{2.29}$$

$$\rho \frac{de}{dt} = \underline{\sigma} : \underline{\dot{\varepsilon}} + W - \text{div}\,\vec{q} = 0\ (III) \tag{2.30}$$

$$\rho \frac{ds}{dt} \geqq \frac{W}{T} - \text{div}\left(\frac{\vec{q}}{T}\right)\ (IV). \tag{2.31}$$

2.5 Behavior Laws

Let us consider $\begin{cases} \vec{x} = \vec{f}\,(\vec{x^0}, t) \\ \underline{F} = \underline{\nabla}_{\vec{x^0}}\, \vec{x} \\ T \end{cases}$ representing the motion law, the geometric transformation gradient and the temperature, respectively.

We generally accept the existence of laws called "constitutive laws" linking the stresses, the internal energy, the heat flux, and the entropy to the state of transformation and temperature at a given time:

$$\underline{\sigma}(t) = \underset{\tau \leq t}{\underline{\Sigma}}\,(\underline{F}(\tau), T(\tau)) \tag{2.32}$$

$$e(t) = \underset{\tau \leqq t}{E}\,(\underline{F}(\tau), T(\tau)) \tag{2.33}$$

$$\vec{q}\,(t) = \underset{\tau \leq t}{\vec{Q}}\,(\underline{F}(\tau), T(\tau)) \tag{2.34}$$

$$s(t) = \underset{\tau \leq t}{S}\,(\underline{F}(\tau), T(\tau)). \tag{2.35}$$

2.5.1 Clausius–Duhem Inequality

A question arises: is it possible to choose any values for the functionals $\underline{\Sigma}$, E , $\overrightarrow{Q}$, S?

Assuming that we take the rate law for $\overrightarrow{V}$ and the history of the transformation gradients $\underline{F}(\tau)$ and temperatures $T(t)$.

$\overrightarrow{V}$ is a known entity, Eq. (2.28) can be used to find ρ.

Equation (2.29) is satisfied when choosing the value of $\overrightarrow{f}$.

Equation (2.30) is satisfied when adopting W.

Equation (2.31) therefore gives us an inequality which must be verified for all the constitutive laws of Eq. (2.30).

We find the value of W and then substitute it into Eq. (2.31) so that

$$\rho T\frac{ds}{dt} \geq \rho\frac{de}{dt} - \underline{\sigma} : \dot{\underline{\varepsilon}} + \text{div}\overrightarrow{q} - T\text{div}\left(\frac{\overrightarrow{q}}{T}\right). \tag{2.36}$$

Noting that $\text{div}(\frac{\overrightarrow{q}}{T}) = \frac{1}{T}div\overrightarrow{q} + \overrightarrow{q}.\overrightarrow{grad}(\frac{1}{T})$, one obtains the dissipation Φ:

$$\Phi = \rho\left(T\frac{ds}{dt} - \frac{de}{dt}\right) + \underline{\sigma} : \frac{d\underline{\varepsilon}}{dt} - \overrightarrow{q}.\frac{\overrightarrow{\text{grad}}T}{T} \geqq 0, \tag{2.37}$$

which is called the Clausius–Duhem inequality. If one introduces the specific-free energy ψ defined by

$$\psi = e - Ts. \tag{2.38}$$

This leads to $T\frac{ds}{dt} - \frac{de}{dt} = -(\frac{d\psi}{dt} + s\frac{dT}{dt})$:

$$\rho\frac{de}{dt} = \underline{\sigma} : \dot{\underline{\varepsilon}} + W - \text{div}\overrightarrow{q}. \tag{2.39}$$

We gather the terms together in two types of dissipation $\Phi = \Phi_i + \Phi_{th}$:
– the intrinsic dissipation:

$$\Phi_i = \underline{\sigma} : \frac{d\underline{\varepsilon}}{dt} - \left(\frac{d\psi}{dt} + s\frac{dT}{dt}\right) \tag{2.40}$$

– the thermal dissipation:

$$\Phi_{th} = -\overrightarrow{q}.\frac{\overrightarrow{\text{grad}}T}{T}. \tag{2.41}$$

Classically, we make the hypothesis of uncoupling between the two types of dissipation (which is absolutely not compulsory), meaning that each of the dissipations separately must comply with the inequality:

$$\begin{cases} -\vec{q}.\overrightarrow{\text{grad}}T \geq 0 \\ \underline{\sigma} : \frac{d\underline{\varepsilon}}{dt} - (\frac{d\psi}{dt} + s\frac{dT}{dt}) \geq 0 \end{cases} \tag{2.42}$$

Interpretation:

• The first inequality implies that the "heat flux" vector $\vec{q}$ must form an obtus angle with the vector $\overrightarrow{gradT}$. This is satisfied for isotropic materials, which are governed by the Fourier law:

$$\vec{q} = -k\overrightarrow{\text{grad}}T. \tag{2.43}$$

If the material is not isotropic:

$$\vec{q} = -\underline{K}.\overrightarrow{\text{grad}}T, \tag{2.44}$$

where $\underline{K}$ constitutes the matrix (3×3) of calorific diffusivity which must be positive definite.

– the second inequality can be interpreted using the local state method.

The local state method postulates that the thermomechanics state of a material medium at a given point and at a given time is entirely defined by the known values, at that time, of a certain number of variables which depend only on the point in question (and on the physical problem at hand). These variables, called state variables s, are the observable variables and internal variables.

– observable variables:

temperature T, strain tensor $\underline{\varepsilon}$

– internal variables:

for dissipative phenomena, the state also depends on the previous history, represented by values at each moment, other variables, called internal variables.

Plasticity and viscoplasticity require the introduction of variables such as plastic deformation $\underline{\varepsilon}^{pl}$ with

$$\underline{\varepsilon} = \underline{\varepsilon}^{\text{el}} + \underline{\varepsilon}^{\text{pl}}, \tag{2.45}$$

where in the small deformations hypothesis, $\underline{\varepsilon}^{\text{el}}$ is the elastic strain tensor and $\underline{\varepsilon}^{\text{pl}}$ the plastic or viscoplastic strain tensor.

The description of the phase transformation will require the introduction of deformation associated with this process known as $\underline{\varepsilon}^{\text{tr}}$ with

$$\underline{\varepsilon} = \underline{\varepsilon}^{\text{el}} + \underline{\varepsilon}^{\text{tr}}. \tag{2.46}$$

One or many variables will be necessary, for example, z the volume fraction of martensite in austenite phase ($z\epsilon$ [0, 1]).

However, there is no objective method to choose the nature of the internal variables which are best adapted for such-and-such phenomenon: it is the necessity to describe the physics that prevails.

Let us consider the phase transformation:

$$\psi = \psi(\underline{\varepsilon}^{\text{el}} = \underline{\varepsilon} - \underline{\varepsilon}^{\text{tr}}, z, T) \tag{2.47}$$

which shows that

$$\frac{\partial \psi}{\partial \underline{\varepsilon}^{\text{el}}} = \frac{\partial \psi}{\partial \underline{\varepsilon}} - \frac{\partial \psi}{\partial \underline{\varepsilon}^{\text{tr}}}. \tag{2.48}$$

One use the Clausius–Duhem inequality with

$$\frac{d\psi}{dt} = \frac{\partial \psi}{\partial \underline{\varepsilon}^{\text{el}}} : \frac{d\boldsymbol{\varepsilon}^{\text{el}}}{dt} + \frac{\partial \psi}{\partial T}\frac{d\psi}{dt} + \frac{\partial \psi}{\partial z}\frac{dz}{dt} \tag{2.49}$$

$$\left[-\frac{\partial \psi}{\partial \underline{\varepsilon}^{\text{el}}}\right] : \frac{d\underline{\boldsymbol{\varepsilon}}^{\text{el}}}{dt} + \underline{\sigma} : \frac{d\underline{\varepsilon}^{tr}}{dt} - \rho\left(s + \frac{\partial \psi}{\partial T}\right)\frac{dT}{dt} - \rho\frac{\partial \psi}{\partial z}\frac{dz}{dt}. \tag{2.50}$$

A classical hypothesis permits to cancel certain terms of the inequality, and obtains

$$\underline{\sigma} = \frac{\partial \psi}{\partial \underline{\varepsilon}^{\text{el}}} = \frac{\partial \psi}{\partial \underline{\varepsilon}} - \frac{\partial \psi}{\partial \underline{\varepsilon}^{\text{tr}}};\; s = -\frac{\partial \psi}{\partial T}. \tag{2.51}$$

If we introduce the variables of "thermodynamic forces" associated with the internal variables, e.g., here

$$\Pi^f = -\rho\frac{\partial \psi}{\partial z}, \tag{2.52}$$

the Clausius–Duhem inequality is reduced to

$$dD = \Pi^f dz. \tag{2.53}$$

We shall see that knowing the specific free energy ψ and two dissipation pseudopotentials ϕ_1 and ϕ_2, dependent of phase transformation rates and of the kinetics of direct transformation A $\Rightarrow$ M and reverse transformation M $\Rightarrow$ A, is entirely sufficient to determine the pseudoelastic behavior of the SMA (see the works of J.J. Moreau about convex analysis Moreau 1970).

For elasto(visco)plasticity in the case of small perturbations:

$$\psi = \psi(\underline{\varepsilon}^{el} = \underline{\varepsilon} - \underline{\varepsilon}^{\text{pl}}, V_K, T), \tag{2.54}$$

where $\mathbf{V}_{\text{K}}$ internal variable replaces the variable z (associated to phase transformation). One can denote $\underline{V_0} = \underline{\varepsilon}^{\text{pl}}$ and the associated variable is

$$\underline{A_0} = \rho\frac{\partial \psi}{\partial \underline{\varepsilon}^{\text{pl}}} = -\underline{\sigma}. \tag{2.55}$$

To work with a potential in terms of stress rather than deformations, one introduce the Gibbs specific free enthalpy G deduced from ψ by partial Legendre transformation:

$$G\left(\underline{\sigma}, V_K, T\right) = sup_{\mathbf{E}}\left[\frac{1}{\rho}\underline{\sigma} : \underline{\varepsilon} - \psi = \psi\left(\underline{\varepsilon}, V_K, T\right)\right]. \tag{2.56}$$

Thus, a second expression of state laws is equivalent to the first:

$$\underline{\varepsilon} = \rho\frac{\partial \mathrm{G}}{\partial \underline{\sigma}},\ s = \frac{\partial G}{\partial T},\ \underline{\mathrm{V}}_{\mathrm{K}} = -\rho\frac{\partial G}{\partial \underline{\mathrm{A}}_{\mathrm{K}}}. \tag{2.57}$$

2.6 Heat Equation

Let us consider the energy conservation equation:

$$\rho\dot{e} = \underline{\sigma} : \underline{\dot{\varepsilon}} + W - \mathrm{div}\,\vec{q}\,, \tag{2.58}$$

and replace $\rho\dot{e}$ by its expression $\rho\dot{e} = \rho\dot{\psi} + \rho\dot{T}s + \rho T\dot{s}$ derived from $e = \psi + Ts$ and $\dot{\psi}$ by its expression as a function of the state variables, e.g.,

$$\rho\dot{e} = \rho\left[\frac{1}{\rho}\underline{\sigma} : \underline{\dot{\varepsilon}} - s\dot{T} + \frac{1}{\rho}\underline{A_K}.\underline{\dot{V}_K}\right] + \rho T\dot{s} + \rho s\dot{T}.$$

One obtains

$$\rho T\dot{s} = W - \mathrm{div}\,\vec{q} - \underline{A_K}.\underline{\dot{V}_K}\ with\ s = -\frac{\partial\psi\left(\underline{\varepsilon}, T, \underline{V_K}\right)}{\partial T}.$$

We can write

$$\dot{s} = -\frac{\partial^2\psi}{\partial\underline{\varepsilon}\partial T} : \underline{\dot{\varepsilon}} - \frac{\partial^2\psi}{\partial T^2}\dot{T} - \frac{\partial^2\psi}{\partial \underline{V_K}\partial T} : \underline{\dot{V}_K} = -\frac{1}{\rho}\frac{\partial\underline{\sigma}}{\partial T} : \underline{\dot{\varepsilon}} + \frac{\partial s}{\partial T}\dot{T} - \frac{1}{\rho}\frac{\partial \underline{\underline{A_K}}}{\partial T} : \underline{\dot{V}_K}.$$

By introducing the specific heat defined by

$$C_v = T\frac{\partial s}{\partial T} = -T\frac{\partial^2\psi}{\partial T^2}$$

and taking into account the Fourier equation, we obtain

$$\rho C_v\dot{T} = k\Delta T + W - \underline{A_K}.\underline{\dot{V}_K} + T\left[\frac{\partial\underline{\sigma}}{\partial T} : \underline{\dot{\varepsilon}} + \frac{\partial \underline{\underline{A_K}}}{\partial T} : \underline{\dot{V}_K}\right]. \tag{2.59}$$

Let recall that the internal variables A_K can be scalar, vectorial, or tensorial (for example, second-order tensor) as we choose to write.

The heat equation reminds us that there is no strictly isothermal tests but the tests where the temperature is kept constant at the boundary of the sample. Similarly, there is no "stricto sensu" adiabatic tests

Besides following the rate solicitation on a mechanical loading, we define two bounds: isothermal and adiabatic for loading very slow and very fast, respectively.

2.7 Synthesis and Chapter Conclusion

The tools required for modeling the nonlinear mechanical behavior of materials were given.

The definitions of vectorial displacement, strain and stress, with exercises, are easy to integrate.

The "thermodynamics of irreversible processes" framework laid down at the outset will help us to write the behavior laws of different materials under complex loading.

It may be difficult to understand for the uninformed reader. But it will clarify, in later chapters, particularly those on linear and nonlinear elastic behaviors.

Chapter 3
Linear Elastic Behavior, Thermoelasticity

Abstract This chapter is devoted to linear elasticity and thermoelasticity in the «small strains» hypothesis. At first, the isotropic behavior is studied and the planar linear elasticity is introduced (Airy function). In the second part, the anisotropic elastic behavior is investigated for composite materials.

3.1 Introduction

The elastic behavior is defined at the beginning by the existence of a domain in the stress space, called «elasticity domain» whose size and geometrical shape depend on temperature, mechanical loading, and obviously on the chosen material (Criteria of «yield elasticity domain» will be defined in Chap. 4).

The material will have a linear or a nonlinear elastic behavior when the working point (stress state for example) is inside the domain.

In a characteristic way, the elastic behavior represented by the curve $\sigma \Longleftrightarrow \varepsilon^{el}$ is shown on the Fig. 3.1 for a tensile test.

Starting from a stress-free state, the mechanical loading produces a reversible increase of the axial deformation. The unloading exactly reproduces the loading path at reversion, returning to the origin O as the applied stress returns to 0.

We can say that the material has the "memory" of a single state, the nondeformed natural state (e.g., unstressed).

In other words, a solid is elastic if its free energy is only function of the elastic strain tensor $\underline{\varepsilon}^{el}$ and the dissipation potential is equal to zero (cf. Chap. 2) (Nguyen 1982).

3.2 Linear Elastostaticity Isotropic and Homogeneous

When resuming the method of «local state » developed in Chap. 2; the state variables are the temperature and the strain. If we consider the isothermal transformation, only

C. Lexcellent, *Linear and Non-linear Mechanical Behavior of Solid Materials*, DOI 10.1007/978-3-319-55609-3_3

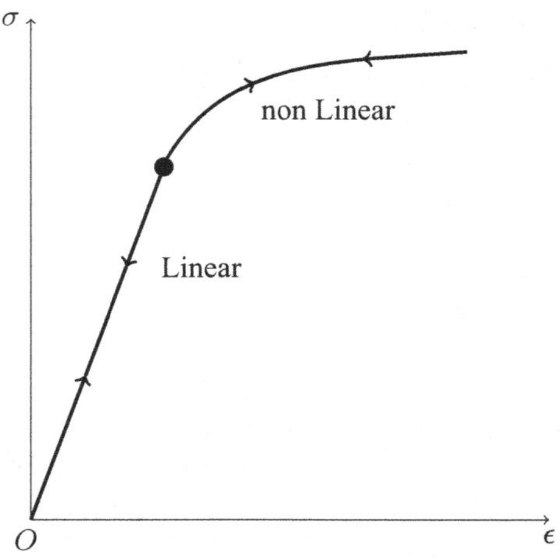

Fig. 3.1 Elastic behavior for simple tension

$\underline{\varepsilon} = \underline{\varepsilon}^{el}$ acts as a state variable; the temperature occurs as a parameter whose elasticity coefficients depend.

In this case, the free energy is written:

$$\rho\psi = \frac{1}{2}\underline{\varepsilon}^{el} : \underline{\underline{C}} : \underline{\varepsilon}^{el} \text{ or again } \rho\psi = \frac{1}{2}\varepsilon_{ij}^{el} C_{ijkl} \varepsilon_{kl}^{el}, \tag{3.1}$$

where $\underline{\underline{C}}$ is the fourth-order tensor of elastic stiffness. By definition of the associated variable, the stress tensor $\underline{\sigma}$ is the derivative of a potential ψ which gives the following state law:

$$\underline{\sigma} = \rho \frac{\partial \psi}{\partial \underline{\varepsilon}^{el}} = \underline{\underline{C}} : \underline{\varepsilon}^{el} \text{ or } \sigma_{ij} = \rho \frac{\partial \psi}{\partial \varepsilon_{ij}^{el}} = C_{ijkl} \varepsilon_{kl}^{el} \tag{3.2}$$

The hypothesis of elastic isotropy (the same mechanical properties in all directions around a point) is well satisfied for polycrystalline materials and ceramics. For single crystals, composites, and wood, particular anisotropy theories will be used. In the case of isotropy, there are only two independent elastic coefficients and the elastic moduli tensor becomes:

$$\underline{\underline{C}} = \lambda \left(\underline{1} \otimes \underline{1}\right) + 2\mu \underline{\underline{1}}$$

In this case, the behavior model is written as:

$$\underline{\varepsilon}^{el} \Longrightarrow \underline{\sigma}\ \underline{\sigma} = \lambda \left(\text{tr}\underline{\varepsilon}^{el}\right) \underline{1} + 2\mu \underline{\varepsilon}^{el}, \tag{3.3}$$

where λ and μ are called Lamé coefficients. They have the same dimensions as the stress σ_{ij}, that are expressed in N.m^{-2} (or in MPa).

The nonlinearity (see Fig. 3.1) may correspond to an energy $\rho\psi$ of an order in $\underline{\varepsilon}^{el}$ more than 2. In general, the situation is more complex and will not be considered here.

The anisotropic linear elastic case will be studied for the composite materials. The number of independent elastic coefficients is completely dependent on the degree of symmetry of the material. A fully anisotropic material has 21 independent constants to determine what is a "sacred" design of experiments to achieve this. These materials exist in nature; in the Vosges as minerals. (see for example "Plagioclase feldspars Elasticity of plagioclase feldspars" J.M. Brown R.J. Angel and N. Ross 2006) (Brown et al. 2006).

By projection of the isotropic laws, stress is written as:

$$\begin{aligned}\sigma_{11} &= \lambda \mathrm{tr}\underline{\varepsilon}^{el} + 2\mu\varepsilon_{11}^{el},\ \sigma_{12} = 2\mu\varepsilon_{12}^{el}\\ \sigma_{22} &= \lambda \mathrm{tr}\underline{\varepsilon}^{el} + 2\mu\varepsilon_{22}^{el},\ \sigma_{23} = 2\mu\varepsilon_{23}^{el}\\ \sigma_{33} &= \lambda \mathrm{tr}\underline{\varepsilon}^{el} + 2\mu\varepsilon_{33}^{el},\ \sigma_{31} = 2\mu\varepsilon_{31}^{el}\end{aligned}$$

We can reverse the relationship and get $\underline{\sigma} \Longrightarrow \underline{\varepsilon}^{el}$ with λ et μ.

In fact:

$$\sigma_{11} + \sigma_{22} + \sigma_{33} = \mathrm{tr}\left(\underline{\sigma}\right) = (3\lambda + 2\mu)\,\mathrm{tr}\left(\underline{\varepsilon}^{el}\right)$$

that gives:

$$\underline{\sigma} \Longrightarrow \underline{\varepsilon}^{el}\ \underline{\varepsilon}^{el} = \frac{\underline{\sigma}}{2\mu} - \frac{\lambda}{2\mu\,(3\lambda + 2\mu)}\mathrm{tr}\left(\underline{\sigma}\right)\underline{1} \tag{3.4}$$

In order to establish the behavior laws with E the Young modulus and ν the Poisson coefficient, elastic constants that are more common for materials mechanics, consider the uniaxial tensile test (along $\overrightarrow{x_1}$). The slope of the initial straight line $\left(\sigma_{11}, \varepsilon_{11}^{el}\right)$ constitutes the Young modulus.

Thus, we write Hooke's Law (1676):

$$\sigma_{11} = E\varepsilon_{11}^{el} \tag{3.5}$$

$$\underline{\sigma} = \begin{bmatrix}\sigma_{11} & 0 & 0\\ 0 & 0 & 0\\ 0 & 0 & 0\end{bmatrix} \Longrightarrow \underline{\varepsilon}^{el} = \begin{bmatrix}\varepsilon_{11}^{el} & 0 & 0\\ 0 & \varepsilon_{22}^{el} & 0\\ 0 & 0 & \varepsilon_{33}^{el}\end{bmatrix}$$

The specimen-sectional reduction is handled by the Poisson's ratio:

$$\nu = -\frac{\varepsilon_{22}^{el}}{\varepsilon_{11}^{el}} = -\frac{\varepsilon_{33}^{el}}{\varepsilon_{11}^{el}} \tag{3.6}$$

These two experimental information suggest introducing a behavior law in the form:

$$\underline{\sigma} \Longrightarrow \underline{\varepsilon}^{el} \; \underline{\varepsilon}^{el} = \frac{1+\nu}{E}\underline{\sigma} - \frac{\nu}{E}\text{tr}\left(\underline{\sigma}\right)\underline{1} \tag{3.7}$$

We verify that «it works» for simple tension test.

The inversion of the relationship is not a problem. Indeed, previous relationships give:

$$\text{tr}\left(\underline{\varepsilon}^{el}\right) = \frac{1-2\nu}{E}\text{tr}\left(\underline{\sigma}\right)$$

So:

$$\underline{\varepsilon}^{el} \Longrightarrow \underline{\sigma} \; \underline{\sigma} = \frac{E}{1+\nu}\left(\underline{\varepsilon}^{el}\right) + E\frac{\nu}{(1+\nu)(1-2\nu)}\text{tr}\left(\underline{\varepsilon}^{el}\right)\underline{1} \tag{3.8}$$

A comparison between the four formulations provides:

$$(E,\nu) \Longrightarrow (\lambda,\mu) \;\; \mu = \frac{E}{2(1+\nu)}, \;\; \lambda = \frac{\nu}{(1+\nu)(1-2\nu)} \tag{3.9}$$

$$(\lambda,\mu) \Longrightarrow (E,\nu) \;\; E = \frac{\mu(3\lambda+2\mu)}{(\lambda+\mu)}, \;\; \nu = \frac{\lambda}{2(\lambda+\mu)} \tag{3.10}$$

Note that μ is called shear modulus (or torsional modulus $G = \mu$). In fact:

$$\underline{\sigma} = \begin{bmatrix} 0 & \sigma_{12} & 0 \\ \sigma_{21} = \sigma_{12} & 0 & 0 \\ 0 & 0 & 0 \end{bmatrix} \Longrightarrow \underline{\varepsilon}^{el} = \begin{bmatrix} 0 & \varepsilon_{12}^{el} & 0 \\ \varepsilon_{21}^{el} = \varepsilon_{12}^{el} & 0 & 0 \\ 0 & 0 & 0 \end{bmatrix}$$

with:

$$\varepsilon_{12}^{el} = \frac{\sigma_{12}}{2\mu} \Longrightarrow \mu = \frac{\sigma_{12}}{2\varepsilon_{12}^{el}} = \frac{\sigma_{12}}{\gamma_{12}^{el}},$$

where $\gamma_{12}^{el} = 2\varepsilon_{12}^{el}$ is the distortion of the right angle $\left(\overrightarrow{x_1}, \overrightarrow{x_2}\right)$. This term γ_{ij}^{el} with $i \neq j$ is often used for composites materials.

Among the four materials parameters (λ, μ, E, ν), only the physical interpretation of λ is not obvious.

A very important fact for the plasticity is to decompose the symmetric tensors $\underline{\sigma}$ and $\underline{\varepsilon}^{el}$ into its isotropic part and its deviatoric one, that allows the introduction of the bulk modulus K.

$$\underline{\sigma} = \frac{1}{3}\text{tr}\left(\underline{\sigma}\right)\underline{1} + \text{dev}\left(\underline{\sigma}\right) \tag{3.11}$$

The partition has been carried out such that $\text{tr}\left(\text{dev}\left(\underline{\sigma}\right)\right) = 0$.

Moreover:

$$\underline{\varepsilon}^{el} = \frac{1}{3}\text{tr}\left(\underline{\varepsilon}^{el}\right)\underline{1} + \text{dev}\left(\underline{\varepsilon}^{el}\right)$$

We obtain:

$$\begin{array}{rl} \text{tr}\left(\underline{\varepsilon}^{el}\right) = \frac{1-2\nu}{E}\text{tr}\left(\underline{\sigma}\right) & \text{tr}\left(\underline{\sigma}\right) = 3K\,\text{tr}\left(\underline{\varepsilon}^{el}\right) \\ \text{dev}\left(\underline{\varepsilon}^{el}\right) = \frac{1+\nu}{E}\text{dev}\left(\underline{\sigma}\right) & \text{dev}\left(\underline{\sigma}\right) = 2\mu\text{dev}\left(\underline{\varepsilon}^{el}\right) \end{array} \tag{3.12}$$

So:

$$K = \frac{E}{3(1-2\nu)} = \frac{3\lambda + 2\mu}{3}$$

As the modulus K must be positive, since the pressure exerted on an elastic sphere causes a concomitant decrease in the radius of the sphere, this means that:

$$0 < \nu < 0.5 \ \textit{with évidently} \ E > 0,\ G = \mu > 0$$

and also $3\lambda + 2\mu > 0$.

Note that the incompressibility corresponds to $\nu = 0.5$.

3.2.1 *General Equations Applied to Linear Elastic Isotropic Behavior*

3.2.1.1 Equilibrium Equations in Terms of Displacement $\vec{u}$ or Navier Equations

We have:

$$\left\{\begin{array}{c} \text{div}\underline{\sigma} + \overrightarrow{f^\star} = \vec{0} \\ \underline{\sigma} = \lambda\left(\text{div}\,\vec{u}\right)\underline{1} + 2\mu\underline{\varepsilon}^{el} \\ \underline{\varepsilon}^{el} = \left(\underline{\nabla_{x^0}\vec{u}}\right)^s \end{array}\right\} \ or \ \left\{\begin{array}{c} \sigma_{ij,j} + f_i^\star = 0 \\ \sigma_{ij} = \lambda\left(\text{div}\,\vec{u}\right)\delta_{ij} + 2\mu\varepsilon_{ij}^{el} \\ \varepsilon_{ij}^{el} = \frac{1}{2}\left(u_{i,j} + u_{j,i}\right) \end{array}\right\}$$

For simplicity, let i = 1 (Then we generalize the result)

$$\sigma_{11,1} + \sigma_{12,2} + \sigma_{13,3} + f_1^\star = 0$$

$$\begin{array}{r} \sigma_{11,1} = \lambda\left(\text{div}\,\vec{u}\right)_{,1} + 2\mu u_{1.11} \\ +\sigma_{12,2} = \mu\left(u_{1,2} + u_{2,1}\right)_{,2} \\ +\sigma_{13,3} = \left(u_{1,3} + u_{3,1}\right)_{,3} \\ \Longrightarrow -f_1^\star = (\lambda + \mu)\left(\text{div}\,\vec{u}\right)_{,1} + \mu\Delta u_1 \end{array}$$

with $\text{div}\,\vec{u} = u_{1,1} + u_{2,2} + u_{3,3} = \text{tr}\left(\underline{\varepsilon}^{el}\right)$.

$$\overrightarrow{f^\star} + (\lambda + \mu)\,\overrightarrow{\text{grad}}\left(\text{div}\,\overrightarrow{u}\right) + \mu\Delta\overrightarrow{u} = \overrightarrow{0} \tag{3.13}$$

Noting that:

$$\Delta\overrightarrow{u} = \overrightarrow{\text{grad}}\left(\text{div}\,\overrightarrow{u}\right) - \mu\overrightarrow{\text{rot}}\left(\overrightarrow{\text{rot}}\,\overrightarrow{u}\right)$$

The equilibrium equations in terms of displacement can also be written as:

$$\overrightarrow{f^\star} + (\lambda + 2\mu)\,\overrightarrow{\text{grad}}\left(\text{div}\,\overrightarrow{u}\right) - \mu\overrightarrow{rot}\left(\overrightarrow{rot}\,\overrightarrow{u}\right) = \overrightarrow{0} \tag{3.14}$$

They can afford to separate displacements at $\text{div}\,\overrightarrow{u} = 0$ (said indivergents) of those at $\overrightarrow{\text{rot}}\left(\overrightarrow{u}\right) = \overrightarrow{0}$, (said irrotationals); but this distinction is most operative in fluid mechanics.

Navier equations are useful when we research a solution of the elasticity problem by identifying $\overrightarrow{u}\,(x_1, x_2, x_3)$.

This method is called the Lamé and Clapeyron method. It will be illustrated in Exercise 3.1.

3.2.2 *Exercise 3.1*

A spherical tank under internal and external pressures (Fig. 3.2) (Mandel 1974).

A hollow sphere (interval between two concentric spheres of radii r_0 and r_1) is subjected to different uniform normal pressures (p_0 *et* p_1) on inside and outside.

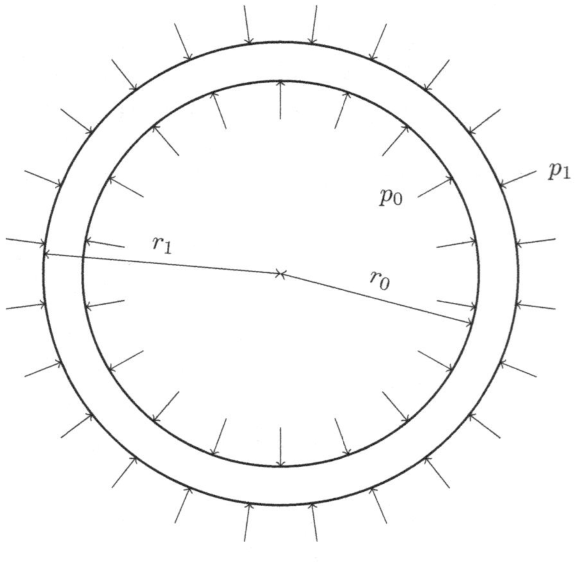

Fig. 3.2 Hollow sphere under internal and external pressures

Volume forces are neglected and the problem is considered as static ($\overrightarrow{\gamma} = \overrightarrow{0}$).
(1) We write the Cartesian coordinates of the displacement $\overrightarrow{u}\,(x_1, x_2, x_3)$ under the form:

$$u_i = g\,(r)\, x_i \ with\ r = \left(x_1^2 + x_2^2 + x_3^2\right)^{\frac{1}{2}}$$

Succinctly justify this choice.

Let us determine the function $g\,(r)$ (in the expression of $g\,(r)$, there are two integrations constants called α and β).

(2) Calculate the strain tensor $\underline{\varepsilon}^{el}$ then the stress is given by $\underline{\sigma}$ (only elastic constants used will be the bulk modulus $K = 3\lambda+2\mu/3$ and the Coulomb modulus μ).

The tensor $\underline{\sigma}$ is of revolution about $\overrightarrow{r}$. Let us find the eigenstress σ_r associated with $\overrightarrow{r}$: eigen direction associated. We can do the calculation at a point $(x_1 = r,\ x_2 = 0,\ x_3 = 0)$ and generalize the result and deduce the values σ_r and σ_θ.

(3) Writing the boundaries conditions, let us determine the two constants α and β.

Correction:

(1) The choice of u_i is natural because of the spherical symmetry of the problem around $\overrightarrow{r}$.

The equilibrium equations give $\overrightarrow{\mathrm{rot}}\,\overrightarrow{u} = \overrightarrow{0}\ \forall g\,(r)$ and deliver the condition $\mathrm{div}\,\overrightarrow{u} = cte$ that we choose equal to 3α.

The equation to solve is $\mathrm{div}\,\overrightarrow{u} = 3g\,(r) + rg^{'}\,(r) = 3\alpha$.

Solution: $g\,(r) = \alpha + \frac{\beta}{r^3}$.

(2) Strain tensor:

$$\underline{\varepsilon}^{el} = \begin{bmatrix} g\,(r) + \frac{x_1^2}{r} g'\,(r) & 0 & 0 \\ 0 & g\,(r) + \frac{x_2^2}{r} g'\,(r) & 0 \\ 0 & 0 & g\,(r) + \frac{x_3^2}{r} g'\,(r) \end{bmatrix}_{x_1, x_2, x_3}$$

becomes:

$$\underline{\varepsilon}^{el} = \begin{bmatrix} g\,(r) + rg'\,(r) = \alpha - \frac{2\beta}{r^3} & 0 & 0 \\ 0 & g\,(r) = \alpha + \frac{\beta}{r^3} & 0 \\ 0 & 0 & g\,(r) = \alpha + \frac{\beta}{r^3} \end{bmatrix}_{x_1 = r, x_2 = 0, x_3 = 0}$$

Let us design σ_{rr} the eigenstress associated to $\overrightarrow{r}$ and $\sigma_{\theta\theta} = \sigma_{\varphi\varphi}$ the eigenstress perpendicular to $\overrightarrow{r}$.

We have:

$$\sigma_{rr} = A - \frac{2B}{r^3} \quad \sigma_{\theta\theta} = \sigma_{\varphi\varphi} = A + \frac{B}{r^3}$$

let us assign:

$$A = 3K\alpha = (3\lambda + 2\mu)\,\alpha, \;\; B = 2\mu\beta$$

(3) The values of A et B are obtained by writing that $\sigma_{rr} = -p_0$ for $r = r_0$ and $\sigma_{rr} = -p_1$ for $r = r_1$.

Radial displacement; $u_r(r) = \alpha r + \frac{\beta}{r^2}$.

3.2.2.1 Compatibility Equations in Stress Terms (Said Beltrami-Mitchell Equations)

The aim is to examine how to transform the compatibility equations in strain terms by introducing behavior laws (and the equilibrium Cauchy equations).

Under contracted form, compatibility equations are:

$$\varepsilon^{el}_{il,jm} - \varepsilon^{el}_{jl,im} - \varepsilon^{el}_{im,jl} + \varepsilon^{el}_{jm,il} = 0 \; \forall \, i, j, l, m = 1, 2, 3 \tag{3.15}$$

One introduces the behavior laws:

$$\varepsilon^{el}_{ij} = \frac{1+\nu}{E}\sigma_{ij} - \frac{\nu}{E}\mathrm{tr}\left(\underline{\sigma}\right)\delta_{ij}$$

with:

$$\sigma_{ij,j} + f_i^{\bigstar} = 0.$$

After calculations, one obtains:

$$\Delta\sigma_{ij} + \frac{1}{1+\nu}\mathrm{tr}\left(\underline{\sigma}\right)_{,ij} = -\delta_{ij}\frac{\nu}{1-\nu}\mathrm{div}\overrightarrow{f^{\bigstar}} - f^{\bigstar}_{i,j} - f^{\bigstar}_{j,i} \tag{3.16}$$

When $\overrightarrow{f^{\bigstar}} = \overrightarrow{C^{te}}$, so:

$$\Delta\sigma_{ij} + \frac{1}{1+\nu}\mathrm{tr}\left(\underline{\sigma}\right)_{,ij} = 0 \tag{3.17}$$

Use is made of these equations when a direct stress calculation is performed (Beltrami method).

Suppose we study the equilibrium of a body "simply convex" and that in any point of the border, are given the surface densities of strength.

The Exercise 3.2 gives the solution of this Beltrami problem as an example, a deformable cube placed on a dimensionally stable frame.

The Exercise 3.2 bis is a variant of the same problem by replacing the cube by a cylinder and the Cartesian coordinates by the cylindrical coordinates.

Fig. 3.3 Cube placed on an nondeformable support

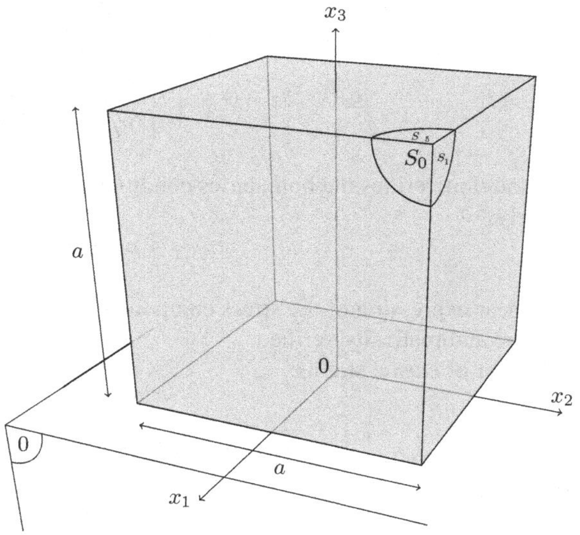

3.2.3 *Exercise 3.2*

Cube deformation under its own weight (Fig. 3.3).

Statement: On a dimensionally stable frame (D), a cube is placed whose side is a. This cube weighting of ρ density is homogeneous, isotropic, linear elastic, of Young's modulus E and Poisson ratio ν. The gravity acceleration is $\overrightarrow{g} = -g\overrightarrow{x_3}$. (Air pressure shall be ignored on the walls).
(1) After making an inventory of external mechanical stresses, let us write the boundary conditions on the faces:

$$S_0\,(x_1 = {}^a\!/2)\,,\ S_1\,(x_1 = -{}^a\!/2)\,,\ S_2\,(x_2 = {}^a\!/2)\,,\ S_3\,(x_2 = -{}^a\!/2)\,,\ S_4\,(x_3 = 0)\ \ et\ S_5\,(x_3 = a)\ .$$

(2) Let us write the equilibrium equations and deduce the stress tensor $\underline{\sigma}$
(3) Let us deduce the elastic strains tensor $\underline{\varepsilon}^{el}$ and the displacement field $\overrightarrow{u}$.
Studying transformed from an initial surface $x_3 = ka\ \ (0 \leq k \leq 1)$.

Correction:

Boundaries conditions: On S_0 : $\overset{x_1}{\overrightarrow{T}}\,({}^a\!/2, x_2, x_3) = \sigma_{11}\,({}^a\!/2, x_2, x_3)\,\overrightarrow{x_1} + \sigma_{21}\,({}^a\!/2, x_2, x_3)\,\overrightarrow{x_2} + \sigma_{31}\,({}^a\!/2, x_2, x_3)\,\overrightarrow{x_3} = \overrightarrow{0}$.

So the surfaces S_i $i = 1, 2, 3, 5$ are free. Only the surface S_4 is stressed (action of the support D on the cube):

$$\overset{-x_3}{\overrightarrow{T}}\,(x_1, x_2, 0) = -\sigma_{13}\,(x_1, x_2, 0)\,\overrightarrow{x_1} - \sigma_{23}\,(x_1, x_2, 0)\,\overrightarrow{x_2} - \sigma_{33}\,(x_1, x_2, 0)\,\overrightarrow{x_3} = \rho g a \overrightarrow{x_3}.$$

We suppose that each point of the cube is under uniaxial compression in the direction $\overrightarrow{x_3}$:

$$\underline{\sigma}(x_1, x_2, x_3) = \begin{bmatrix} 0 & 0 & 0 \\ 0 & 0 & 0 \\ 0 & 0 & \rho g (x_3 - a) \leq 0 \end{bmatrix}$$

This solution verifies the boundaries conditions, the equilibrium equations and particularly:

$$\sigma_{33,3} - \rho g = 0$$

The linear expression of the stress component σ_{33} indicates that the Beltrami equations are automatically verified.
(3) Tensor of elastic strains:

$$\underline{\varepsilon}^{el} = \begin{bmatrix} -\frac{\nu\rho g}{E}(x_3 - a) & 0 & 0 \\ 0 & -\frac{\nu\rho g}{E}(x_3 - a) & 0 \\ 0 & 0 & \frac{\rho g}{E}(x_3 - a) \end{bmatrix}$$

The rotation is equal to 0 at the point O (0, 0, 0) e.g.:

$$\vec{\alpha}(0, 0, 0) = \vec{0} \iff \omega_{ij}(0, 0, 0) = 0 \; \forall i, j$$

We obtain:

$$\underline{\omega} \begin{Bmatrix} \omega_{12} = -\omega_{21} = 0 \\ \omega_{23} = -\omega_{32} = -\frac{\nu\rho g}{E} x_2 \\ \omega_{31} = -\omega_{13} = \frac{\nu\rho g}{E} x_1 \end{Bmatrix}$$

with the condition $\vec{u}(0, 0, 0) = \vec{0}$.

$$\vec{u} \begin{Bmatrix} u_1 = -\frac{\nu\rho g}{E} x_1 (x_3 - a) \\ u_2 = -\frac{\nu\rho g}{E} x_2 (x_3 - a) \\ u_3 = \frac{\rho g}{2E}\left(\left(x_3^2 - 2ax_3\right) + \nu\left(x_1^2 + x_2^2\right)\right) \end{Bmatrix}.$$

Transformed of the initial surface $x_3 = ka \;\; (0 \leq k \leq 1)$.

For a steel cube: $\rho = 8000\,\text{kg/m}^3$; g = 10 m/s^2; E = 2.10^{11}N/m^{2}and $\nu = 0.3$ $\epsilon = \frac{\nu\rho g}{E} a$.
Let:

$$x_i' = x_i + u_i$$

$$\begin{aligned} x_1' &= x_1 (1 - \epsilon \{k - 1\}) \\ x_2' &= x_2 (1 - \epsilon \{k - 1\}) \\ x_3' &= a \left\{k + \frac{\epsilon (k^2 - 2k)}{2} + \frac{\epsilon\upsilon}{2} \frac{(x_1^2 + x_2^2)}{a^2}\right\} \end{aligned}$$

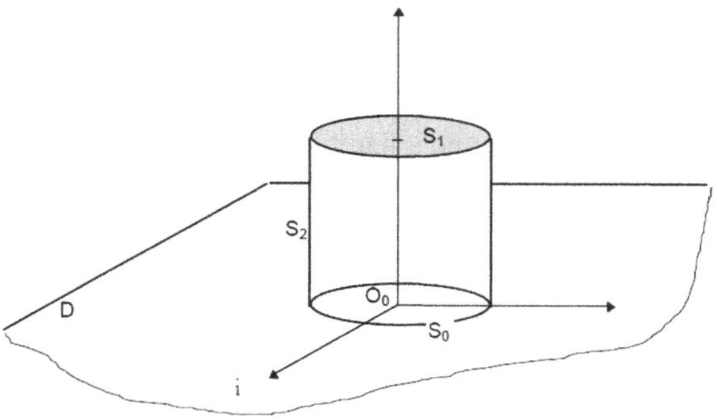

Fig. 3.4 Weighting cylinder of height h and of radius R placed on a nondeformable support

A right section $x_3 = ka$ becomes:

$$x_3^{'} = ka\left(1-\epsilon\left(1-\frac{k}{2}\right)\right)+\frac{\epsilon\nu}{2}\left(x_1^2+x_2^2\right) \text{ at the first order.}$$

This equation represents a paraboloïd of $\overrightarrow{x_3}$ axis and indicates that the solution is true in the agreement with the Saint-Venant's principle.

3.2.4 *Exercise 3.2 Bis*

Cylinder deformation under its own weight (Fig. 3.4).

It constitutes the same exercise in cylindric coordinates.

Comment: the same physical problem is to solve with cylindrical coordinates instead of Cartesian ones.

$$\underline{\sigma}\left(M\left(r,\theta,z\right)\right)=\begin{bmatrix}0\;0 & 0\\ 0\;0 & 0\\ 0\;0\;\rho g\left(z-h\right)\le 0\end{bmatrix}$$

$$\underline{\varepsilon}^{el}\left(M\left(r,\theta,z\right)\right)=\begin{bmatrix}-\frac{\nu\rho g}{E}\left(h-z\right) & 0 & 0\\ 0 & -\frac{\nu\rho g}{E}\left(h-z\right) & 0\\ 0 & 0 & \frac{\rho g}{E}\left(h-z\right)\end{bmatrix}$$

$$\vec{u}\left(M\left(r,\theta,z\right)=\left\{\begin{matrix}u_r=-\frac{\nu\rho g}{E}r\left(h-z\right)\\ u_\theta=0\\ u_z=\frac{\rho g}{2E}\left[\left(z^2-2hz\right)+\nu r^2\right]\end{matrix}\right\}\right.$$

3.3 Plane Linear Elasticity

There are two extreme geometries with loads "ad hoc" which justify this section: a very long cylinder and a flat thin plate.

The cylinder is said «infinite» (free ends and thus without boundary conditions) and is loaded in the plane Mx_1x_2 in the same way whatever x_3.

In this case, the displacement is plane:

$$\vec{u}\,(x_1, x_2)\begin{Bmatrix} u_1 = u_1\,(x_1, x_2) \\ u_2 = u_2\,(x_1, x_2) \\ u_3 = 0 \end{Bmatrix}$$

This results in a plane strain condition:

$$\underline{\varepsilon}^{el} = \begin{bmatrix} \varepsilon_{11}^{el} = u_{1,1} & \varepsilon_{12}^{el} = \frac{1}{2}\left(u_{1,2} + u_{2,1}\right) & 0 \\ \varepsilon_{21}^{el} = \varepsilon_{12} & \varepsilon_{22}^{el} = u_{2,2} & 0 \\ 0 & 0 & 0 \end{bmatrix}$$

but the reverse is not true.

The relation $\varepsilon_{13}^{el} = \varepsilon_{23}^{el} = \varepsilon_{33}^{el} = 0$ gives a displacement outside of the plane Mx_1x_2:

$$\vec{u}\begin{Bmatrix} u_1 = -x_3 w_{0,1}\,(x_1, x_2) + u_1^0\,(x_1, x_2) \\ u_2 = -x_3 w_{0,2}\,(x_1, x_2) + u_2^0\,(x_1, x_2) \\ u_3 = w_0\,(x_1, x_2) \end{Bmatrix} \;\forall\, w_0\,(x_1, x_2)$$

A flat plate whose thickness 2h is small compared to the lateral dimensions, loaded in the directions $\vec{x_1}$ et $\vec{x_2}$.

In this case, one can call of a state of stress (mean) plane.

Recall that behavior laws in tridimensional situation are valid and can therefore be used in bidimensional situation.

They lead, in general, to a plane strain condition which does not imply a plane stress state (and vice versa).

Let:

$$\underline{\varepsilon}^{el}\,(x_1, x_2) = \begin{bmatrix} \varepsilon_{11} & \varepsilon_{12} & 0 \\ \varepsilon_{21} & \varepsilon_{22} & 0 \\ 0 & 0 & 0 \end{bmatrix} \Rightarrow \underline{\sigma}\,(x_1, x_2) = \begin{bmatrix} \sigma_{11} & \sigma_{12} & 0 \\ \sigma_{21} & \sigma_{22} & 0 \\ 0 & 0 & \sigma_{33} \end{bmatrix}$$

$$\varepsilon_{33}^{el} = 0 \Rightarrow \sigma_{33} = \nu\,(\sigma_{11} + \sigma_{22})\,.$$

With $\underline{\sigma} = \lambda\left(\varepsilon_{11}^{el} + \varepsilon_{22}^{el}\right) + 2\mu\underline{\varepsilon}^{el}$

$$\sigma_{33} = 0 \; if\ and\ only\ if \; \varepsilon_{11}^{el} + \varepsilon_{22}^{el} = 0$$

Likewise $\underline{\sigma}\,(x_1, x_2) = \begin{bmatrix} \sigma_{11} & \sigma_{12} & 0 \\ \sigma_{21} & \sigma_{22} & 0 \\ 0 & 0 & 0 \end{bmatrix} \Rightarrow \underline{\varepsilon}^{el}\,(x_1, x_2) = \begin{bmatrix} \varepsilon_{11}^{el} & \varepsilon_{12}^{el} & 0 \\ \varepsilon_{21}^{el} & \varepsilon_{22}^{el} & 0 \\ 0 & 0 & \varepsilon_{33}^{el} \end{bmatrix}$

With:

$$\underline{\varepsilon}^{el} = \frac{1+\nu}{E}\underline{\sigma} - \frac{\nu}{E}\mathrm{tr}\left(\underline{\sigma}\right)\underline{1} \Rightarrow \varepsilon_{33}^{el} = -\frac{\nu}{E}\left(\sigma_{11} + \sigma_{22}\right)$$

$$\varepsilon_{33}^{el} = 0 \; if \; and \; only \; if \; \sigma_{11} + \sigma_{22} = 0$$

Let us examine the compatibility equations.

For plane strain conditions, they are reduced to:

$$2\varepsilon_{12,12}^{el} = \varepsilon_{11,22}^{el} + \varepsilon_{22,11}^{el}$$

with: $\underline{\varepsilon}^{el} = \frac{1+\nu}{E}\underline{\sigma} - \frac{\nu}{E}\mathrm{tr}\left(\underline{\sigma}\right)\underline{1}$

One obtains:

$$2\sigma_{12,12} = \left(\left(1-\nu\right)\sigma_{11} - \nu\sigma_{22}\right)_{,22} + \left(-\nu\sigma_{11} + \left(1-\nu\right)\sigma_{22}\right)_{,11}$$

Recall the equilibrium equations in terms of stress:

$$\begin{Bmatrix} \sigma_{11,1} + \sigma_{12,2} + f_1^{\star} = 0 \\ \sigma_{21,1} + \sigma_{22,2} + f_2^{\star} = 0 \end{Bmatrix} \Rightarrow \begin{Bmatrix} \sigma_{11,11} + \sigma_{12,21} + f_{1,1}^{\star} = 0 \\ \sigma_{21,12} + \sigma_{22,22} + f_{2,2}^{\star} = 0 \end{Bmatrix}$$

$$\Rightarrow 2\sigma_{12,12} = -\sigma_{11,11} - \sigma_{22,22} - \left(f_{1,1}^{\star} + f_{2,2}^{\star}\right)$$

$$\Rightarrow \Delta\left(\sigma_{11} + \sigma_{22}\right) = -\frac{\mathrm{div}\overrightarrow{f^{\star}}}{\left(1-\nu\right)} \; with \; \mathrm{div}\overrightarrow{f^{\star}} = f_{1,1}^{\star} + f_{2,2}^{\star}. \tag{3.18}$$

In plane stress conditions, an identical calculation gives:

$$\Delta\left(\sigma_{11} + \sigma_{22}\right) = -\left(1+\nu\right)\mathrm{div}\overrightarrow{f^{\star}} \tag{3.19}$$

with

$$\varepsilon_{33}^{el} = -\frac{\nu}{E}\left(\sigma_{11} + \sigma_{22}\right) \Rightarrow \varepsilon_{33,12}^{el} = -\frac{\nu}{E}\left(\sigma_{11} + \sigma_{22}\right)_{,12} = 0$$

that means that $(\sigma_{11} + \sigma_{22})$ must be linear in x_1 and x_2.

It follows that a plane stress state, in general, does not exist.

Suffice it to an average plane stress state (by calculation of mean values on the thickness of the plate and reasonable assumptions).

3.3.1 Airy Function (or Stress Function)

Considering once again the equilibrium equations:

$$\begin{Bmatrix} \sigma_{11,1} + \sigma_{12,2} + f_1^\star = 0 \\ \sigma_{21,1} + \sigma_{22,2} + f_2^\star = 0 \end{Bmatrix}.$$

Let us consider the functions G_1 and G_2 defined by the expressions:

$$\begin{matrix} G_{1'1} = f_1^\star \\ G_{2,2} = f_2^\star \end{matrix} \Rightarrow \begin{matrix} (\sigma_{11} + G_1)_{,1} + \sigma_{12,2} = 0 \\ \sigma_{21,1} + (\sigma_{22} + G_2)_{,2} = 0 \end{matrix}$$

These relationships are identically verified if we find two functions $\psi_1$and ψ_2 satisfying:

$$\begin{matrix} \sigma_{11} + G_1 = \psi_{1,2}\ \sigma_{12} = -\psi_{1,1} \\ \sigma_{22} + G_2 = -\psi_{2,1}\ \sigma_{21} = \psi_{2,2} \end{matrix}$$

Leading to the equation:

$$\psi_{1,1} = -\psi_{2,2}$$

which in turn is satisfied if there exists a function $A(x_1, x_2)$ called Airy function checking:

$$\psi_1 = A_{,2}\ and\ \psi_2 = -A_{,1}$$

We obtain:

$$\begin{Bmatrix} \sigma_{11} = A_{,22} - G_1 \\ \sigma_{22} = A_{,11} - G_2 \\ \sigma_{12} = -A_{,12} \end{Bmatrix} \tag{3.20}$$

If $G_1 = G_2 = G$ that means $\overrightarrow{f^\star} = \vec{\nabla} G$ so:

$$\begin{matrix} \Delta\Delta A = \frac{1-2\nu}{1-\nu}\Delta G\ in\ DP \\ \Delta\Delta A = -(1+\nu)\Delta G\ in\ CP \end{matrix} \tag{3.21}$$

So if $\vec{\gamma} = \vec{0}$ and $\vec{f} = \overrightarrow{Cte}$:

$$\begin{matrix} \sigma_{11} = A_{,22} \\ \sigma_{22} = A_{,11}\ et\ \Delta\Delta A = 0 \\ \sigma_{12} = -A_{,12} \end{matrix} \tag{3.22}$$

In cylindrical coordinates, the equilibrium equations are written:

$$\begin{cases} \frac{\partial \sigma_{rr}}{\partial r} + \frac{1}{r}\frac{\partial \sigma_{r\theta}}{\partial \theta} + \frac{1}{r}\left(\sigma_{rr} - \sigma_{\theta}\right) + f_r^{\star} = 0 \\ \frac{\partial \sigma_{r\theta}}{\partial r} + \frac{1}{r}\frac{\partial \sigma_{\theta\theta}}{\partial \theta} + \frac{1}{r}\sigma_{r\theta} + f_{\theta}^{\star} = 0 \end{cases}$$

Let us consider that $\overrightarrow{f^{\star}} = Cte$, the solution is therefore:

$$\begin{cases} \sigma_{rr} = \frac{1}{r}\frac{\partial A}{\partial r} + \frac{1}{r^2}\frac{\partial^2 A}{\partial \theta^2} \\ \sigma_{\theta\theta} = \frac{\partial^2 A}{\partial r^2} \\ \sigma_{r\theta} = -\frac{\partial}{\partial r}\left(\frac{1}{r}\frac{\partial A}{\partial \theta}\right) \end{cases} \tag{3.23}$$

with: $\Delta\Delta A = 0$.

Thus, a dictionary of biharmonics functions was performed and an inventory of mechanical engineering plans shaped media problems can be solved.

Note that the Airy function is mainly used in cylindrical coordinates.

Special case: axisymmetric problem $A = A(r)$

$$\Delta\Delta A = \frac{d^4 A}{dr^4} + \frac{2}{r}\frac{d^3 A}{dr^3} - \frac{1}{r^2}\frac{d^2 A}{dr^2} + \frac{1}{r^3}\frac{dA}{dr} = 0.$$

This Euler equation is solved by the variable change $r = e^u$:

$$\Rightarrow \frac{d^4 A}{du^4} - 4\frac{d^3 A}{du^3} + 4\frac{d^2 A}{du^2} = 0$$

Let us consider $\phi(u) = \frac{d^2 A}{du^2}$; ϕ verifies the equation: $\frac{d^2\phi}{du^2} - 4\frac{d\phi}{du} + 4\phi = 0$

We seek the solution in the form $\phi(u) = e^{\lambda u} \Longrightarrow \lambda^2 - 4\lambda + 4 = (\lambda - 2)^2 = 0 \Longrightarrow \lambda = 2$ twice root.

The solution is written as: $\phi(u) = (\alpha u + \beta)\exp(2u) \Longrightarrow A(u) = (Au + B)\exp(2u) + Cu + D$

Hence finally the Airy function is written as:

$$A(r) = (A \ln r + B) r^2 + C \ln r + D \tag{3.24}$$

3.3.2 *Exercise 3.3*

Let us consider an infinite plate (P) assumed to be linear elastic and perforated by a circular hole centered at the origin O whose coordinates are (Ox_1, Ox_2). The radius of the hole is a. (Fig. 3.5).

The plate is subjected to a stress state (average) plane. In addition, body forces are zero.

It is proposed to determine the stresses and displacements obtained by introducing within the hole, a disk of radius $(a + \eta)$ by force with η small (in reference of a).

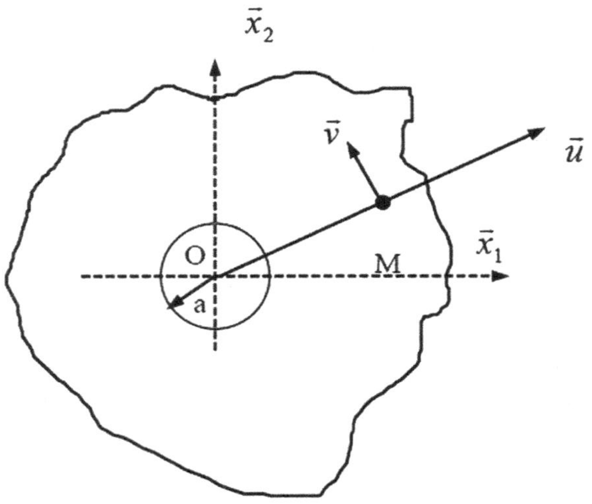

Fig. 3.5 Insertion of a disk in a perforated plate

The contact between the plate and the disk is assumed frictionless. Let E_1, ν_1 (respectively E_2, ν_2) the elastic modulus of the plate (P) (respectively of the disc (D)).

(1) Determine the stresses and displacements in a similar plate (P) submitted at its inner contour radius a at an uniform pressure $p_1 \vec{u}$ (p_1 positive constant).

To do this:

(a) Show that the stresses can be calculated with the help of an Airy function:

$$A(r) = \alpha r^2 + \beta r^2 \ln r + \gamma \ln r + \delta.$$

(b) Calculate the constants of integration: we write in particular that the stress tensor must tend towards 0 at infinity.

(c) Determine the displacement vector at each point. System revolution symmetry will be used and we neglect the rigid solid movement.

(2) Determine the stresses and displacements for the circular disk of radius R submitted to an uniform pressure $-p_2 \vec{u}$ on its contour (p_2 positive constant). As above, we cancel the rigid solid movement.

(3) Let us introduce by means force of the disk of radius $(a + \eta)$ in the plate (P).

Let us calculate from the above results, the contact pressure p between (P) and (D) as function of η, a and the elastic constants of the two solids.

(4) Numerical application:

E_1 =7000 daN/mm^2, E_2=21000 daN/mm^2,a=50 mm

$\nu_1 = \nu_2$=0.3.

Let us calculate of p/η.

Correction:

(1a) To find the solution of a problem axisymmetric:

$$A(r) = \alpha r^2 + \beta r^2 \ln r + \gamma \ln r + \delta.$$

The solution $\Delta\Delta A(r) = 0$ can be searched under the form $A(r) = r^\alpha$ and using the following theorem: "If α multiple root of order p of the characteristic equation, the corresponding solution is the product of r^α by a polynom of degree $(p-1)$ in $\ln r$." Here, evidently $\alpha = 0$ *and* $\alpha = 2$ double roots.
(1b) Stress calculations:

$$\begin{cases} \sigma_{rr} = \frac{1}{r}A_{,r} = 2\alpha \ln r + 2\beta + \alpha + \frac{\gamma}{r^2} \\ \sigma_{\theta\theta} = A_{,rr} = 2\alpha \ln r + 2\beta + 3\alpha - \frac{\gamma}{r^2} \\ \sigma_{r\theta} = 0 \end{cases}$$

When $r \to +\infty$: $\sigma_{rr} \to 0 \implies \alpha = \beta = 0.$

$$\begin{cases} \sigma_{rr} = \frac{\gamma}{r^2} \\ \sigma_{\theta\theta} = -\frac{\gamma}{r^2} \end{cases}$$

Boundary condition for $r = a$:

$$\overrightarrow{T}^{-r}(r = a) = p_1 \overrightarrow{u}\ , \gamma = -p_1 a^2$$

$$\begin{cases} \sigma_{rr} = -p_1 \frac{a^2}{r^2} \\ \sigma_{\theta\theta} = p_1 \frac{a^2}{r^2} \end{cases}$$

(1c) Strains and displacement calculations:
$\underline{\underline{\varepsilon}}^{el} = \frac{1+\nu}{E}\underline{\sigma} - \frac{\nu}{E}\mathrm{tr}\left(\underline{\sigma}\right)\underline{1}$ *with* $\mathrm{tr}\left(\underline{\sigma}\right) = 0$

$$\begin{cases} \varepsilon_{rr}^{el} = -\frac{1+\nu_1}{E_1} p_1 \frac{a^2}{r^2} = u_{r,r} \\ \varepsilon_{\theta\theta}^{el} = \frac{1+\nu_1}{E_1} p_1 \frac{a^2}{r^2} = \frac{u_r}{r} \\ \varepsilon_{r\theta}^{el} = 0 \end{cases}$$

The symmetry of revolution generates $u_\theta = 0$, hence:

$$u_r^{(1)} = \frac{1+\nu_1}{E_1} p_1 \frac{a^2}{r} \ \ and\ \ u_\theta^{(1)} = 0$$

(2) Uniformal compression of disk:

$$\underline{\sigma} = -p_2 \begin{bmatrix} 1 & 0 \\ 0 & 1 \end{bmatrix} \implies \mathrm{tr}\left(\underline{\sigma}\right) = -2p_2$$

and

$$\begin{cases} \varepsilon_{rr}^{el} = -\frac{1-\nu_2}{E_2} P_2 = u_{r,r} \\ \varepsilon_{\theta\theta}^{el} = -\frac{1-\nu_2}{E_2} P_2 = \frac{u_r}{r} \\ \varepsilon_{r\theta}^{el} = 0 \end{cases}$$

That gives:

$$u_r^{(2)} = -\frac{1-\nu_2}{E_2} P_2 r \ and \ u_\theta^{(2)} = 0.$$

(3) Let us denote r_0 the equilibrium radius, one has:

$$\begin{cases} u_r^{(1)} (r = a) = r_0 - a \\ u_r^{(2)} (r = a + \eta) = r_0 - (a + \eta) \end{cases}$$

$$\Rightarrow u_r^{(1)} (r = a) - u_r^{(2)} (r = a + \eta) = \eta \Rightarrow \frac{1+\nu_1}{E_1} P_1 a + \frac{1-\nu_2}{E_2} P_2 (a + \eta) = \eta.$$

At the mechanical equilibrium: $p_1 = p_2 = p \ for \ \eta \ll a$

$$p \simeq \frac{\eta}{a} \left[\frac{1+\nu_1}{E_1} + \frac{1-\nu_2}{E_2} \right]^{-1}$$

(4) Numerical application:
$E_1 = 7000$ daN/mm^2, $E_2 = 21000$ daN/mm^2, a = 50 mm
$\nu_1 = \nu_2 = 0.3$.
$\Longrightarrow {}^{p}/_{\eta} = 91.3$daN/mm

3.3.3 Exercise 3.4

Let us study a plate with a hole (Fig. 3.6): by using of the superposition principle (Fig. 3.7).

It is proposed to determine the stress state in a perforated plate subjected to the following simple tensile test in the direction $\overrightarrow{x_1}$. The hole has a very small radius compared to the size of the plate. The thickness is small compared to the dimensions in the plane $\left(\overrightarrow{x_1}, \overrightarrow{x_2}\right)$. The body forces are neglected.

(1) Show that the problem is the superposition of two problems:
The problem n° 1 is symmetrical while the problem n° 2 depends upon the angle θ.
(2) For the problem n°1, let us calculate the stress state by using the Airy function $A(r)$ (See Exercise 3.3)
(3) For the problem n°2, let us calculate the stress state by using the Airy function under the form:

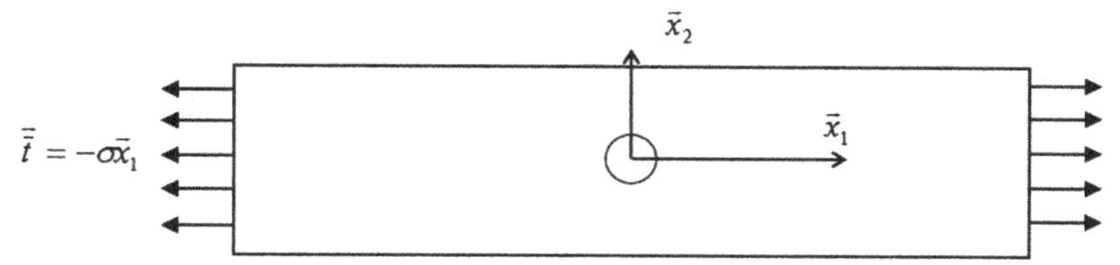

Fig. 3.6 Drawing of the perforated plate

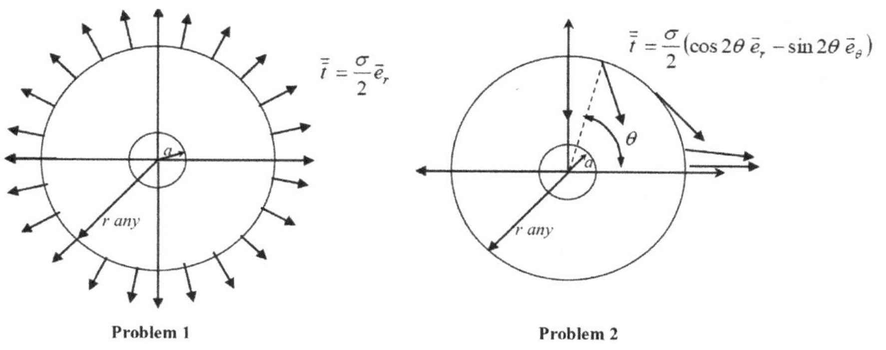

Fig. 3.7 Decomposition of the overall problem into two problems

$$A\,(r,\theta) = \varphi\,(r)\cos\,(2\theta)$$

and apply the superposition principle.
(4) Draw the evolution of $\sigma_{\theta\theta}$ near the hole. What do you observe?

Correction:

(1) Although the rectangular geometry of the plate is put into cylindrical coordinates. Away from the hole (which is a local disturbance), the stress tensor is written as:

$$\underline{\sigma}_{(x_1,x_2)} = \begin{bmatrix} \sigma & 0 \\ 0 & 0 \end{bmatrix} \Longrightarrow \underline{\sigma}_{(r,\theta)} = \begin{bmatrix} \frac{\sigma}{2}\,(1+\cos 2\theta) & -\frac{\sigma}{2}\sin 2\theta \\ -\frac{\sigma}{2}\sin 2\theta & \frac{\sigma}{2}\,(1-\cos 2\theta) \end{bmatrix}$$

The boundary condition in stress terms for $r = R$ is written:

$$\overrightarrow{T}^{e_r}\,(r = R) = \sigma_{rr}\,\overrightarrow{e_r} + \sigma_{r\theta}\,\overrightarrow{e_\theta} = \frac{\sigma}{2}\,(1+\cos 2\theta)\,\overrightarrow{e_r} - \frac{\sigma}{2}\sin 2\theta\,\overrightarrow{e_\theta}$$

$$\overrightarrow{T}^{e_r}\,(r = R) = \overrightarrow{T_1}^{e_r}\,(r = R) + \overrightarrow{T_2}^{e_r}\,(r = R)$$

with:

$$\begin{cases} \overset{e_r}{\overrightarrow{T_1}}(r=R) = \frac{\sigma}{2}\overrightarrow{e_r} \\ \overset{e_r}{\overrightarrow{T_2}}(r=R) = \frac{\sigma}{2}\cos 2\theta\,\overrightarrow{e_r} - \frac{\sigma}{2}\sin 2\theta\,\overrightarrow{e_\theta} \end{cases}$$

What justifies the separation into two problems n°1 and n° 2.

Problem n°1:

The classical resolution delivers:
with the boundaries conditions:

$$\begin{matrix} \sigma^1_{rr}(r=a) = 0 \\ \sigma^1_{r\theta}(r=a) = 0 \end{matrix} \quad and \quad \begin{matrix} \sigma^1_{rr}(r\to\infty) = \frac{\sigma}{2} \\ \sigma^1_{r\theta}(r\to\infty) = 0 \end{matrix}$$

$$\begin{cases} \sigma^1_{rr} = \frac{\sigma}{2}\left(1-\frac{a^2}{r^2}\right) \\ \sigma^1_{\theta\theta} = \frac{\sigma}{2}\left(1+\frac{a^2}{r^2}\right) \\ \sigma^1_{r\theta} = 0 \end{cases}$$

Problem n° 2: more "calculative" with $A(r,\theta) = \varphi(r)\,cos\,(2\theta)$

$$\begin{cases} \sigma^{(2)}_{rr} = \frac{1}{r}\frac{\partial A}{\partial r} + \frac{1}{r^2}\frac{\partial^2 A}{\partial \theta^2} = \left(\frac{\varphi'}{r} - \frac{4\varphi}{r^2}\right)\cos 2\theta \\ \sigma^{(2)}_{\theta\theta} = \frac{\partial^2 A}{\partial r^2} = \varphi''\cos 2\theta \\ \sigma^{(2)}_{r\theta} = -\frac{\partial}{\partial r}\left(\frac{1}{r}\frac{\partial A}{\partial \theta}\right) = \left(\frac{2\varphi'}{r} - \frac{2\varphi}{r^2}\right)\sin 2\theta \end{cases}$$

To be identified $\varphi(r)$ by means of the biharmonic equation $\Delta\Delta A = 0$ one obtains the global solution («Annales ensmp Paris; corrigé 2007 »):

$$\begin{cases} \sigma_{rr} = \frac{\sigma}{2}\left(1-\frac{a^2}{r^2}\right) + \frac{\sigma}{2}\left(1+\frac{3a^4}{r^4}-\frac{4a^2}{r^2}\right)\cos 2\theta \\ \sigma_{\theta\theta} = \frac{\sigma}{2}\left(1+\frac{a^2}{r^2}\right) - \frac{\sigma}{2}\left(1+\frac{3a^4}{r^4}\right)\cos 2\theta \\ \sigma_{r\theta} = -\frac{\sigma}{2}\left(1-\frac{3a^4}{r^4}+\frac{2a^2}{r^2}\right)\sin 2\theta \end{cases}$$

Note that we can consider $\sigma = \sigma^\infty$ (dimension a very small hole in regard to the dimensions of the plate).

Moreover:

$$\sigma_{\theta\theta}(r=a,\theta) = \sigma\,(1-2\cos 2\theta)$$

In simple traction, the orthoradial stress is maximal in $\theta = \pm\pi/2$ and is three times the applied stress. The interest of the problem consists in the discovery of a stress concentration on a holed piece.

However, the calculations are more elegant when using the complex potential as suggested by F. Sidoroff in his solid mechanics course at «Ecole Centrale de Lyon» (Sidoroff 1980).

3.4 Linear Isotropic Thermoelasticity

We will examine the effect of temperature on the elastic behavior of a material.

The temperature change $(T - T_0)$ will be reasonable so that the crystalline structure of the material does not change (for example: solid–solid phase change) or even of state (solid\liquid or gas) and as a first approximation, the elastic constants(E, G, ν) are not changed.

From a temperature change in a nondeformed natural state at T_0, to a state at T, without external mechanical loading, it can be obtained a purely thermal deformation following the isotropy of the material:

$$\underline{\underline{\varepsilon}}^{th} = \alpha\,(T - T_0)\,\underline{\underline{1}}\ with\ \underline{\underline{\sigma}} = \underline{\underline{0}}$$

α is the linear expansion coefficient of the material, its unit is K^{-1}.

If a bar is subjected to a combined mechanical and thermal loadings, then the total deformation $\underline{\underline{\varepsilon}}$ is the sum of two terms:

– $\underline{\underline{\varepsilon}}^{el}$ due to external mechanical forces and governed by the Hooke' law:

$$\underline{\underline{\varepsilon}}^{el} = \frac{1+\nu}{E}\underline{\underline{\sigma}} - \frac{\nu}{E}\mathrm{tr}\left(\underline{\underline{\sigma}}\right)\underline{\underline{1}}$$

– $\underline{\underline{\varepsilon}}^{th}$ due to temperature variation:

$$\underline{\underline{\varepsilon}}^{th} = \alpha\,(T - T_0)\,\underline{\underline{1}}.$$

Thus, with:

$$\underline{\underline{\varepsilon}} = \underline{\underline{\varepsilon}}^{el} + \underline{\underline{\varepsilon}}^{th}$$

One obtains the Hooke–Duhamel's law:

$$\underline{\underline{\sigma}} \Longrightarrow \underline{\underline{\varepsilon}}\ \underline{\underline{\varepsilon}} = \frac{1+\nu}{E}\underline{\underline{\sigma}} - \frac{\nu}{E}\mathrm{tr}\left(\underline{\underline{\sigma}}\right)\underline{\underline{1}} + \alpha\,(T - T_0)\,\underline{\underline{1}}. \tag{3.25}$$

The reversal of laws directly gives:

$$\underline{\underline{\varepsilon}} \Longrightarrow \underline{\underline{\sigma}} = \frac{E}{1+\nu}\left(\underline{\underline{\varepsilon}}\right) + E\frac{\nu}{(1+\nu)\,(1-2\nu)}\mathrm{tr}\left(\underline{\underline{\varepsilon}}\right)\underline{\underline{1}} - 3K\alpha\,(T - T_0)\,\underline{\underline{1}}. \tag{3.26}$$

Similarly, we can write:

$$\underline{\underline{\varepsilon}} \Longrightarrow \underline{\underline{\sigma}}\ \underline{\underline{\sigma}} = \lambda\left(\mathrm{tr}\underline{\underline{\varepsilon}}\right)\underline{\underline{1}} + 2\mu\underline{\underline{\varepsilon}} - (3\lambda + 2\mu)\,\alpha\,(T - T_0)\,\underline{\underline{1}}. \tag{3.27}$$

and:

$$\underline{\underline{\sigma}} \Longrightarrow \underline{\underline{\varepsilon}}\ \underline{\underline{\varepsilon}} = \frac{\underline{\underline{\sigma}}}{2\mu} - \frac{\lambda}{2\mu\,(3\lambda + 2\mu)}\mathrm{tr}\left(\underline{\underline{\sigma}}\right)\underline{\underline{1}} + \alpha\,(T - T_0)\,\underline{\underline{1}}. \tag{3.28}$$

The expressions of the equilibrium equations of Cauchy and compatibility in terms of deformation remain unchanged.

However, in terms of stress, compatibility equations are slightly modified:

$$\Delta\sigma_{ij} + \frac{1}{1+\nu}\mathrm{tr}\left(\underline{\sigma}\right), ij = -\delta_{ij}\frac{\nu}{1-\nu}\mathrm{div}\overrightarrow{f^{\star}} - f^{\star}_{i,j} - f^{\star}_{j,i} - \frac{E}{1+\nu}\alpha T_{,ij} \quad (3.29)$$

To solve a thermoelasticity problem, we must first determine the temperature field. This is done by solving the heat equation (see Chap. 2). It reduces to:

$$\rho C_v \frac{\partial T}{\partial t} = k\Delta T + W + T\left[\frac{\partial \underline{\sigma}}{\partial T} : \underline{\dot{\varepsilon}}\right] \quad (3.30)$$

When the term of mechanical coupling $T\left[\frac{\partial \underline{\sigma}}{\partial T} : \underline{\dot{\varepsilon}}\right]$ is negligible and if there is no heat internal source $W = 0$, we obtain the wave heat equation under the form:

$$\rho C_v \frac{\partial T}{\partial t} = k\Delta T$$

If we examine the steady state, the equation reduces to:

$$\Delta T = 0$$

For example, for a purely thermal loading $\left(\underline{\sigma} = \underline{0}\right)$, what temperature field is allowable? Compatibility equations are checked if and only if:

$$T\left(x_1, x_2, x_3\right) = ax_1 + bx_2 + cx_3 + d$$

This will be proved in the context of the specific Exercise 3.5 devoted to linear thermoelasticity.

3.4.1 Exercise 3.5

Let us consider a circular cylinder of radius R, thickness $2h$, and axis ox_3..

The origin o is taken at the center of gravity of the straight cylinder and the upper and lower sides comply with the conditions, respectively $x_3 = h$ and $x_3 = -h$.

It is assumed that within the cylinder there is a state of elastic deformation:

$$\underline{\varepsilon}^{el} = \alpha k\left(x_1, x_2, x_3\right)\underline{1}$$

with $k\left(x_1, x_2, x_3 = h\right) = k_1$ and $k\left(x_1, x_2, x_3 = -h\right) = k_2$

(1) Show that the compatibility of the strain tensor with the continuity of the medium to suggest that k must be of the form:

$$k(x_1, x_2, x_3) = a_1x_1 + a_2x_2 + a_3x_3 + a_4$$

Let us determine $a_1,\ a_2,\ a_3,\ a_4$.
(2) Let calculate the vectorial displacement $\vec{u}$; integration constants will be determined by assuming zero displacement and rotation in o.

Correction:

(1) Compatibilities equations in terms of strains are reduced to:

$$\left\{\begin{array}{l} 2\varepsilon^{el}_{12,12} = \varepsilon^{el}_{11,22} + \varepsilon^{el}_{22,11} = 0 \Longrightarrow k_{,22} + k_{,11} = 0 \\ 2\varepsilon^{el}_{23,23} = \varepsilon^{el}_{22,33} + \varepsilon^{el}_{33,22} = 0 \Longrightarrow k_{,33} + k_{,22} = 0 \\ 2\varepsilon^{el}_{31,31} = \varepsilon^{el}_{33,11} + \varepsilon^{el}_{11,33} = 0 \Longrightarrow k_{,11} + k_{,33} = 0 \end{array}\right\} \Longrightarrow k_{,11} = k_{,22} = k_{,33} = 0$$

$$\left\{\begin{array}{l} \varepsilon^{el}_{11,23} = 0 \Longrightarrow k_{,23} = 0 \\ \varepsilon^{el}_{22,31} = 0 \Longrightarrow k_{,31} = 0 \\ \varepsilon^{el}_{33,12} = 0 \Longrightarrow k_{,12} = 0 \end{array}\right\}$$

It follows that $k_{,ij} = 0\ \forall i, j = 1, 2, 3$ and so the function $k(x_1, x_2, x_3)$ is linear in x_i:

$$k(x_1, x_2, x_3) = a_1x_1 + a_2x_2 + a_3x_3 + a_4$$

Boundary conditions:

$$for\ x_1^2 + x_2^2 \leq R^2\ \forall x_1, x_2 \left\{\begin{array}{l} a_1x_1 + a_2x_2 + a_3h + a_4 = k_1 \\ a_1x_1 + a_2x_2 - a_3h + a_4 = k_2 \end{array}\right\} \Longrightarrow$$

$$a_1 = a_2 = 0,\ a_3 = \frac{k_1 - k_2}{2h},\ a_4 = \frac{k_1 + k_2}{2}$$

(2) Calculations of $\omega_{ij}\ with\ \omega_{ij,l} = \varepsilon^{el}_{il,j} - \varepsilon^{el}_{jl,i}$

$$\left\{\begin{array}{c} \omega_{12} = -\omega_{21} = 0 \\ \omega_{23} = -\omega_{32} = \alpha a_3 x_2 \\ \omega_{31} = -\omega_{13} = -\alpha a_3 x_1 \end{array}\right\}$$

Calculations of $u_i\ with\ u_{i,j} = \varepsilon^{el}_{ij} + \omega_{ij}$

$$\vec{u}\ \left\{\begin{array}{c} u_1 = \alpha x_1 (a_3x_3 + a_4) \\ u_2 = \alpha x_2 (a_3x_3 + a_4) \\ u_3 = \alpha \left(a_3 \left(\frac{x_3^2}{2} - \left(\frac{x_1^2 + x_2^2}{2}\right)\right) + a_4x_3\right) \end{array}\right\}$$

In this practical problem, we can evaluate the movement of a wall subjected to external and internal temperatures and why not generated stresses.

3.5 Elastic Linear and Anisotropic Mediums

3.5.1 *Behavior Laws*

The law is written as:

$$\underline{\sigma} = \underline{\underline{C}} : \underline{\varepsilon}^{el} \ or \ \sigma_{ij} = C_{ijhk}\varepsilon^{el}_{hk}, \tag{3.31}$$

where $\underline{\underline{C}}$ is the fourth-order elastic tensor. Normally, there are $9\times9 = 81$ independent coefficients C_{ijkl}.

The respective symmetries of the stress and strains tensors permit to reduce the number of independent coefficients to $6 \times 6 = 36$.

The existence of a deformation potential conjugated to use with the reciprocity theorem of Betti–Maxwell (meaning that, the work of the stress field $\underline{\sigma}^{(1)}$ in the strain field $\underline{\varepsilon}^{(2)}$ is equal to the work of the stress field $\underline{\sigma}^{(2)}$ in the strain field $\underline{\varepsilon}^{(1)}$) causes the symmetry of the tensor of compliances.

So, there are no more than 21 independent coefficients (to be determined) for anisotropic elastic materials that exist in nature (see Brown et al. 2006).

The degree of symmetry of material determines the number of independent coefficients to be identified.

If the material is invariant towards the transformation defined by the $\underline{P}$ matrix, the change of reference frame defined by $\underline{P}$ does not modify the behavior law, which must be written with the same $\underline{\underline{C}}$. From $\underline{\sigma} = \underline{P}^{-1}\underline{\sigma}\,\underline{P}$ and $\underline{\varepsilon} = \underline{P}^{-1}\underline{\varepsilon}\,\underline{P}$, one deduces $\underline{\underline{C}} = \underline{P}^{-1}\underline{P}^{-1}\underline{\underline{C}}\,\underline{P}\,\underline{P}$ (Cailletaud et al. 2011).

In order to simplify the notations, one writes:

$$\begin{bmatrix} \sigma_{11} \\ \sigma_{22} \\ \sigma_{33} \\ \sigma_{23} \\ \sigma_{31} \\ \sigma_{12} \end{bmatrix} = \begin{bmatrix} C_{11} & C_{12} & C_{13} & C_{14} & C_{15} & C_{16} \\ C_{12} & C_{22} & C_{23} & C_{24} & C_{25} & C_{26} \\ C_{13} & C_{23} & C_{33} & C_{34} & C_{35} & C_{36} \\ C_{14} & C_{24} & C_{34} & C_{44} & C_{45} & C_{46} \\ C_{15} & C_{25} & C_{35} & C_{45} & C_{55} & C_{56} \\ C_{16} & C_{26} & C_{36} & C_{46} & C_{56} & C_{66} \end{bmatrix} \begin{bmatrix} \varepsilon^{el}_{11} \\ \varepsilon^{el}_{22} \\ \varepsilon^{el}_{33} \\ \gamma^{el}_{23} = 2\varepsilon^{el}_{23} \\ \gamma^{el}_{31} = 2\varepsilon^{el}_{31} \\ \gamma^{el}_{12} = 2\varepsilon^{el}_{12} \end{bmatrix} \tag{3.32}$$

(1) Symmetry with respect to a coordinate plane $x_3 = 0$
There are 13 independent coefficients:

$$\underline{\underline{C}} = \begin{bmatrix} C_{11} & C_{12} & C_{13} & 0 & 0 & C_{16} \\ C_{12} & C_{22} & C_{23} & 0 & 0 & C_{26} \\ C_{13} & C_{23} & C_{33} & 0 & 0 & C_{36} \\ 0 & 0 & 0 & C_{44} & C_{45} & 0 \\ 0 & 0 & 0 & C_{45} & C_{55} & 0 \\ C_{16} & C_{26} & C_{36} & 0 & 0 & C_{66} \end{bmatrix}$$

(2) Symmetry with respect to two orthogonal planes $x_1 = 0$ et $x_3 = 0$, this is the case of orthotropic materials

The matrix is written as:

$$\underline{\underline{C}} = \begin{bmatrix} C_{11} & C_{12} & C_{13} & 0 & 0 & 0 \\ C_{12} & C_{22} & C_{23} & 0 & 0 & 0 \\ C_{13} & C_{23} & C_{33} & 0 & 0 & 0 \\ 0 & 0 & 0 & C_{44} & 0 & 0 \\ 0 & 0 & 0 & 0 & C_{55} & 0 \\ 0 & 0 & 0 & 0 & 0 & C_{66} \end{bmatrix} \tag{3.33}$$

We can introduce $\underline{\underline{S}}$ called "the compliance tensor" in the orthotropic case under the form:

$$\underline{\varepsilon}^{el} = \underline{\underline{S}} : \underline{\sigma} \ with \ \underline{\underline{S}} = \underline{\underline{C}}^{-1}$$

$$\begin{bmatrix} \varepsilon_{11}^{el} \\ \varepsilon_{22}^{el} \\ \varepsilon_{33}^{el} \\ \gamma_{23}^{el} = 2\varepsilon_{23}^{el} \\ \gamma_{31}^{el} = 2\varepsilon_{31}^{el} \\ \gamma_{12} = 2\varepsilon_{12}^{el} \end{bmatrix} = \begin{bmatrix} S_{11} & S_{12} & S_{13} & 0 & 0 & 0 \\ S_{12} & S_{22} & S_{23} & 0 & 0 & 0 \\ S_{13} & S_{23} & S_{33} & 0 & 0 & 0 \\ 0 & 0 & 0 & S_{44} & 0 & 0 \\ 0 & 0 & 0 & 0 & S_{55} & 0 \\ 0 & 0 & 0 & 0 & 0 & S_{66} \end{bmatrix} \begin{bmatrix} \sigma_{11} \\ \sigma_{22} \\ \sigma_{33} \\ \sigma_{23} \\ \sigma_{31} \\ \sigma_{12} \end{bmatrix} \tag{3.34}$$

A tensile test in the direction x_1 allows us to measure the Young modulus E_1 and the two Poisson coefficients ν_{12} et ν_{13}:

$$\begin{cases} \varepsilon_{11}^{el} = \varepsilon_1^{el} = S_{11}\sigma_{11} = S_{11}\sigma_1 \Rightarrow S_{11} = \frac{1}{E_1} \\ \varepsilon_2^{el} = S_{12}\sigma_{11} = S_{12}\sigma_1 \Rightarrow \varepsilon_{22}^{el} = \varepsilon_2^{el} = \frac{S_{12}}{S_{11}}\varepsilon_1^{el} = -\nu_{12}\varepsilon_1^{el} \\ \varepsilon_3^{el} = S_{13}\sigma_{11} = S_{13}\sigma_1 \Rightarrow \varepsilon_{33}^{el} = \varepsilon_3^{el} = \frac{S_{12}}{S_{11}}\varepsilon_1^{el} = -\nu_{13}\varepsilon_1^{el} \end{cases}$$

Let us consider the following notations specific to composites materials: $\sigma_I = \sigma_{9-(i+j)}$ and $\varepsilon_I = 2\varepsilon_{9-(i+j)}^{el}$ for $i \neq j$.

For the pure shear test: $\sigma_{23} = \sigma_4 \neq 0$, the others $\sigma_I = 0$:

$$2\varepsilon_{23}^{el} = \varepsilon_4^{el} = S_{44}\sigma_4 \Rightarrow S_{44} = \frac{1}{G_{23}}$$

Thus:

$$\begin{bmatrix} \varepsilon_1^{el} \\ \varepsilon_2^{el} \\ \varepsilon_3^{el} \\ \varepsilon_4^{el} \\ \varepsilon_5^{el} \\ \varepsilon_6^{el} \end{bmatrix} = \begin{bmatrix} \frac{1}{E_1} & \frac{-\nu_{12}}{E_1} & \frac{-\nu_{13}}{E_1} & 0 & 0 & 0 \\ -\frac{\nu_{21}}{E_2} & \frac{1}{E_2} & -\frac{\nu_{23}}{E_2} & 0 & 0 & 0 \\ -\frac{\nu_{31}}{E_3} & -\frac{\nu_{32}}{E_3} & \frac{1}{E_3} & 0 & 0 & 0 \\ 0 & 0 & 0 & \frac{1}{G_{23}} & 0 & 0 \\ 0 & 0 & 0 & 0 & \frac{1}{G_{13}} & 0 \\ 0 & 0 & 0 & 0 & 0 & \frac{1}{G_{12}} \end{bmatrix} \begin{bmatrix} \sigma_1 \\ \sigma_2 \\ \sigma_3 \\ \sigma_4 \\ \sigma_5 \\ \sigma_6 \end{bmatrix} \tag{3.35}$$

Considering the symmetry of $\underline{\underline{S}} \Longrightarrow$

$$\frac{\nu_{ij}}{E_i} = \frac{\nu_{ji}}{E_j} \; with \; i \neq j.$$

Case of plane stresses in orthotropic material:

In this case $\sigma_3 = \sigma_4 = \sigma_5 = 0$. This corresponds to plane materials (fabrics laminates). In the orthotropic axis $(o\overline{x_1}\,\overline{x_2}\,\overline{x_3})$, we have:

$$\begin{bmatrix} \overline{\varepsilon_1^{el}} \\ \overline{\varepsilon_2^{el}} \\ \overline{\varepsilon_6^{el}} \end{bmatrix} = \begin{bmatrix} \overline{S_{11}} & \overline{S_{12}} & 0 \\ \overline{S_{12}} & \overline{S_{22}} & 0 \\ 0 & 0 & \overline{S_{66}} \end{bmatrix} \begin{bmatrix} \overline{\sigma_1} \\ \overline{\sigma_2} \\ \overline{\sigma_6} \end{bmatrix}$$

with: $\overline{\varepsilon_3^{el}} = \overline{S_{13}}\sigma_1 + \overline{S_{23}}\sigma_2$

$$\begin{cases} \overline{S_{11}} = \frac{1}{\overline{E_1}}\; \overline{S_{22}} = \frac{1}{\overline{E_2}}\; \overline{S_{12}} = -\frac{\overline{\nu_{12}}}{\overline{E_1}} = -\frac{\overline{\nu_{21}}}{\overline{E_2}} \\ \overline{S_{66}} = \frac{1}{\overline{G_{12}}}\; \overline{S_{13}} = -\frac{\overline{\nu_{13}}}{\overline{E_1}} = -\frac{\overline{\nu_{31}}}{\overline{E_3}} \\ \overline{S_{23}} = -\frac{\overline{\nu_{23}}}{\overline{E_2}} = -\frac{\overline{\nu_{32}}}{\overline{E_3}} \end{cases}$$

The inversion of these relationships gives:

$$\begin{bmatrix} \overline{\sigma_1} \\ \overline{\sigma_2} \\ \overline{\sigma_6} \end{bmatrix} = \begin{bmatrix} \overline{Q_{11}} & \overline{Q_{12}} & 0 \\ \overline{Q_{12}} & \overline{Q_{22}} & 0 \\ 0 & 0 & \overline{Q_{66}} \end{bmatrix} \begin{bmatrix} \overline{\varepsilon_1^{el}} \\ \overline{\varepsilon_2^{el}} \\ \overline{\varepsilon_6^{el}} \end{bmatrix}$$

with:

$$\begin{cases} \overline{Q_{11}} = \frac{\overline{E_1}}{1-\overline{\nu_{12}}\,\overline{\nu_{21}}}\; \overline{Q_{22}} = \frac{\overline{E_2}}{1-\overline{\nu_{12}}\,\overline{\nu_{21}}} \\ \overline{Q_{12}} = \frac{\overline{\nu_{12}}\,\overline{E_2}}{1-\overline{\nu_{12}}\,\overline{\nu_{21}}}\; \overline{Q_{66}} = \overline{G_{12}} \end{cases}$$

(3) Equivalence of two axis of symmetry (for example 1 and 2), it is a tetragonal case (tetragonal crystal), six independent coefficients:

$$C_{11} = C_{22},\; C_{13} = C_{23},\; C_{44} = C_{55}$$

$$\underline{\underline{C}} = \begin{bmatrix} C_{11} & C_{12} & C_{13} & 0 & 0 & 0 \\ C_{12} & C_{11} & C_{13} & 0 & 0 & 0 \\ C_{13} & C_{13} & C_{33} & 0 & 0 & 0 \\ 0 & 0 & 0 & C_{44} & 0 & 0 \\ 0 & 0 & 0 & 0 & C_{44} & 0 \\ 0 & 0 & 0 & 0 & 0 & C_{66} \end{bmatrix}$$

(4) Equivalence of the three axes of symmetry: it is the case of cubic symmetry three independent coefficients:

$$C_{11} = C_{33},\; C_{13} = C_{12},\; C_{44} = C_{66}$$

$$\underline{\underline{C}} = \begin{bmatrix} C_{11} & C_{12} & C_{12} & 0 & 0 & 0 \\ C_{12} & C_{11} & C_{12} & 0 & 0 & 0 \\ C_{12} & C_{12} & C_{11} & 0 & 0 & 0 \\ 0 & 0 & 0 & C_{44} & 0 & 0 \\ 0 & 0 & 0 & 0 & C_{44} & 0 \\ 0 & 0 & 0 & 0 & 0 & C_{44} \end{bmatrix}$$

(5) Transverse isotropic relationship. The stiffness matrix for an orthotropic material is therefore:

$$\underline{\underline{C}} = \begin{bmatrix} C_{11} & C_{12} & C_{13} & 0 & 0 & 0 \\ C_{12} & C_{22} & C_{23} & 0 & 0 & 0 \\ C_{13} & C_{23} & C_{33} & 0 & 0 & 0 \\ 0 & 0 & 0 & C_{44} & 0 & 0 \\ 0 & 0 & 0 & 0 & C_{55} & 0 \\ 0 & 0 & 0 & 0 & 0 & C_{66} \end{bmatrix}$$

It is assumed that a material is transverse isotropic when the plane $o\overline{x_1}\,\overline{x_2}$ is a plane of isotropy, meaning that $\underline{\underline{C}}$ is invariant under any rotation of α around the axis $\overrightarrow{x_3}$.

We deduce the transverse isotropic relationship (Casimir and Chevalier 2014):

$$C_{66} = \frac{C_{11} - C_{12}}{2}$$

3.5.2 *Unidirectional Composites with Long Fibers*

Consider an elementary layer consisting of unidirectional fibers oriented along $\overline{x_1}$ (Fig. 3.8).

Let us denote E_m the matrix Young modulus, E_f the fibers Young modulus and V_m, V_f the volume proportions of the matrix and the fibers $\left(V_m + V_f = 1\right)$.

In the case of a material where the fibers are continuous, it is reasonable to assume that the approximation "in parallel", in which the deformations are uniform from one phase to the other, is respected. This permits to assess the Young's modulus $\overline{E_1}$ corresponding to a tensile test in the direction of the fibers with a uniform deformation approximation.

It comes:

$$\overline{E_1} = E_m V_m + E_f V_f \tag{3.36}$$

If the sollicitation is perpendicular to the fibers, the phases will be considered "in series" when applying the uniform stress approximation (Fig. 3.8).

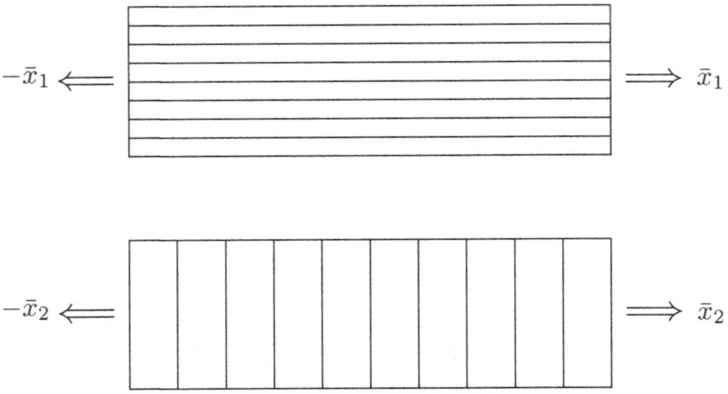

Fig. 3.8 Measurement of $\overline{E_1}$ and $\overline{E_2}$

On a plate, one obtains $\overline{E_2}$ such that:

$$\frac{1}{\overline{E_2}} = \frac{V_m}{E_m} + \frac{V_f}{E_f} \tag{3.37}$$

with a traction along $\overline{x_1}$, the sum of lateral deformations of each phase gives:

$$\overline{\varepsilon_2^{el}} = \overline{\varepsilon_{2m}^{el}} V_m + \overline{\varepsilon_{2f}^{el}} V_f$$

where

$$\overline{\varepsilon_{2m}^{el}} = -\overline{\nu_{12m}}\, \overline{\varepsilon_1^{el}}\ , \ \overline{\varepsilon_{2f}} = -\overline{\nu_{12f}}\, \overline{\varepsilon_1^{el}}$$

$\Longrightarrow \overline{\varepsilon_2^{el}} = -\left(\overline{\nu_{12m}} V_m + \overline{\nu_{12f}} V_f\right) \overline{\varepsilon_1^{el}}$ what gives the equivalent Poisson coefficient:

$$\overline{\nu_{12}} = \overline{\nu_{12m}} V_m + \overline{\nu_{12f}} V_f \tag{3.38}$$

For the transverse shear term, the most realistic assumption is to assume that the shear stress is the same in the matrix and in the fiber. The average is therefore applied to the reverse of the modulus:

$$\frac{1}{\overline{G_{12}}} = \frac{V_m}{G_m} + \frac{V_f}{G_f} \tag{3.39}$$

Note that these results from the mixtures laws do not take into account the nature of multiaxial loading. In addition, when a material is heterogeneous, it leads to a multiaxial field of internal stresses produced by the deformation incompatibilities.

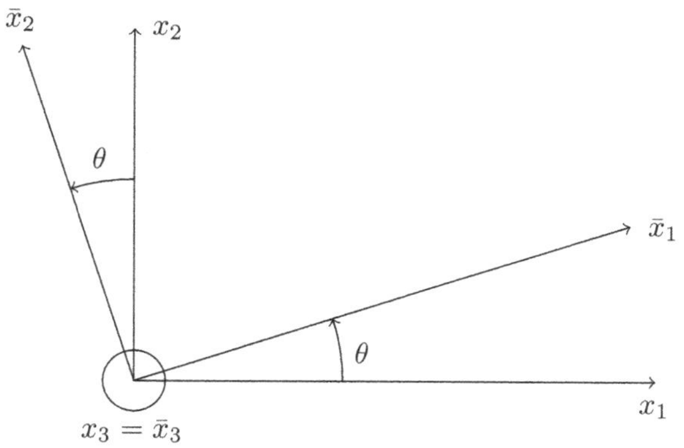

Fig. 3.9 Material frame R becomes orthotropic frame $\overline{R}$ by rotation of axis $\overrightarrow{x_3}$ and angle θ

3.5.3 *Expression of the Elastic Constants in a Reference Frame Linked to the Geometry of the Structure*

In the previous paragraph, material and orthotropy directions are the same. This is not the case for layers with different fiber orientations or tubes with different winding angles of 0–90° (Fig. 3.9).

Let us denote $(ox_1x_2x_3)$ the reference frame R linked to the geometry of the structure, and $(o\overline{x_1x_2x_3})$ the orthotropic reference frame $\overline{R}$.

The state of stress is naturally given in R and the behavior laws are expressed in $\overline{R}$.

$$R \Longrightarrow \overline{R}\ par\ rot\left(\theta, \overrightarrow{x_3} = \overrightarrow{\overline{x_3}}\right)\ ou\ \overline{R} \Longrightarrow R\ rot\left(-\theta, \overrightarrow{x_3} = \overrightarrow{\overline{x_3}}\right)$$

Let us denote $c = \cos\theta$ and $s = \sin\theta$

For example, let us calculate the components $\overline{\sigma_i}$ as a function of σ_i:

$$\begin{bmatrix} \overline{\sigma_1} \\ \overline{\sigma_2} \\ \overline{\sigma_3} \\ \overline{\sigma_4} \\ \overline{\sigma_5} \\ \overline{\sigma_6} \end{bmatrix} = \begin{bmatrix} c^2 & s^2 & 0 & 0 & 0 & 2sc \\ s^2 & c^2 & 0 & 0 & 0 & -2sc \\ 0 & 0 & 1 & 0 & 0 & 0 \\ 0 & 0 & 0 & c & -s & 0 \\ 0 & 0 & 0 & s & c & 0 \\ -sc & sc & 0 & 0 & 0 & c^2 - s^2 \end{bmatrix} \begin{bmatrix} \sigma_1 \\ \sigma_2 \\ \sigma_3 \\ \sigma_4 \\ \sigma_5 \\ \sigma_6 \end{bmatrix}$$

In plane stress,one can reduce expression to lines 1, 2, and 6:

$$\begin{bmatrix} \overline{\sigma_1} \\ \overline{\sigma_2} \\ \overline{\sigma_6} \end{bmatrix} = \begin{bmatrix} c^2 & s^2 & 2sc \\ s^2 & c^2 & -2sc \\ -sc & sc & c^2 - s^2 \end{bmatrix} \begin{bmatrix} \sigma_1 \\ \sigma_2 \\ \sigma_6 \end{bmatrix}$$

For convenience, let us denote $\underline{T}$ the 3×3 matrix base change:

$$\underline{T} = \begin{bmatrix} c^2 & s^2 & -2sc \\ s^2 & c^2 & 2sc \\ sc & -sc & c^2 - s^2 \end{bmatrix}$$

We easily obtain $\underline{T}^{-1}$ considering a change of the rotation angle from $\underline{T}$ by $\theta \Longrightarrow -\theta$ e.g., $s \Longrightarrow -s$ et $c \Longrightarrow c$, what gives:

$$\begin{bmatrix} \overline{\sigma_1} \\ \overline{\sigma_2} \\ \overline{\sigma_6} \end{bmatrix} = \underline{T}^{-1} \begin{bmatrix} \sigma_1 \\ \sigma_2 \\ \sigma_6 \end{bmatrix} \quad and \quad \begin{bmatrix} \overline{\varepsilon_1^{el}} \\ \overline{\varepsilon_2^{el}} \\ \overline{\varepsilon_6^{el}/2} \end{bmatrix} = \underline{T}^{-1} \begin{bmatrix} \varepsilon_1^{el} \\ \varepsilon_2^{el} \\ \varepsilon_6^{el}/2 \end{bmatrix}$$

by introducing the Reuter matrix $\underline{R}$:

$$\begin{bmatrix} \varepsilon_1^{el} \\ \varepsilon_2^{el} \\ \varepsilon_3^{el} \\ \varepsilon_4^{el} \\ \varepsilon_5^{el} \\ \varepsilon_6^{el} \end{bmatrix} = \begin{bmatrix} 1&0&0&0&0&0 \\ 0&1&0&0&0&0 \\ 0&0&1&0&0&0 \\ 0&0&0&2&0&0 \\ 0&0&0&0&2&0 \\ 0&0&0&0&0&2 \end{bmatrix} \begin{bmatrix} \varepsilon_{11}^{el} \\ \varepsilon_{22}^{el} \\ \varepsilon_{33}^{el} \\ \varepsilon_{23}^{el} \\ \varepsilon_{31}^{el} \\ \varepsilon_{12}^{el} \end{bmatrix} \quad \left(\varepsilon_i^{el}\right) = \underline{R} \left(\varepsilon_{ij}^{el}\right)$$

we obtain in the case of plane strains:

$$\begin{bmatrix} \varepsilon_1^{el} \\ \varepsilon_2^{el} \\ \varepsilon_6^{el} \end{bmatrix} = \begin{bmatrix} 1&0&0 \\ 0&1&0 \\ 0&0&2 \end{bmatrix} \begin{bmatrix} \varepsilon_{11}^{el} \\ \varepsilon_{22}^{el} \\ \varepsilon_{12}^{el} \end{bmatrix}$$

and therefore, we have in $\overline{R}$:

$$\begin{bmatrix} \overline{\sigma_1} \\ \overline{\sigma_2} \\ \overline{\sigma_6} \end{bmatrix} = \underline{\overline{Q}} \begin{bmatrix} \overline{\varepsilon_1^{el}} \\ \overline{\varepsilon_2^{el}} \\ \overline{\varepsilon_6^{el}} \end{bmatrix} \quad with\ \underline{\overline{Q}} = \begin{bmatrix} \overline{Q_{11}} & \overline{Q_{12}} & 0 \\ \overline{Q_{12}} & \overline{Q_{22}} & 0 \\ 0 & 0 & \overline{Q_{66}} \end{bmatrix}$$

Finally:

$$\underline{\sigma} = \underline{T}^{-1} \underline{\bar{Q}} \underline{R} \underline{T} \underline{R}^{-1} \underline{\varepsilon}^{el}$$

If the plane stress conditions are not satisfied, $\underline{\overline{Q}}$ can be replaced by $\underline{\overline{C}}$:

$$\begin{bmatrix} \sigma_1 \\ \sigma_2 \\ \sigma_6 \end{bmatrix} = Q \begin{bmatrix} \varepsilon_1^{el} \\ \varepsilon_2^{el} \\ \varepsilon_6^{el} \end{bmatrix} = \begin{bmatrix} Q_{11} & Q_{12} & Q_{16} \\ Q_{12} & Q_{22} & Q_{26} \\ Q_{16} & Q_{26} & Q_{66} \end{bmatrix} \begin{bmatrix} \varepsilon_1^{el} \\ \varepsilon_2^{el} \\ \varepsilon_6^{el} \end{bmatrix} \tag{3.40}$$

$$\begin{cases} Q_{11} = c^4\overline{Q_{11}} + s^4\overline{Q_{22}} + 2c^2s^2\left(\overline{Q_{12}} + 2\overline{Q_{66}}\right) \\ Q_{22} = s^4\overline{Q_{11}} + c^4\overline{Q_{22}} + 2c^2s^2\left(\overline{Q_{12}} + 2\overline{Q_{66}}\right) \\ Q_{66} = c^2s^2\left(\overline{Q_{11}} + \overline{Q_{22}} - 2\overline{Q_{12}} - 2\overline{Q_{66}}\right) + \left(c^4+s^4\right)\overline{Q_{66}} \\ Q_{12} = c^2s^2\left(\overline{Q_{11}} + \overline{Q_{22}} - 4\overline{Q_{66}}\right) + \left(c^4+s^4\right)\overline{Q_{12}} \\ Q_{16} = sc^3\left(\overline{Q_{11}} - \overline{Q_{12}} - 2\overline{Q_{66}}\right) + s^3c\left(\overline{Q_{12}} - \overline{Q_{22}} + 2\overline{Q_{66}}\right) \\ Q_{26} = s^3c\left(\overline{Q_{11}} - \overline{Q_{12}} - 2\overline{Q_{66}}\right) + s^3c\left(\overline{Q_{12}} - \overline{Q_{22}} + 2\overline{Q_{66}}\right) \end{cases} \tag{3.41}$$

Expression of compliance tensors components:

$$\left(\underline{\varepsilon^{el}}\right) = \underline{RT^{-1}R^{-1}\overline{S}T}\left(\underline{\sigma}\right)$$

after calculation, the following relationships can be found:

$$\begin{cases} S_{11} = c^4\overline{S_{11}} + s^4\overline{S_{22}} + s^2c^2\left(2\overline{S_{12}} + \overline{S_{66}}\right) \\ S_{22} = s^4\overline{S_{11}} + c^4\overline{S_{22}} + s^2c^2\left(2\overline{S_{12}} + \overline{S_{66}}\right) \\ S_{66} = 2s^2c^2\left(2\overline{S_{11}} + 2\overline{S_{22}} - 4\overline{S_{12}} - \overline{S_{66}}\right) + \left(c^4+s^4\right)\overline{S_{66}} \\ S_{12} = s^2c^2\left(\overline{S_{11}} + \overline{S_{22}} - \overline{S_{66}}\right) + \left(c^4+s^4\right)\overline{S_{12}} \\ S_{16} = sc^3\left(2\overline{S_{11}} - 2\overline{S_{12}} - \overline{S_{66}}\right) - s^3c\left(2\overline{S_{22}} - 2\overline{S_{12}} - \overline{S_{66}}\right) \\ S_{26} = s^3c\left(2\overline{S_{11}} - 2\overline{S_{12}} - \overline{S_{66}}\right) - sc^3\left(2\overline{S_{22}} - 2\overline{S_{12}} - \overline{S_{66}}\right) \end{cases} \tag{3.42}$$

with engineering constants:

$$\begin{cases} \overline{S_{11}} = \frac{1}{\overline{E_1}}\ \overline{S_{22}} = \frac{1}{\overline{E_2}} \\ \overline{S_{12}} = -\frac{\overline{\nu_{12}}}{\overline{E_1}} = -\frac{\overline{\nu_{21}}}{\overline{E_2}}\ \overline{S_{66}} = \frac{1}{\overline{G_{12}}} \end{cases}$$

One obtains:

$$\begin{cases} \frac{1}{E_1} = \frac{c^4}{\overline{E_1}} + \frac{s^4}{\overline{E_2}} + s^2c^2\left(\frac{1}{\overline{G_{12}}} - \frac{2\overline{\nu_{12}}}{\overline{E_1}}\right) \\ \frac{1}{E_2} = \frac{s^4}{\overline{E_1}} + \frac{c^4}{\overline{E_2}} + s^2c^2\left(\frac{1}{\overline{G_{12}}} - \frac{2\overline{\nu_{12}}}{\overline{E_1}}\right) \\ -\frac{\nu_{12}}{E_1} = s^2c^2\left(\frac{1}{\overline{E_1}} + \frac{1}{\overline{E_2}} - \frac{1}{\overline{G_{12}}}\right) - \left(c^4+s^4\right)\frac{\overline{\nu_{12}}}{\overline{E_1}} \\ \frac{1}{G_{12}} = 2s^2c^2\left(\frac{2}{\overline{E_1}} + \frac{2}{\overline{E_2}} + 4\frac{\overline{\nu_{12}}}{\overline{E_1}} - \frac{1}{\overline{G_{12}}}\right) + \left(c^4+s^4\right)\frac{1}{\overline{G_{12}}} \end{cases} \tag{3.43}$$

Even if they are not exploited, components S_{16} and S_{26} are not equal to zero. We only need four equations to obtain the elastic constants $(E_1, E_2, \nu_{12}, G_{12})$ in R, from the knowledge of the elastic constants $\left(\overline{E_1}, \overline{E_2}, \overline{\nu_{12}}, \overline{G_{12}}\right)$ in the anisotropic frame $\overline{R}$ and angle θ of fibers orientation.

3.5.4 Exercises 3.6 et 3.7

3.5.4.1 Exercise 3.6

Statement:

Let us consider a unidirectional ply "composite consisting of high-strength carbon/epoxy resin."

What is the volume fraction of fiber to obtain a Young's modulus in the longitudinal sense similar to the one of duralumin (AU4G-2024)?

Numerical application: high-strength carbon: $E_f = 230000$ MPa, epoxy resin $E_m = 4500$ MPa, duralumin $E_{AU4G} = 75000$ MPa.

Correction:

$$\begin{cases} E_L = \overline{E_1} = E_m V_m + E_f V_f = E_{AU4G} \\ 1 = V_m + V_f \end{cases}$$

$$\Longrightarrow E_L = \left(1 - V_f\right) E_m + V_f E_f \Longrightarrow V_f = \frac{E_{AU4G} - E_m}{E_f - E_m}$$

A.N.: $V_f \simeq 0.31$.

Exercise 3.7 (Berthelot 1999).

Statement:

An unidirectional ply is subject to the following strain state in its material plane $(x_1,\ x_2)$:

$$\begin{cases} \varepsilon_1^{el} = 10^{-2} \\ \varepsilon_2^{el} = -5 \cdot 10^{-3} \\ \varepsilon_6^{el} = 2 \cdot 10^{-2} \end{cases}$$

The angle θ between x_1 *and* $\overline{x_1}$ is equal to 30°. The elastic constants of the composite material are equal to:

$\overline{E_1} = 40$ GPa, $\overline{E_2} = 10$ GPa, $\overline{\nu_{12}} = 0.32$, $\overline{G_{12}} = 4.5$ GPa.

Considering that the ply in plane stress state, let us determine:

1. the stresses $\sigma_1,\ \sigma_2,\ \sigma_6$ in R.
2. the stresses $\overline{\sigma_1},\ \overline{\sigma}_2,\ \overline{\sigma_6}$ in $\overline{R}$.

Correction:

In order to determine the stiffness matrix $\underline{Q}$ in the frame R of the sample, one must determine $\overline{\underline{Q}}$ in $\overline{R}$ e.g. $\overline{Q_{ij}}$ and Q_{ij} in GPa.

$$\begin{cases} \overline{Q_{11}} = \frac{\overline{E_1}}{1-\overline{\nu_{12}\nu_{21}}} = 41.0151 \\ \overline{Q_{22}} = \frac{\overline{E_2}}{1-\overline{\nu_{12}\nu_{21}}} = 10.263 \\ \overline{Q_{12}} = \frac{\overline{\nu_{12}E_2}}{1-\overline{\nu_{12}\nu_{21}}} = 3.284 \\ \overline{Q_{66}} = \overline{G_{12}} = 4.5 \end{cases} \qquad \overline{Q_{16}} = \overline{Q_{26}} = 0$$

$$\begin{cases} Q_{11} = c^4\overline{Q_{11}} + s^4\overline{Q_{22}} + 2c^2s^2\left(\overline{Q_{12}} + 2\overline{Q_{66}}\right) = 28.339 \\ Q_{22} = s^4\overline{Q_{11}} + c^4\overline{Q_{22}} + 2c^2s^2\left(\overline{Q_{12}} + 2\overline{Q_{66}}\right) = 12.945 \\ Q_{66} = c^2s^2\left(\overline{Q_{11}} + \overline{Q_{22}} - 2\overline{Q_{12}} - 2\overline{Q_{66}}\right) + \left(c^4 + s^4\right)\overline{Q_{66}} = 9.515 \\ Q_{12} = c^2s^2\left(\overline{Q_{11}} + \overline{Q_{22}} - 4\overline{Q_{66}}\right) + \left(c^4 + s^4\right)\overline{Q_{12}} = 8.299 \\ Q_{16} = sc^3\left(\overline{Q_{11}} - \overline{Q_{12}} - 2\overline{Q_{66}}\right) + s^3c\left(\overline{Q_{12}} - \overline{Q_{22}} + 2\overline{Q_{66}}\right) = 9.561 \\ Q_{26} = s^3c\left(\overline{Q_{11}} - \overline{Q_{12}} - 2\overline{Q_{66}}\right) + s^3c\left(\overline{Q_{12}} - \overline{Q_{22}} + 2\overline{Q_{66}}\right) = 3.770 \end{cases}$$

$$\begin{bmatrix} \sigma_1 \\ \sigma_2 \\ \sigma_6 \end{bmatrix} = \underline{Q} \begin{bmatrix} \varepsilon_1^{el} \\ \varepsilon_2^{el} \\ \varepsilon_6^{el} \end{bmatrix}$$

It comes σ_i and $\overline{\sigma}_i$ in MPa:

$$\begin{cases} \sigma_1 = 433 \\ \sigma_2 = 94 \\ \sigma_6 = 267 \end{cases}$$

$$\begin{bmatrix} \overline{\sigma_1} \\ \overline{\sigma_2} \\ \overline{\sigma_6} \end{bmatrix} = \begin{bmatrix} c^2 & s^2 & 2sc \\ s^2 & c^2 & -2sc \\ -sc & sc & c^2 - s^2 \end{bmatrix} \begin{bmatrix} \sigma_1 \\ \sigma_2 \\ \sigma_6 \end{bmatrix} = \begin{bmatrix} 580 \\ -53 \\ -13.5 \end{bmatrix}$$

3.5.5 *Theory of Laminates*

A laminate plate consists of a stack of several unidirectional layers (Fig. 3.10). These layers are usually oriented differently (typically 0°, 45°, 90°, −45°).

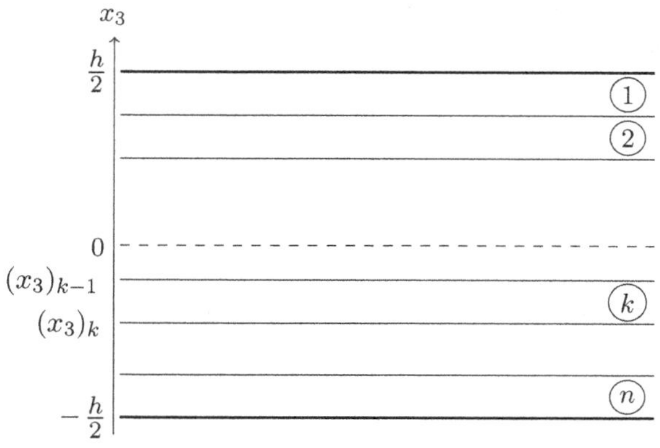

Fig. 3.10 Classification of a multilayer in the thickness

In design, it is important to comply with to the "mirror symmetry", meaning that the stacking sequence, of both sides of the mean plane, is symmetrical. Indeed, if the plate does not have this symmetry, it is likely to "cure" during manufacture due to differences in expansion or in service as a result of the tensile shear coupling.

Naturally, the axis $o\overrightarrow{x_3}$ will be chosen as perpendicular to the plane of the laminated plate.

It is a defined nomenclature.

For example, the notation $[0_3/90_2/45/-45]_{s\ ou\ T}$ means: S symmetric or T antisymmetric and:
−3 Layers oriented at 0°
−2 Layers oriented at 90°
−1 Layer oriented at 45°
−1 Layer oriented at −45°.
If we examine the bending of laminated plates, the classical theory of plates gives:

$$\vec{u}\,(M_0) = \vec{u}\,(H_0) + \vec{\beta} \wedge \overrightarrow{HM_0}$$

with $M_0\,(x_1, x_2, x_3)$ et $H_0\,(x_1, x_2, 0)$.
by clarifying:

$$\begin{bmatrix} u_1\,(x_1, x_2, x_3) \\ u_2\,(x_1, x_2, x_3) \\ u_3\,(x_1, x_2, x_3) \end{bmatrix} = \begin{bmatrix} u_1\,(x_1, x_2, 0) \\ u_2\,(x_1, x_2, 0) \\ u_3\,(x_1, x_2, 0) \end{bmatrix} + \begin{bmatrix} \frac{\partial u_3}{\partial x_2}\,(x_1, x_2, 0) \\ -\frac{\partial u_3}{\partial x_1}\,(x_1, x_2, 0) \\ 0 \end{bmatrix} \wedge \begin{bmatrix} 0 \\ 0 \\ x_3 \end{bmatrix}$$

Let us denote: $\begin{cases} u_1\,(x_1, x_2, 0) = u_1^0\,(x_1, x_2) \\ u_2\,(x_1, x_2, 0) = u_2^0\,(x_1, x_2) \\ u_3\,(x_1, x_2, 0) = u_3^0\,(x_1, x_2) \end{cases}$.

We make the following standard assumptions:
– Straight sections before deformation remain straight after deformation.
– Continuity of deformation at the interfaces.
– The interaction of membrane and bending effects due to the large displacement is neglected.

$$\begin{cases} u_1 = u_1^0 - x_3 \frac{\partial u_3^0}{\partial x_1} \\ u_2 = u_2^0 - x_3 \frac{\partial u_3^0}{\partial x_2} \\ u_3 = u_3^0 \end{cases}$$

Strains calculations:

This is a problem of large displacements and small strains.

$$\varepsilon_{ij}^{el} = \frac{1}{2}\left(\frac{\partial u_i}{\partial x_j} + \frac{\partial u_j}{\partial x_i}\right)$$

$$\begin{cases} \varepsilon_{11}^{el} = \frac{\partial u_1}{\partial x_1} = \frac{\partial u_1^0}{\partial x_1} - x_3 \frac{\partial^2 u_3^0}{\partial x_1^2} \\ \varepsilon_{22}^{el} = \frac{\partial u_2}{\partial x_2} = \frac{\partial u_2^0}{\partial x_2} - x_3 \frac{\partial^2 u_3^0}{\partial x_2^2} \\ \varepsilon_{12}^{el} = \frac{1}{2}\left(\frac{\partial u_1}{\partial x_2} + \frac{\partial u_2}{\partial x_1}\right) = \frac{1}{2}\left(\frac{\partial u_1^0}{\partial x_2} + \frac{\partial u_2^0}{\partial x_1} - 2x_3 \frac{\partial^2 u_3^0}{\partial x_1 \partial x_2}\right) \end{cases} \qquad \varepsilon_{13}^{el} = \varepsilon_{23}^{el} = \varepsilon_{33}^{el} = 0$$

Note that the assumption of small perturbations generates a plane strain problem in this case.

It can be formulated as:

$$\begin{bmatrix} \varepsilon_1^{el} \\ \varepsilon_2^{el} \\ \varepsilon_6^{el} \end{bmatrix} = \begin{bmatrix} \varepsilon_1^0 \\ \varepsilon_2^0 \\ \varepsilon_6^0 \end{bmatrix} + x_3 \begin{bmatrix} \rho_1 = -\frac{\partial^2 u_3}{\partial x_1^2} \\ \rho_2 = -\frac{\partial^2 u_3}{\partial x_2^2} \\ \rho_6 = -2\frac{\partial^2 u_3}{\partial x_1 \partial x_2} \end{bmatrix} \tag{3.44}$$

The same relations on the $\overline{\varepsilon_i}$ can be obtained in the orthotropic frame $\overline{R}$.

The stresses can be written in the kth layer:

$$\begin{bmatrix} \overline{\sigma_1} \\ \overline{\sigma_2} \\ \overline{\sigma_6} \end{bmatrix} = \begin{bmatrix} \overline{Q_{11}} & \overline{Q_{12}} & 0 \\ \overline{Q_{12}} & \overline{Q_{22}} & 0 \\ 0 & 0 & \overline{Q_{66}} \end{bmatrix} \left[\begin{bmatrix} \overline{\varepsilon_1^0} \\ \overline{\varepsilon_2^0} \\ \overline{\varepsilon_6^0} \end{bmatrix} + x_3 \begin{bmatrix} \overline{\rho_1} \\ \overline{\rho_2} \\ \overline{\rho_6} \end{bmatrix} \right] \tag{3.45}$$

We consider an external mechanical loading.

The generalized forces can be written on one layer as follows:

$$\begin{bmatrix} N_1 & N_2 & N_{12} \\ M_1 & M_2 & M_{12} \end{bmatrix} = \int_{-\frac{h}{2}}^{+\frac{h}{2}} \begin{bmatrix} 1 \\ x_3 \end{bmatrix} \left[\sigma_{11}, \sigma_{22}, \sigma_{12} \right] dx_3 \tag{3.46}$$

For a multilayer, each integral from N_1 to N_{12} is the result of a sum of integrals over the thickness of each layer.

$$\begin{bmatrix} N_1 & N_2 & N_{12} \\ M_1 & M_2 & M_{12} \end{bmatrix} = \sum_{k=1}^{k=n} \int_{(x_3)_{k-1}}^{(x_3)_k} \begin{bmatrix} 1 \\ x_3 \end{bmatrix} \left[\sigma_{11}^{(k)}, \sigma_{22}^{(k)}, \sigma_{12}^{(k)} \right] dx_3$$

$$\begin{bmatrix} N_1 \\ N_2 \\ N_{12} \end{bmatrix} = \sum_{k=1}^{k=n} \int_{(x_3)_{k-1}}^{(x_3)_k} \left[Q_{ij} \right]_k \left[\begin{bmatrix} \varepsilon_1^0 \\ \varepsilon_2^0 \\ \varepsilon_6^0 \end{bmatrix} + x_3 \begin{bmatrix} \rho_1 \\ \rho_2 \\ \rho_6 \end{bmatrix} \right] dx_3$$

$$\begin{bmatrix} M_1 \\ M_2 \\ M_{12} \end{bmatrix} == \sum_{k=1}^{k=n} \int_{(x_3)_{k-1}}^{(x_3)_k} \left[Q_{ij} \right]_k \left[x_3 \begin{bmatrix} \varepsilon_1^0 \\ \varepsilon_2^0 \\ \varepsilon_6^0 \end{bmatrix} + x_3^2 \begin{bmatrix} \rho_1 \\ \rho_2 \\ \rho_6 \end{bmatrix} \right] dx_3$$

Let us recall that for all practical purposes, ε_i and ρ_i are not functions of the coordinate x_3, and also the elastic constants of each layer k: $\left[Q_{ij} \right]_k$.

It follows that:

$$\begin{bmatrix} N_1 \\ N_2 \\ N_{12} \end{bmatrix} = \begin{bmatrix} A_{11} & A_{12} & A_{16} \\ A_{12} & A_{22} & A_{26} \\ A_{16} & A_{26} & A_{66} \end{bmatrix} \begin{bmatrix} \varepsilon_1^0 \\ \varepsilon_2^0 \\ \varepsilon_6^0 \end{bmatrix} + \begin{bmatrix} B_{11} & B_{12} & B_{16} \\ B_{12} & B_{22} & B_{26} \\ B_{16} & B_{26} & B_{66} \end{bmatrix} \begin{bmatrix} \rho_1 \\ \rho_2 \\ \rho_6 \end{bmatrix} \tag{3.47}$$

$$\begin{bmatrix} M_1 \\ M_2 \\ M_{12} \end{bmatrix} = \begin{bmatrix} B_{11} & B_{12} & B_{16} \\ B_{12} & B_{22} & B_{26} \\ B_{16} & B_{26} & B_{66} \end{bmatrix} \begin{bmatrix} \varepsilon_1^0 \\ \varepsilon_2^0 \\ \varepsilon_6^0 \end{bmatrix} + \begin{bmatrix} D_{11} & D_{12} & D_{16} \\ D_{12} & D_{22} & D_{26} \\ D_{16} & D_{26} & D_{66} \end{bmatrix} \begin{bmatrix} \rho_1 \\ \rho_2 \\ \rho_6 \end{bmatrix} \tag{3.48}$$

with:

$$\begin{cases} A_{ij} = \sum_{k=1}^{k=n} \left[Q_{ij}\right]_k \left((x_3)_k - (x_3)_{k-1}\right) \\ B_{ij} = \frac{1}{2} \sum_{k=1}^{k=n} \left[Q_{ij}\right]_k \left((x_3)_k^2 - (x_3)_{k-1}^2\right) \\ D_{ij} = \frac{1}{3} \sum_{k=1}^{k=n} \left[Q_{ij}\right]_k \left((x_3)_k^3 - (x_3)_{k-1}^3\right) \end{cases} \tag{3.49}$$

Note that the tensile force in the mean plane of the plate generates the bending (Fig. 3.11).
The coupling is associated with from the term $\underline{B}$.
To eliminate this coupling, we need to cancel the terms B_{ij}.
By doing this, the mirror symmetry condition is met.

For example:

layer at 45°		
layer at 0°		Al
	or "sandwich"	époxy resin
layer at 0°		Al
layer at 45°		

Plane coupling:

The terms A_{16} *and* A_{26} show the existence of a coupling between normal force angular distortion (Fig. 3.12).

To eliminate this coupling, A_{16} *and* A_{26} must be removed in the matrix $\underline{A}$. Only if there exists in the thickness of the plate, the thickness of ply is equal to Q_{16} (Q_{26}) that cancel in pairs.

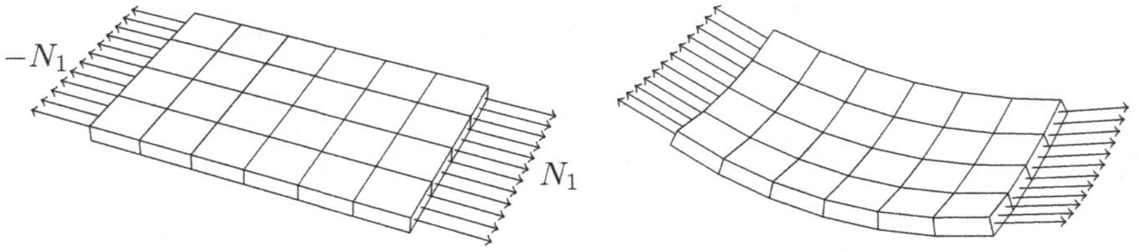

Fig. 3.11 tensile force in the mean plane of the plate generates the bending

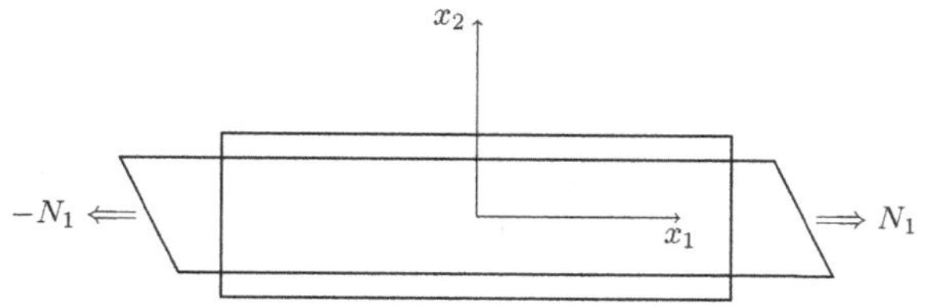

Fig. 3.12 A "rectangle" can generate a parallelogram

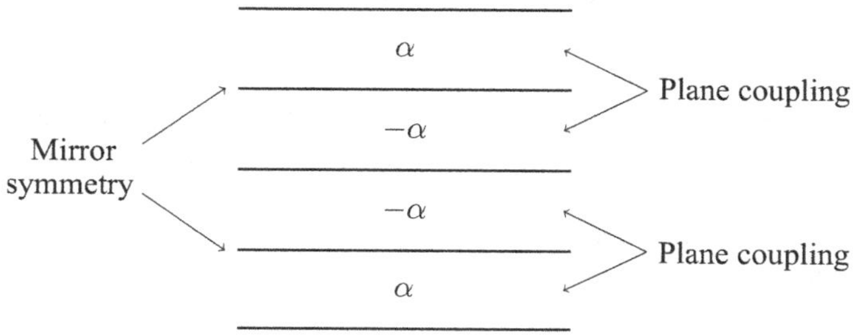

Fig. 3.13 Example of multilayer with "mirror symmetry" 2 layers oriented at α and at $-\alpha$

To eliminate membrane-bending coupling and coupling plane, a multilayer is chosen with "mirror symmetry" and n layers oriented at α and n layers oriented at $-\alpha$. (Fig. 3.13)

Resolution:

To solve a problem of plates, it will get a satisfying solution:
(1) the equilibrium equations,
(2) kinematic conditions,
(3) the boundaries conditions in terms of stress and (or) displacement.
For plates having mirror symmetry loaded in their plane, only constraints existing are σ_{11}, σ_{22}, σ_{12}.

The equations of static equilibrium:

$$\begin{cases} \sigma_{11,1} + \sigma_{12,2} + f_1 = 0 \\ \sigma_{21,1} + \sigma_{22,2} + f_2 = 0 \end{cases}$$

The integrations from $-\frac{h}{2}$ to $+\frac{h}{2}$ give:

$$\begin{cases} N_{1,1} + N_{12,2} + P_1 = 0 \\ N_{12,1} + N_{2,2} + P_2 = 0 \end{cases}$$

with:

$$\begin{cases} N_1 = \sum_{k=1}^{k=n} Q_{1j}^k \left(x_k - x_{k-1}\right) \varepsilon_j^0 \\ N_2 = \sum_{k=1}^{k=n} Q_{2j}^k \left(x_k - x_{k-1}\right) \varepsilon_j^0 \\ N_{12} = \sum_{k=1}^{k=n} Q_{6j}^k \left(x_k - x_{k-1}\right) \varepsilon_j^0 \end{cases}$$

3.5.6 Exercise 3.8 on Laminated Media

Statement (Berthelot 1999):

Consider a laminate of two layers of unidirectional composite. The lower layer 1 of 3 mm thick is oriented at 45° from the reference material laminate R. The upper layer 2 is oriented at 0° and a thickness of 5 mm.

The composite material consists of two materials «resin epoxy» and «glass fibers» with respective elastic constants:

$\overline{E_1} = 46$ GPa, $\overline{E_2} = 10$ GPa, $\overline{G_{12}} = 4.6$ GPa, $\overline{\nu_{12}} = 0.31$

Let us write the constitutive equation of the laminate.

Correction:

(1) Calculation of $\overline{Q_{ij}}$ in GPa: $\overline{\nu_{21}} = \frac{\overline{E_2}}{\overline{E_1}} \overline{\nu_{12}}$

$$\begin{cases} \overline{Q_{11}} = \frac{\overline{E_1}}{1 - \overline{\nu_{12}}\,\overline{\nu_{21}}} = 46.892 \quad \overline{Q_{22}} = \frac{\overline{E_2}}{1 - \overline{\nu_{12}}\,\overline{\nu_{21}}} = 10.213 \\ \overline{Q_{12}} = \frac{\overline{\nu_{12}}\,\overline{E_2}}{1 - \overline{\nu_{12}}\,\overline{\nu_{21}}} = 3.166 \quad \overline{Q_{66}} = \overline{G_{12}} = 4.6 \end{cases}$$

(2) Stiffness matrices of each layer expressed in the axis of the laminate.
Layer at 0°:

$$\underline{\underline{Q_{0^\circ}}} = \begin{bmatrix} 46.892 & 3.166 & 0 \\ 3.166 & 10.213 & 0 \\ 0 & 0 & 4.6 \end{bmatrix} \text{GPa}$$

Layer at 45°:

$$\begin{cases} Q_{11} = \left(\overline{Q_{11}} + \overline{Q_{22}} + 2\left(\overline{Q_{12}} + 2\overline{Q_{66}}\right)\right) cos^4 45^\circ = 20.482 \text{ GPa} \\ Q_{22} = \left(\overline{Q_{11}} + \overline{Q_{22}} + 2\left(\overline{Q_{12}} + 2\overline{Q_{66}}\right)\right) cos^4 45^\circ = Q_{11} = 20.482 \text{ GPa} \\ Q_{66} = \left(\overline{Q_{11}} + \overline{Q_{22}} - 2\overline{Q_{12}}\right) cos^4 45^\circ = 12.716 \text{ GPa} \\ Q_{12} = \left(\overline{Q_{11}} + \overline{Q_{22}} - 4\overline{Q_{66}} + 2\overline{Q_{12}}\right) cos^4 45^\circ = 11.282 \text{ GPa} \\ Q_{16} = \left(\overline{Q_{11}} - \overline{Q_{22}}\right) cos^4 45^\circ = 9.192 \text{ GPa} \\ Q_{26} = Q_{16} = 9.192 \text{ GPa} \end{cases}$$

$$\underline{Q_{45^\circ}} = \begin{bmatrix} 20.482 & 11.282 & 9.192 \\ 11.282 & 20.482 & 9.192 \\ 9.192 & 9.192 & 12.716 \end{bmatrix} \text{GPa}$$

(3) Matrix A: A_{ij} in 10^6 Nm^{-1}

$$\underline{A} = \left[3\underline{Q_{45^\circ}} + 5\underline{Q_{0^\circ}}\right] 10^{-3}$$

$$\underline{A} = \begin{bmatrix} 296.35 & 49.676 & 25.576 \\ 49.676 & 112.51 & 27.576 \\ 25.576 & 25.576 & 61.147 \end{bmatrix}$$

(4) Matrix B:

$$B_{ij} = \frac{1}{2}\sum_{k=1}^{k=n}\left[Q_{ij}\right]_k \left((x_3)_k^2 - (x_3)_{k-1}^2\right)$$

$$\underline{B} = 7.5\left[-\underline{Q_{45^\circ}} + \underline{Q_{0^\circ}}\right] 10^{-6}$$

(5) Matrix D:

$$D_{ij} = \frac{1}{3}\sum_{k=1}^{k=n}\left[Q_{ij}\right]_k \left((x_3)_k^3 - (x_3)_{k-1}^3\right)$$

$$\underline{D} = \left[21\underline{Q_{45^\circ}} + \frac{65}{3}\underline{Q_{0^\circ}}\right] 10^{-9}$$

and finally:

$$\begin{bmatrix} N_1 \\ N_2 \\ N_{12} \\ M_1 \\ M_2 \\ M_{12} \end{bmatrix} = \begin{bmatrix} A & B \\ B & D \end{bmatrix} \begin{bmatrix} \varepsilon_1^0 \\ \varepsilon_2^0 \\ \varepsilon_6^0 \\ \rho_1 \\ \rho_2 \\ \rho_6 \end{bmatrix}$$

In summary, the calculations in composites, sometimes tedious, need to be well structured.

Chapter 4
Yield Elasticity Criteria

Abstract The establishment of behavior laws requires the definition of an elasticity domain, at least in the stress space for which there is no plastic or viscoplastic flow. In a general case, this domain must be convex. One introduces the independence from the hydrostatic pressure (Von Mises or Tresca criterium). The asymmetry between tension and compression is integrated and also the material anisotropy.

4.1 Introduction and Tools

The establishment of behavior laws requires the definition of an elasticity domain, at least in the stress space, for which there are no plastic or viscoplastic flow. For a tensile/compression test, the trace of this field is limited to a straight line, for example, on the axis of the uniaxial stress σ_{11}. Already is a dichotomy between the symmetries of the elastic boundary between compression and tension or not. In addition, some models are able of representing a maximum stress tolerable by the material.

In order to approach the study of multiaxial loadings, it is necessary to define such tridimensional limits.

An elasticity domain C is defined in the stress space σ_{ij}. When the frontier is regular, it is expressed as an inequality:

$$C = \left\{ \underline{\sigma} \parallel f\left(\underline{\sigma}\right) \leq 0 \right\}, \tag{4.1}$$

where the convex function $f\left(\underline{\sigma}\right)$, called «plasticity criterion » allowed to clarify the yield plasticity stress by equality $f\left(\underline{\sigma}\right) = 0$.

One function f in the stress space $\underline{\sigma}$ is said convex if

$$\begin{gathered} f\left(\xi \underline{\sigma_1} + (1-\xi)\,\underline{\sigma_2}\right) \leq \xi f\left(\underline{\sigma_1}\right) + (1-\xi)\, f\left(\underline{\sigma_2}\right) \\ \forall \left(\underline{\sigma_1}, \underline{\sigma_2}\right) \in \underline{\sigma},\ \forall \xi \in [0, 1]\,. \end{gathered} \tag{4.2}$$

In addition, a regular function is a differentiable function without singularity points (points where the derivative has not the same value to the left and right of the singular point in question). Of course, a monotonic function is regular. The maximum shear

C. Lexcellent, *Linear and Non-linear Mechanical Behavior of Solid Materials*, DOI 10.1007/978-3-319-55609-3_4

criterion called Tresca constitutes a special case as it cannot correspond to a smooth function.

Experimental data shows that, for most materials, the area of initial elasticity is convex (this is particularly true for the metals deformed by crystallographic slip). The initial domain is the first mechanical loading of a "virgin" material.

As in the case of the study of elasticity tensor (see Chap. 3), the material symmetries are respected. This implies in particular that f in the case of an isotropic material is a symmetrical function of the eigenstress $\sigma_1,\ \sigma_2,\ \sigma_3$, or which is equivalent to three invariants of the stress tensor:

$$\begin{cases} \Theta_1 = \sigma_1 + \sigma_2 + \sigma_3 = \sigma_{11} + \sigma_{22} + \sigma_{33} = \mathrm{tr}\left(\underline{\sigma}\right) \\ \Theta_2 = \sigma_1\sigma_2 + \sigma_2\sigma_3 + \sigma_3\sigma_1 = \sigma_{11}\sigma_{22} - \sigma_{12}\sigma_{21} + \sigma_{22}\sigma_{33} - \sigma_{23}\sigma_{32} + \sigma_{33}\sigma_{11} - \sigma_{31}\sigma_{13} \\ \Theta_3 = \sigma_1\sigma_2\sigma_3 = \det\left(\underline{\sigma}\right). \end{cases}$$

4.2 Criteria Integrating the Independence from Hydrostatic Pressure of the Mechanical Behavior of the Material

This leads to consider as critical variable, either the stress tensor $\underline{\sigma}$ or the deviatoric tensor of $\underline{\sigma}$, e.g., $\underline{s} = \mathrm{dev}\left(\underline{\sigma}\right)$ defined by

$$\underline{s} = \underline{\sigma} - {}^1\!/\!_3\left(\mathrm{tr}\left(\underline{\sigma}\right)\right)\underline{1}. \tag{4.3}$$

The deviatoric tensor is defined by

$$\mathrm{tr}\left(\mathrm{dev}\underline{\sigma}\right) = 0.$$

The three invariants of $\underline{s}$ are

$$\begin{cases} J_1 = \mathrm{tr}\left(\underline{s}\right) = 0 \\ J_2 = s_1 s_2 + s_2 s_3 + s_2 s_3 = s_{11}s_{22} - s_{12}s_{21} + s_{22}s_{33} - s_{23}s_{32} + s_{33}s_{11} - s_{31}s_{13} \\ J_3 = s_1 s_2 s_3 = \det\left(\underline{s}\right). \end{cases} \tag{4.4}$$

Note that the values of f are homogeneous to stress, facilitating the comparison with the experimental results.

Let us introduce $J = \sqrt{3J_2}$ which can be expressed in terms of $\sigma_1, \sigma_2, \sigma_3$ and reduce to stress σ_1 for tensile test:

$$J = \left(\frac{3}{2}s_{ij}s_{ij}\right)^{\frac{1}{2}} = \left(\frac{1}{2}\left((\sigma_1 - \sigma_2)^2 + (\sigma_2 - \sigma_3)^2 + (\sigma_3 - \sigma_1)^2\right)\right)^{\frac{1}{2}} = \overline{\sigma}, \tag{4.5}$$

that is to say that when $\sigma_1 \geq 0,\ \sigma_2 = \sigma_3 = 0$. So $J = \sigma_1$.

Let us notice that $J = \overline{\sigma}$ is the Huber-von Mises «equivalent stress ».

4.2.1 *Exercise 4.1*

Statement:

The equivalent stress is similar to the shear stress in the octahedral plane. The octahedral planes are those of normal $\overrightarrow{\nu}: [1/\sqrt{3}, 1/\sqrt{3}, 1/\sqrt{3}]$ in the main frame $\overrightarrow{X_1}, \overrightarrow{X_2}, \overrightarrow{X_3}$. Let us calculate the stress vector $\overrightarrow{T}\overrightarrow{\nu}$ and its normal component $N = \sigma_{oct}$ and tangential one $T = \tau_{oct}$.

Correction:

$$\overrightarrow{T}\overrightarrow{\nu} = \begin{bmatrix} \sigma_1 & 0 & 0 \\ 0 & \sigma_2 & 0 \\ 0 & 0 & \sigma_3 \end{bmatrix} \begin{bmatrix} \frac{\sqrt{3}}{3} \\ \frac{\sqrt{3}}{3} \\ \frac{\sqrt{3}}{3} \end{bmatrix} = \frac{\sqrt{3}}{3} \begin{bmatrix} \sigma_1 \\ \sigma_2 \\ \sigma_3 \end{bmatrix}$$

$$\sigma_{oct} = \overrightarrow{\nu}.\overrightarrow{T}.\overrightarrow{\nu} = \frac{1}{3}(\sigma_1 + \sigma_2 + \sigma_3) = \frac{1}{3}\Theta_1 \tag{4.6}$$

$$\tau_{oct} = \left(\left(\overrightarrow{T}\overrightarrow{\nu}\right)^2 - (\sigma_{oct})^2\right)^{\frac{1}{2}} = \frac{1}{3}\left((\sigma_1 - \sigma_2)^2 + (\sigma_2 - \sigma_3)^2 + (\sigma_3 - \sigma_1)^2\right)^{\frac{1}{2}} = \frac{\sqrt{2}}{3}\overline{\sigma}. \tag{4.7}$$

–Huber-von Mises criterion:

In the function f, only the second invariant of the stress deviator tensor is included:

$$f\left(\underline{\sigma}\right) = \overline{\sigma} - \sigma_y, \tag{4.8}$$

where σ_y represents the elastic yield in tension.

Let us notice that this criterion does not take into account an eventual asymmetry between tension and compression.

–Tresca criterion:

It corresponds to the maximal shearing stress of the material. A simple examination of the Mohr tri-circle is sufficient for its definition. If in each point, we classify the eigenstress $\sigma_1 \geq \sigma_2 \geq \sigma_3$ so

$$\tau_{max} = \frac{\sigma_1 - \sigma_3}{2}$$

and

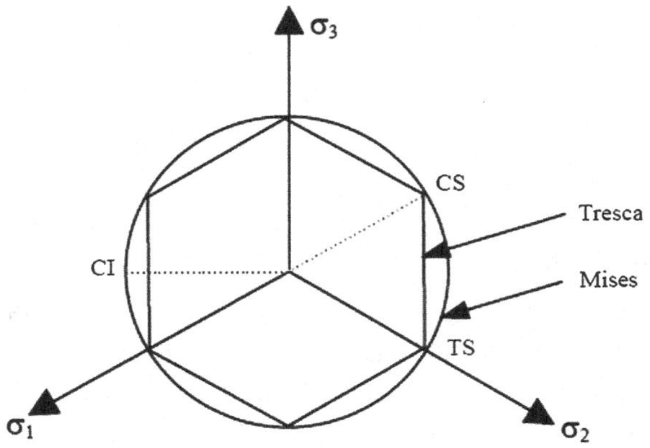

Fig. 4.1 Criteria representation in the deviatoric plane: Tresca; Huber-von Mises

$$f\left(\underline{\sigma}\right) = \sigma_1 - \sigma_3 - \sigma_y. \tag{4.9}$$

In the deviatoric plan, the Huber-von Mises criterion is represented by a circle and the Tresca criterion by an hexagon. (See Fig. 4.1).

4.2.2 Exercise 4.2: Comparison of Tresca Criteria and Huber-von Mises Criteria for Tension–Torsion and Biaxial Tension Tests (Fig. 4.2)

Statement 1: Let us consider the stress state of a thin tube in tension (compression)–torsion:

$$\underline{\sigma}\left(M\left(r,\theta,z\right)\right) = \begin{bmatrix} 0 & 0 & 0 \\ 0 & 0 & \sigma_{z\theta} = \tau \\ 0 & \sigma_{\theta z} = \tau & \sigma_{zz} = \sigma \end{bmatrix}.$$

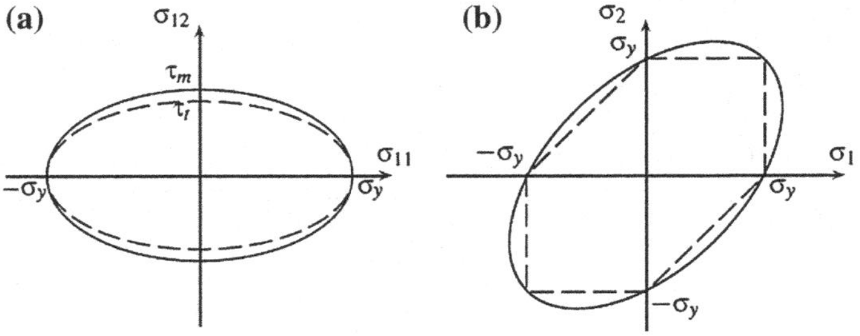

Fig. 4.2 Comparison between Tresca (*dotted*) and Huber-von Mises (*in solid lines*) criteria: **a** tension (compression)–torsion: Huber-von Mises $\tau_m = \sigma_y/\sqrt{3}$, Tresca $\tau_t = \sigma_y/2$, **b** biaxial tension

Give the expressions of Huber-von Mises and Tresca criteria.
Response 1:

–Huber-von Mises: $f(\sigma, \tau) = \left(\sigma^2 + 3\tau^2\right)^{\frac{1}{2}} - \sigma_y$

–Tresca: $f(\sigma, \tau) = \left(\sigma^2 + 4\tau^2\right)^{\frac{1}{2}} - \sigma_y$

Statement 2: Let us consider a thin plate under biaxial tension:

$$\underline{\sigma} M(x_1, x_2, x_3) = \begin{bmatrix} \sigma_1 & 0 & 0 \\ 0 & \sigma_2 & 0 \\ 0 & 0 & 0 \end{bmatrix}.$$

Give the expressions of Huber-von Mises and Tresca criteria.
Response 2:

–Huber-von Mises: $f(\sigma_1, \sigma_2) = \left(\sigma_1^2 + \sigma_2^2 - \sigma_1\sigma_2\right)^{\frac{1}{2}} - \sigma_y$

–Tresca:

$$f(\sigma_1, \sigma_2) = \sigma_2 - \sigma_y \; if \; 0 \le \sigma_1 \le \sigma_2$$
$$f(\sigma_1, \sigma_2) = \sigma_1 - \sigma_y \; if \; 0 \le \sigma_2 \le \sigma_1$$
$$f(\sigma_1, \sigma_2) = \sigma_1 - \sigma_2 - \sigma_y \; if \; \sigma_2 \le 0 \le \sigma_1$$

4.3 Criteria Involving Hydrostatic Pressure

They reflect the fact that the material is dependent on any isotropic action of the stress tensor. Porous media, especially soils, are sensitive to hydrostatic pressure, in contrast to metal alloys and also to shape memory alloys. These criteria include, by construction, a tension–compression asymmetry.

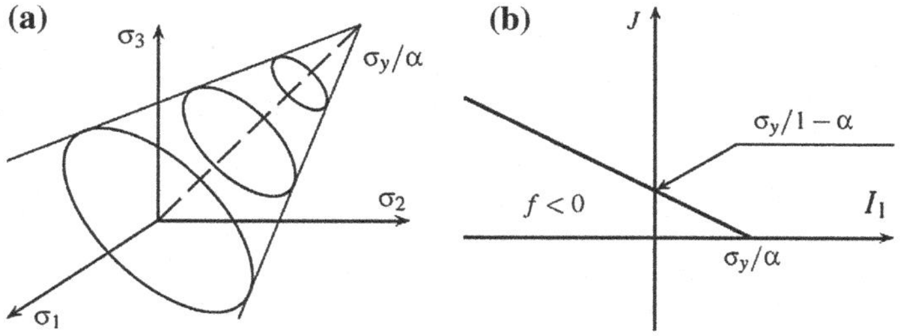

Fig. 4.3 Drucker–Prager criterion representation: **a** in the space of eigenstresses, **b** in the plane $I_1 - J$ (Cailletaud et al. 2011)

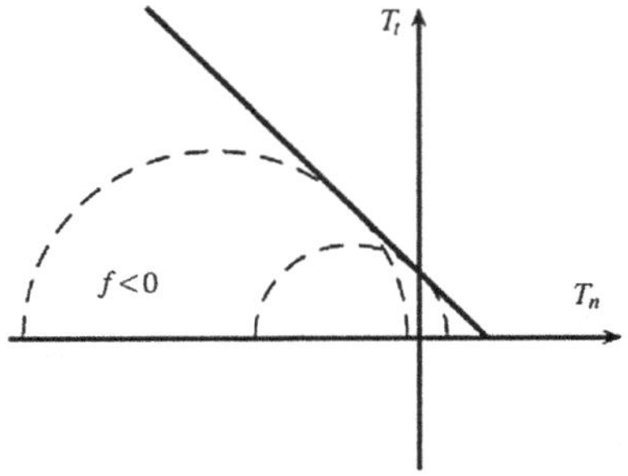

Fig. 4.4 Mohr–Coulomb criterion representation (Cailletaud et al. 2011)

4.3.1 Drucker–Prager Criterion

It consists of a linear combination of the first invariant of the stress tensor $\Theta_1 = I_1$ and the Huber-von Mises equivalent stress $J = \overline{\sigma}$. As said Cailletaud et al. (2011), it is always a circle whose radius depends on "elevation" on the axis trisector $\sigma_1, \sigma_2, \sigma_3$ of the eigenstresses (Fig. 4.3):

$$f\left(\underline{\sigma}\right) = \alpha I_1 + (1-\alpha)\,J - \sigma_y. \tag{4.10}$$

The tensile yield strength is σ_y and the one in compression $-\sigma_y/(1-2\alpha)$. The value of coefficient α depends on the chosen material and is found as Huber-von Mises criterion for $\alpha = 0$.

4.3.2 Mohr–Coulomb Criterion

For materials other than metals, Mohr (1900) worked on the idea that plasticity is generated by sliding and proposed a generalization of the Tresca criterion as

$$f\left(\underline{\sigma}\right) = \sigma_1 - \sigma_3 + (\sigma_1 + \sigma_3)\, sin\varphi - 2C cos\varphi \; with \; \sigma_3 \le \sigma_2 \le \sigma_1.$$

This criterion incorporates the concept of friction and assumes that the maximum shear that can undergo the material (T_t in Fig. 4.4) is even larger than the normal compressive stress (T_n) is great. The admissible limit is "an intrinsic curve" in the Mohr diagram. The preceding formula is obtained with a linear friction rule:

$$\mid T_t \mid < -tan\varphi\; T_n + C. \tag{4.11}$$

4.3.3 Gurson Criterion for the Porous Materials Behavior

Their behavior is rendered sensitive to hydrostatic pressure because of their porosity p:

$$f\left(\overline{\sigma},\sigma_m\right)=\frac{\overline{\sigma}^2}{\sigma_y^2}+2p\cosh\left(\frac{3\sigma_m}{2\sigma_y}\right)-1-p^2, \tag{4.12}$$

with $\sigma_m=\frac{1}{3}\mathrm{tr}\left(\underline{\sigma}\right)$. For $p=0$, we find again the Huber-von Mises criterion.

4.3.4 Criteria Integrating the Second and the Third Deviatoric Stress Invariants

We study materials insensitive to pressure which exhibit an asymmetry in the tension and compression behaviors.

This is true for certain metallic materials including shape memory alloys, so the criterion can be written in the following form:

$$f\left(\overline{\sigma},J_3\right)=0. \tag{4.13}$$

To simplify the writing, we introduce a parameter y which will be a "third invariant normalized" in the form

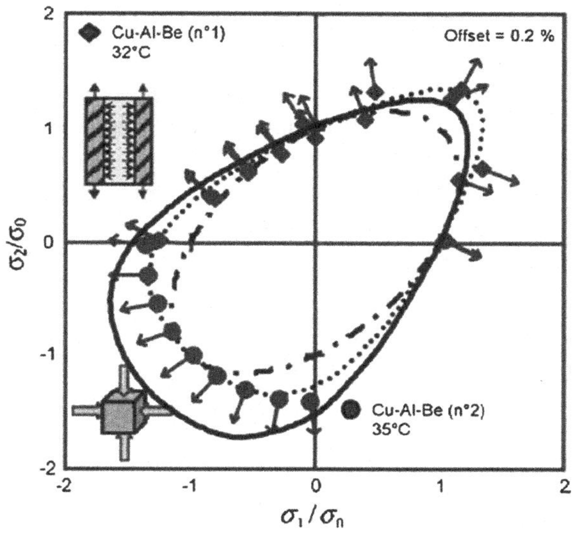

Fig. 4.5 Boundary curve of the elastic field of austenite obtained by biaxial mechanical testing on the Cu–Al–Be. In spaced lines, prediction of Huber-von Mises. In *solid lines*, Eqs. (4.14) and (4.16). *Dotted*, the experimental curve (Bouvet et al. 2002)

$$y = \frac{27J_3}{2J^3} = \frac{27\det\left(\underline{s}\right)}{2\overline{\sigma}^3}, \tag{4.14}$$

where y is a dimensional parameter such that $-1 \leq y \leq 1$. The calculations give $y = 1$ for pure tension, $y = -1$ for pure compression, and $y = 0$ for pure shearing.

We can formulate the criterion in the form given by Raniecki and Lexcellent: (Raniecki and Lexcellent 1998):

$$f\left(\underline{\sigma}\right) = \overline{\sigma}g\left(y\right) - \sigma_y, \tag{4.15}$$

with

$$g\left(y\right) = 1 + ay. \tag{4.16}$$

The convexity of f may be shown if and only if $0 \leq a \leq 1/8$.

A second formulation was introduced by Bouvet et al. in 2002 (Bouvet et al. 2002) under the form

$$g\left(y\right) = \cos\left(1/3\arccos\left(1 - a\left(1 - y\right)\right)\right), \tag{4.17}$$

which ensures the convexity of f if and only if $0 \leq a \leq 1$.

Figure 4.5 is meaningful; beyond the elastic domain, the phase transformation (austenite$\Longrightarrow$ martensite) occurs. Let us notice the normal flow rule that is verified experimentally, and is a classic result for elastoplastic materials.

4.4 Exercise 4.3

Statement:

Consider a Ni–Ti material, tensile (compression)–torsion tests made by Taillard—Laverne et al. (Laverhne Taillard et al. 2008) proposed the following values on the border of the elastic area (see Fig. 4.6):

(1) Draw the curve of border area in the axes $\left(\sigma_{zz}, \sqrt{3}\sigma_{z\theta}\right)$.

(2) Model the curve using the following function:

$$f\left(\underline{\sigma}\right) = \overline{\sigma}g\left(y\right) - \sigma_c$$

with

$$g\left(y\right) = \cos\left(1/3\arccos\left(1 - a\left(1 - y\right)\right)\right)$$

or

$$g\left(y\right) = b_1 y + b_2.$$

$$\text{NiTi}: \sigma_i = \begin{pmatrix} 0 & 0 & 0 \\ 0 & 0 & \sigma_{z\theta} \\ 0 & \sigma_{z\theta} & \sigma_{zz} \end{pmatrix}, \ 1 \leq i \leq n = 15,$$

i	1	2	3	4	5	6	7	8	9	10	11	12	13	14	15
σ_{zz}	400	390	310	220	0	-210	-370	-520	-530	-520	-430	-200	30	250	370
$\sigma_{z\theta}$	0	150	240	340	440	420	330	220	0	-150	-340	-400	-370	-300	-150

Fig. 4.6 Border point measurements of the elastic range in tension (compression)– torsion on a Ni–Ti tube (Laverhne Taillard et al. 2008)

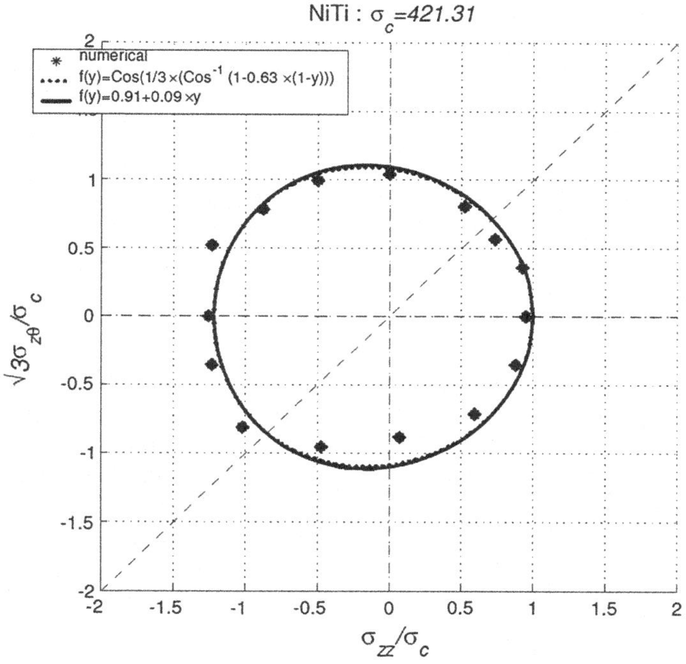

Fig. 4.7 Modeling of the « yield elastic surface » in the case of a Ni–Ti alloy (Laydi and Lexcellent 2010)

Optimize values of a and σ_c.

Correction:

$a = 0.63$, $\sigma_c = 421$ MPa and $b_1 = 0.91$, $b_2 = 0.09$ (Fig. 4.7).

4.5 Criteria Integrating the Three Invariants of the Stress Tensor I_1, J, *and* J_3

As suggested by Cailletaud et al. (2011), in the expression (10)

$$f\left(\underline{\sigma}\right) = \alpha I_1 + (1-\alpha)\, J - \sigma_y,$$

we replace J by t

$$t = \frac{J}{2}\left(1 + \frac{1}{K} - \left(1 - \frac{1}{K}\right)\left(\frac{S}{j}\right)^3\right) \quad with\ S = \left(\det\left(\text{dev}\underline{\sigma}\right)\right)^{\frac{1}{3}} = (J_3)^{\frac{1}{3}}, \quad (4.18)$$

where K is a material-dependent parameter. One must have $0.778 \leq K \leq 1$ in order to comply with the convexity of the criterion (Cailletaud et al. 2011).

4.6 Anisotropic Criteria

Let us recall that the materials are inherently anisotropic due to manufacturing conditions, such as long fiber composites, for example. We can say that the criterion is based on six components of the stress tensor in a given base. However, the shape chosen must be intrinsic, which requires that the result must be invariant under changes in the reference frame.

The most generally suggested solution constitutes an extension of the Huber-von Mises criterion using instead of $J(s) = \overline{\sigma}$, and the corresponding expression is

$$J_B(s) = \left(\underline{s} : \underline{\underline{B}} : \underline{s}\right)^{\frac{1}{2}}, \quad (4.19)$$

which involves the tensor of the fourth-order $\underline{\underline{B}}$. As in elasticity, considerations of symmetry may reduce the number of independent components of $\underline{\underline{B}}$.

If one combines the plastic incompressibility to the existence of three perpendicular planes of symmetry, there are only six components. One found classical expression (Hill criterion) in the coordinate system $\overline{R}$:

$$\begin{aligned} f\left(\underline{\sigma}\right) &= (F(\overline{\sigma}_{11} - \overline{\sigma}_{22})^2 + G(\overline{\sigma}_{22} - \overline{\sigma}_{33})^2 + H(\overline{\sigma}_{33} - \overline{\sigma}_{11})^2 \\ &\quad + 2L\overline{\sigma}_{12}^2 + 2M\overline{\sigma}_{23}^2 + 2N\overline{\sigma}_{31}^2)^{\frac{1}{2}} - \sigma_y = f_H\left(\underline{s}\right) = 0. \end{aligned} \quad (4.20)$$

Representing the tensor of the four order as a 6X6 matrix, the terms of $\underline{\underline{B}}$ are written as

$$\begin{bmatrix} F+H & -F & -H & 0 & 0 & 0 \\ -F & G+F & -G & 0 & 0 & 0 \\ -H & -G & H+G & 0 & 0 & 0 \\ 0 & 0 & 0 & 2L & 0 & 0 \\ 0 & 0 & 0 & 0 & 2M & 0 \\ 0 & 0 & 0 & 0 & 0 & 2N \end{bmatrix}. \tag{4.21}$$

One can verify that the same test can also be expressed in terms of the components of the stress deviator $\underline{s}$, with $J_{B^0}\left(\underline{s}\right) = \left(\underline{s} : \underline{\underline{B}}^0 : \underline{s}\right)^{\frac{1}{2}}$:

$$\underline{\underline{B}}^0 = \begin{bmatrix} 2F-G+2H & 0 & 0 & 0 & 0 & 0 \\ 0 & 2F+2G-H & 0 & 0 & 0 & 0 \\ 0 & 0 & -F+2G+2H & 0 & 0 & 0 \\ 0 & 0 & 0 & 2L & 0 & 0 \\ 0 & 0 & 0 & 0 & 2M & 0 \\ 0 & 0 & 0 & 0 & 0 & 2N \end{bmatrix}. \tag{4.22}$$

If we want to represent the asymmetry between tension and compression, we reintroduced in addition a linear form in the criterion and thus appear the Hoffman criterion:

$$\begin{gathered} \left(F\left(\overline{\sigma}_{11} - \overline{\sigma}_{22}\right)^2 + G\left(\overline{\sigma}_{22} - \overline{\sigma}_{33}\right)^2 + H\left(\overline{\sigma}_{33} - \overline{\sigma}_{11}\right)^2 \right. \\ \left. +2L\overline{\sigma}_{12}^2 + 2M\overline{\sigma}_{23}^2 + 2N\overline{\sigma}_{31}^2 + P\overline{\sigma}_{11} + Q\overline{\sigma}_{22} + R\overline{\sigma}_{33}\right)^{\frac{1}{2}} - \sigma_y = 0. \end{gathered} \tag{4.23}$$

4.7 Conclusion

Providing a detailed bibliographic data of all yield criteria is a challenge! Instead, I here tried to describe the outline.

However, the determination of these functions is important for elastoplastic modeling.

Chapter 5
Multiaxial Plasticity

Abstract The formulation of the elastic-plastic behavior of a deformable continuous medium is considered under the assumption of «small perturbations». As plasticity is defined as a «time-independent» behavior, the constitutive law is incremental. A distinction is made between materials with «work hardening» or without (perfect plasticity). A particular attention is paid to the normal strain rate vector of the elasticity domain. Proportional and non-proportional loadings are investigated via an exercise on thin tubes under tension-torsion.

5.1 Introduction

The formulation of the elastic–plastic behavior of a deformable continuous medium is considered under the assumption of infinitesimal transformation. This behavior model eliminates any effect of aging of the material or viscosity.

Let us recall that in elasticity, the relationship between stress and strain is bi-univocal. In plasticity, the constitutive law is INCREMENTAL.

The material behavior is entirely determined by the evolution of the local stress state, defined by a load path for the Cauchy stress tensor in space R^6.

As defined in Chap. 4, "initial elasticity domain" is generated by the set of all the load paths, from the original natural state $\left(\underline{\sigma} = \underline{0} \; and \; \underline{\varepsilon} = \underline{0}\right)$, along which the behavior of the material is continuously elastic. As discussed in Chap. 4, this domain is usually convex. When the load path comes out for the first time of "initial elasticity domain," then a new phenomenon is superimposed on the elastic deformation. This phenomenon, which corresponds to the plastic deformation, is "irreversible"; it is only activated if the load of the material element keeps in inversing; in the event of discharge, only the elastic deformation evolves.

So far, let's recall that the "work hardening" of a metal is a result of its plastic deformation (metallurgical modification of its internal structure, occurrence of line defects such as dislocations).

Some materials do not present "work hardening," that is to say when the corresponding stresses at the elastic border is reached, they begin to flow plastically

C. Lexcellent, *Linear and Non-linear Mechanical Behavior of Solid Materials*, DOI 10.1007/978-3-319-55609-3_5

"indefinitely" without additional stress increment. The corresponding behavior model is elastic and "perfectly plastic."

The elastic domain is unique: it is the «initial elastic domain». In contrast, the flow rate rule leaves the plastic deformation "indetermined" by a scalar factor "arbitrary nonnegative."

For materials having strain hardening, one defined for a given state of loading, the «current elasticity domain» of the material. The domain is also usually convex. It serves as a reference for the load concepts material element: "outgoing" path which corresponds to an arc load "increasing" and unloading ("decreasing" path).

The evolution of the elasticity domain, along the arcs of increasing load paths, translates the physical phenomenon of hardening.

The mathematical description of the elasticity domain is obtained through the introduction of "a loading function" which is a scalar function of the stress tensor, set by the hardening state.

It defines, for a given hardening state, "the yield surface" equation of the current elasticity boundary, while loading and unloading are distinguished by the sign of the rate change of this function.

The second brick consists in the definition of "plastic flow rule." It means to define the expression of "plastic deformation rate" according to the strain rate in considered loaded state: the elastic–plastic constitutive law is, in essence, incremental.

The "principle of maximum plastic work" is a characteristic property of the end plastic dissipation. It implies the convexity of the current range of elasticity and the normality of plastic deformation rate vis-à-vis this area.

We will see that the plastic strain rate is a scalar dependence in relation to the stress state.

The evolution of the hardening work is concomitant with the plastic deformation. It is the subject of the "hardening rule" consistent with the loading function.

Plasticity is defined as a "time-independent" behavior. This means that the stress-strain curve is the same regardless of the strain rate (or stress rate). The temperature range is in general $T < 0, 3T_f$ where T_f is the melting temperature of the material in K.

The elastic–plastic material does not deform over time under a dead load. The stress relaxation is not possible if its state of deformation is kept constant at a given time. However, it goes without saying that creep and relaxation are completely dependent on hold time; very long time may show creep deformations and stress relaxation.

Note that plasticity is one of the first manifestation of the nonlinear behavior of materials.

Construction of nonlinear behavioral models of materials includes the study of rheological properties and the formatting of equations to solve three-dimensional problems. Rheology: "study of flow" is used to associate springs (for elastic behavior), shock absorbers (to represent the viscosity), pads (for the management of the yield strength). These elements combined together to form rheological models. The assembly of these elements gives a schematic representation of the material behavior without any actual physical meaning. The lack of viscosity of a material is the field of elastoplasticity.

In the next chapters, materials characterized by a “time-dependent” behavior will be considered. These materials exhibit creep and relaxation, e.g., the viscoelastic and viscoplastic behaviors.

5.2 Uniaxial Plasticity

5.2.1 *Elastic Perfectly Plastic Model*

One-dimensional case (1D).

The series association of a spring and a pad produces a perfect elastoplastic behavior shown in Fig. 5.1.

The system cannot support a stress whose absolute value is greater than σ_y (elastic limit).

The loading function $f(\sigma)$ is defined by:

$$f(\sigma) = |\sigma| - \sigma_y \tag{5.1}$$

This is the one-dimensional version of the Huber–Von Mises criterion (Chap. 4).

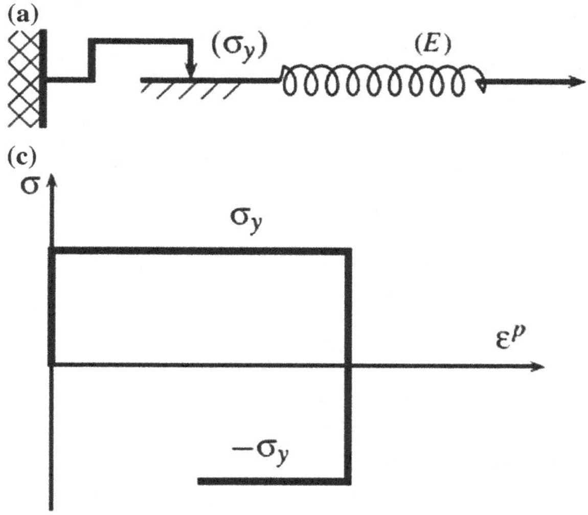

Fig. 5.1 Rheological model perfectly elastic-plasticity (**a**) stress curve - plastic deformation (**c**) (Cailletaud et al. 2011)

The behavior is summarized by the following equations:

$$\begin{cases} \textit{Elasticity domain if} \ : f < 0 \ \left(\dot{\varepsilon} = \dot{\varepsilon}^{el} = \dot{\sigma}/E\right) \\ \textit{Elastic unloading if} \ : f < 0 \ \textit{and} \ \dot{f} < 0 \ \left(\dot{\varepsilon} = \dot{\varepsilon}^{el} = \dot{\sigma}/E\right) \\ \textit{Plastic flow if} \ : f = 0 \ \textit{and} \ \dot{f} = 0 \ \left(\dot{\varepsilon} = \dot{\varepsilon}^{p}\right) \end{cases} \tag{5.2}$$

In elastic mode, there is no plastic deformation and therefore no rate $\dot{\varepsilon}^p$ associated; $\dot{\varepsilon}^{el}$ becoming null during plastic flow.

The model is «without hardening» because the stress level does not change so beyond the elastic domain: $|\sigma| = \sigma_y$.

In the thermodynamical frame, the internal variable is ε^p. As in Chap. 2, one writes a dissipation potential $D(\dot{\varepsilon}, \dot{\varepsilon}^p)$ differentiable to write this relationship as:

$$\begin{cases} \sigma^{ir} = \frac{\partial D}{\partial \dot{\varepsilon}} \\ A = \frac{\partial D}{\partial \dot{\varepsilon}^p} \end{cases} \tag{5.3}$$

with a separation of the stress into a reversible component and an irreversible one:

$$\sigma = \sigma^r + \sigma^{ir} \tag{5.4}$$

and also the separation between the elastic and the plastic rates:

$$\dot{\varepsilon} = \dot{\varepsilon}^{el} + \dot{\varepsilon}^p \tag{5.5}$$

One admits for an elastoplastic solid that the irreversible stress is equal to zero: $\left(\sigma^{ir} = 0\right)$ that gives us $D = A\dot{\varepsilon}^p$.

The free energy ψ and the dissipation D are written as:

$$\begin{cases} \psi(\varepsilon, \varepsilon^p) = \frac{1}{2}E(\varepsilon - \varepsilon^p)^2 \\ D(\dot{\varepsilon}^p) = \sigma_y \mid \dot{\varepsilon}^p \mid \end{cases} \tag{5.6}$$

with:

$$\begin{cases} A = E(\varepsilon - \varepsilon^p) = \sigma \\ \mid A \mid \leq \sigma_y. \end{cases} \tag{5.7}$$

Natural extension: the «isotropic hardening» case.

The physical aspects are not addressed here. One can find some answers in Jaoul (1965) or in Burlet et al. (1988) and obviously in Friedel (1964) concerning the dislocations. In the model shown in Fig. 5.2a, we replace the pad threshold σ_y by $R(p)$ with p representing «the cumulated plastic strain» with the following definition:

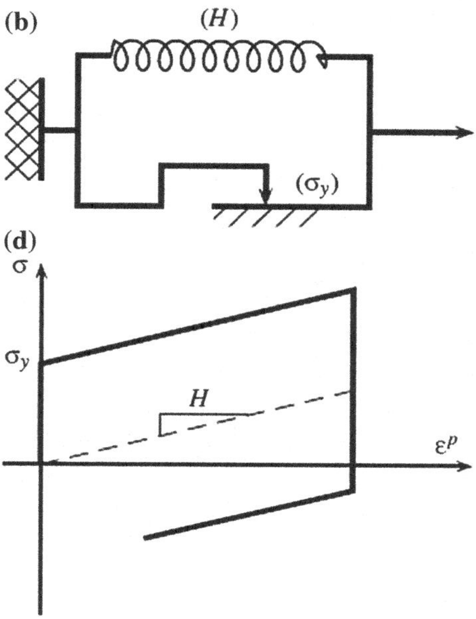

Fig. 5.2 Rheological model inelastoplasticity with hardening (**b**) Curve stress-plastic strain (**d**) (Cailletaud et al. 2011)

$$p\,(t) = \int_0^t \left(\frac{2}{3}\dot{\underline{\varepsilon}}^p : \dot{\underline{\varepsilon}}^p\right)^{\frac{1}{2}} d\tau. \tag{5.8}$$

In three-dimensional plasticity, we can choose to write:

$$R\,(p) = R_s + \left(\sigma_y - R_s\right) \exp\left(-bp\right) \tag{5.9}$$

to comply with the natural conditions $R\,(p = 0) = \sigma_y$ et $R\,(p \rightarrow +\infty) = R_s$.

5.2.2 Prager Model

The parallel combination of Fig. 5.2b corresponds to the behavior shown in Fig. 5.2d.

In this case, the model presents work hardening. We observe a «linear kinematic hardening» because it is linear-dependent on the plastic deformation. As shown in Fig. 5.2b, the model is rigid-plastic, simply it is sufficient to add a serial spring to become elastoplastic.

One calls X the stress exerted on the spring. It is called "kinematic stress": $X = H\varepsilon^p$.

A new loading function f is written:

$$f\,(\sigma, X) = \mid \sigma - X \mid - \sigma_y. \tag{5.10}$$

There will be plastic flow when: $f = 0$ and $\dot{f} = 0$. This leads to the following condition:

$$\frac{\partial f}{\partial \sigma}\dot{\sigma} + \frac{\partial f}{\partial X}\dot{X} = 0 \tag{5.11}$$

Hence:

$$sign\,(\sigma - X)\,\dot{\sigma} - sign\,(\sigma - X)\,\dot{X} = 0 \tag{5.12}$$

and partially:

$$\dot{\sigma} = \dot{X} \text{ and finally } \dot{\varepsilon}^p = {}^{\dot{\sigma}}/_{H} \text{ or } \dot{\varepsilon}^p = \frac{E}{E+H}\dot{\varepsilon} \tag{5.13}$$

From the thermodynamical point of view:

internal variable ε^p

free energy ψ and dissipation potential D:

$$\begin{cases} \psi\,(\varepsilon, \varepsilon^p) = \frac{1}{2}E\,(\varepsilon - \varepsilon^p)^2 + \frac{1}{2}H\,(\varepsilon^p)^2 \\ D\,(\dot{\varepsilon}^p) = \sigma_y \mid \dot{\varepsilon}^p \mid \end{cases} \tag{5.14}$$

with:

$$\begin{cases} \sigma = \frac{\partial \psi}{\partial \varepsilon} = E\,(\varepsilon - \varepsilon^p) \\ A = -\frac{\partial \psi}{\partial \varepsilon^p} \end{cases} \tag{5.15}$$

$$\begin{cases} A = -\frac{\partial \psi}{\partial \varepsilon} = -H\varepsilon^p + E\,(\varepsilon - \varepsilon^p) = \sigma - H\varepsilon^p \\ \mid A \mid \leq \sigma_y. \end{cases} \tag{5.16}$$

These expressions are prior to a three-dimensional version of the model.

Natural extension: the kinematical hardenings case.

Instead of choosing an isotropic variable $R\,(p)$ (that is to say scalar), we choose one or more second order tensors said internal stress tensors $\underline{X}$ which constitutes a three-dimensional extension of the scalar variable X. As for isotropic hardening, the reader will find information in the works of Jaoul (1965) Burlet et al. (1988).

5.3 Tridimensional Plasticity

5.3.1 Introduction

Without considering into the variety of behavior laws and especially the wide variety of criteria and evolution laws, the general framework will be exposed through limiting to the "small perturbations" formulation.

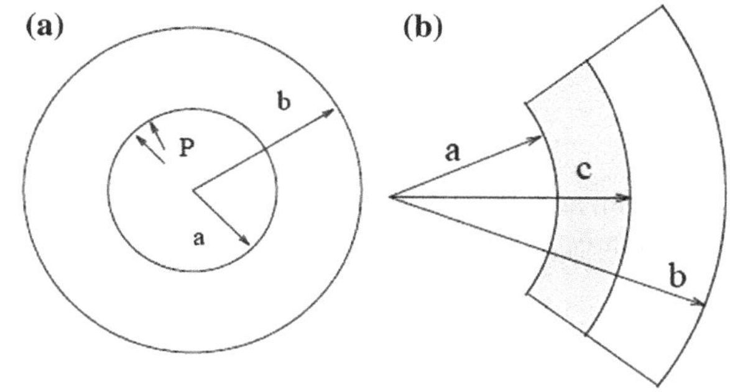

Fig. 5.3 **a** Spherical shell geometry and mechanical loading; **b** Plastic zone progression starting from the inner surface (Cailletaud et al. 2011)

5.3.1.1 Strain Decomposition

The symmetric strain tensor is divided into three parts:
an elastic part $\underline{\varepsilon}^{el}$ *with* $\underline{\sigma_i}$ representing the stress tensor in its initial state, which is most of the time null except in geotechnics:

$$\underline{\varepsilon}^{el} = \underline{\underline{S}} : \left(\underline{\sigma} - \underline{\sigma_i}\right) \ with\ \underline{\underline{S}} = \underline{\underline{C}}^{-1} \tag{5.17}$$

a thermal part $\underline{\varepsilon}^{th}$ function of the initial temperature T_0 and the current temperature T and the second order tensor $\underline{\alpha}$ which is optionally temperature-dependent and which is spherical in the case of isotropic materials (cf. Chap. 2):

$$\underline{\varepsilon}^{th} = \underline{\alpha}\,(T - T_0) \tag{5.18}$$

a plastic strain $\underline{\varepsilon}^p$.

Hence the total strain $\underline{\varepsilon}$ is written as:

$$\underline{\varepsilon} = \underline{\varepsilon}^{el} + \underline{\varepsilon}^{th} + \underline{\varepsilon}^{p} = \underline{\underline{S}} : (\underline{\sigma} - \underline{\sigma_i}) + \underline{\alpha}\,(T - T_0) + \underline{\varepsilon}^{p}. \tag{5.19}$$

5.3.2 *Example of Perfect Elastoplastic Problem (Exercise 5.1)*

Thick spherical shell subjected to internal pressure p (Fig. 5.3).

Let's first recall the elastic solution developed in Exercise 3.1 by Mandel (1974).

The elastic solution corresponds to spherical coordinates:

$$u_r(r) = \alpha r + \tfrac{\beta}{r^2},\ \sigma_{rr} = A - \tfrac{2B}{r^3},\ \ \sigma_{\theta\theta} = \sigma_{\varphi\varphi} = A + \tfrac{B}{r^3}\,. \tag{5.20}$$

The other stress tensor components are equal to zero. And:

$$A = p\frac{1}{(b/a)^3 - 1}\ \ B = Ab^3/2\ \ \alpha = A/(3\lambda + 2\mu)\ \ \beta = B/2\mu$$

If one accepts that the material is elastic, perfectly plastic and obeys the criterion of Tresca, the plasticity condition is written in terms of eigenstress:

$$Max_{i,j} \mid \sigma_i - \sigma_j \mid \leq 2k. \tag{5.21}$$

To be compatible with the Huber–von Mises criterion, we take $2k = \sigma_y$.

Explicitly, the elastic solution is:

$$\sigma_{rr} = -\frac{a^3}{b^3 - a^3}\left(\frac{b^3}{r^3} - 1\right)p \tag{5.22}$$

$$\sigma_{\theta\theta} = \sigma_{\varphi\varphi} = \frac{a^3}{b^3 - a^3}\left(\frac{b^3}{2r^3} + 1\right)p \tag{5.23}$$

$$u_r(r) = \frac{a^3}{b^3 - a^3}\left((1 - 2\nu)\,r + (1 + \nu)\,\frac{b^3}{2r^2}\right)\frac{p}{E}. \tag{5.24}$$

As the sphere remains elastic, we have:

$$\sigma_{rr} - \sigma_{\theta\theta} = -\frac{3}{2}\frac{a^3}{b^3 - a^3}\left(\frac{b^3}{r^3}\right)p. \tag{5.25}$$

The plasticity criterion is reached when $(\sigma_{rr} - \sigma_{\theta\theta})$ decreasing function of p becomes equal to the elastic limit $(-\sigma_y)$ for simple compression. The beginning of plastification appears for $r = a$ when the pressure reaches the value p_e, which is the initial elasticity limit of a spherical shell under pressure:

$$p_e = \frac{2}{3}\left(1 - \frac{a^3}{b^3}\right)\sigma_y. \tag{5.26}$$

Elastoplastic resolution:

To solve the problem, we write the continuity of the stress vector in $r = c$, that is to say: $\sigma^I_{rr}(r = c) = \sigma^{II}_{rr}(r = c)$ with the elastic zone $(I:\ c \leq r \leq b)$ and the plastic zone $(II:\ a \leq r \leq c)$. Without forgetting that the inner surface of equation $(r = a)$ is subjected to the pressure $\left(\sigma^{II}_{rr}(r = a) = -p\right)$ and that the outer surface $(r = b)$ is free from any external mechanical loading: $(\sigma^I_{rr}(r = b) = 0)$.

Practically, the stresses are given by the above equations in which one replaces a by c and p by $\left(-\sigma^I_{rr}(r = c)\right)$.

$$\sigma^I_{rr} = -\frac{2}{3}\frac{c^3}{b^3}\left(\frac{b^3}{r^3} - 1\right)\sigma_y \tag{5.27}$$

$$\sigma^I_{\theta\theta} = \frac{2}{3}\frac{c^3}{b^3}\left(\frac{b^3}{2r^3} + 1\right)\sigma_y \tag{5.28}$$

$$u_r^I(r) = \frac{3}{2E}\frac{c^3}{b^3}\left((1-2\nu)r + (1+\nu)\frac{b^3}{2r^2}\right)\sigma_y \tag{5.29}$$

Let's now consider the plastic zone II. To determine stresses, one has to consider the equilibrium equations (expressed in spherical coordinates obviously Sidoroff 1980) and the plasticity criterion:

$$\frac{d\sigma_{rr}^{II}}{dr} + \frac{2}{r}\left(\sigma_{rr}^{II} - \sigma_{\theta\theta}^{II}\right) = 0 \tag{5.30}$$

$$\left(\sigma_{rr}^{II} - \sigma_{\theta\theta}^{II}\right) = -\sigma_y. \tag{5.31}$$

That gives:

$$\frac{d\sigma_{rr}^{II}}{dr} - \frac{2\sigma_y}{r} = 0 \Longrightarrow \sigma_{rr}^{II} = 2\sigma_y \ln(r) + C$$

By using the continuity of σ_{rr} for $r = c$ one obtains the constant C by:

$$2\sigma_y \ln(c) + C = -\frac{2}{3}\frac{c^3}{b^3}\left(\frac{b^3}{c^3} - 1\right)\sigma_y.$$

Finally, one finds:

$$\sigma_{rr}^{II} = -\frac{2}{3}\sigma_y\left(1 + 3\ln\left(\frac{c}{r}\right) - \frac{c^3}{b^3}\right) \tag{5.32}$$

$$\sigma_{\theta\theta}^{II} = \frac{2}{3}\sigma_y\left(\frac{1}{2} - 3\ln\left(\frac{c}{r}\right) + \frac{c^3}{b^3}\right) \tag{5.33}$$

The stress are dependent on parameter c, hence the evolution of p as a function of parameter c is needed.

For $r = a$, we have:

$$\sigma_{rr}^{II}(r = a) = -p \Longrightarrow p = \frac{2}{3}\sigma_y\left(1 + 3\ln\left(\frac{c}{a}\right) - \frac{c^3}{b^3}\right)$$

In the event of "small strain" used in these calculations, a and b are considered as constants:

$$\frac{dp}{dc} = \frac{2\sigma_y}{c}\left(1 - \frac{c^3}{b^3}\right) \geq 0. \tag{5.34}$$

The outer radius of the plastic zone reaches the value b when p reaches the value p_p:

$$p_p = 2\sigma_y \ln\left(\frac{b}{a}\right). \tag{5.35}$$

Cailletaud et al. (2011) completed the solution by determining the plastic strains and their velocities. They particularly give the radial displacement in the plastic zone:

$$u_r^{II} = \frac{\sigma_y}{E} r \left[(1-\nu) \frac{c^3}{r^3} - \frac{2}{3}(1-2\nu)\left(1 + 3\ln\left(\frac{c}{r}\right) - \frac{c^3}{b^3}\right)\right]. \qquad (5.36)$$

As said Nguyen Quoc Son (1982) about this standard problem, he also treated; "in this simple example, the stress field in the plastic zone could be determined from the equilibrium equations and plasticity conditions, meaning that the problem is statically determined. This is a very special case, the elastoplastic evolution is often complex and in the literature, we only know some examples of explicit solution."

5.4 Three-Dimensional Elasto-Plasticity (in "Small Perturbations")

5.4.1 Elasto-Plasticity Working Hypotheses (Maugin 1992)

From all the properties and experimental facts, we may conclude that the plasticity of metals, with or without hardening, at room or low temperature is:

(H1) a thermodynamically irreversible phenomenon (existence of plastic strains),
(H2) which is practically independent of the strain rate,
(H3) and of volume changes,
(H4) implying the notion of threshold or plastic-yield surface (see Chap. 4),
(H5) represented by a convex hypersurface in the stress space,
(H6) possibly representing angular points (nonuniqueness of the exterior normal),
(H7) and possibly dependent upon the plastic strain history (hardening),
(H8) moreover, for an isotropic medium, with an identical response in traction-compression, the yield surface in the stress space admits a section with symmetry of order 6 (hexagonal). (See Chap. 4 which includes also the possibility to take into account the asymmetry between traction and compression).

5.4.2 Reminder of the Thermomechanical Formulation

We will go from a scalar formulation to a tensor writing.

First of all:

$$\underline{\sigma} = \underline{\sigma_{rev}} + \underline{\sigma_{ir}} \qquad (5.37)$$

N'Guyen Quoc Son (1982) admits that the irreversible stress part is equal to zero: $\underline{\sigma_{ir}} = \underline{0}$, the free energy is written as $\psi\left(\underline{\varepsilon}, \underline{\alpha}, T\right)$, that gives us the dissipation function:

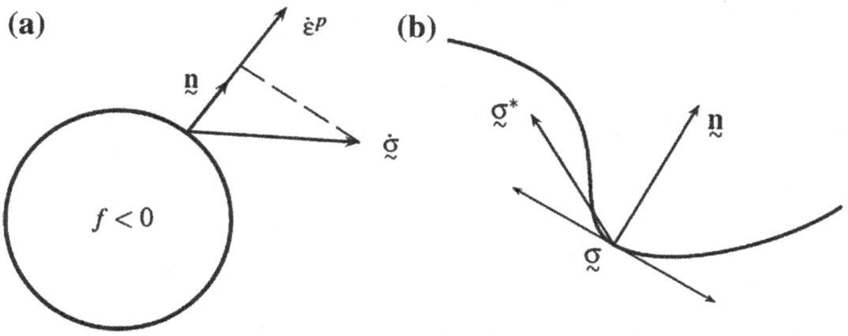

Fig. 5.4 Consequence of the principle of maximum work: **a** illustration of normality rule **b** convexity of f (Cailletaud et al. 2011)

$D =< \underline{A}, \dot{\underline{\alpha}} >$(<> scalar product).

In the space of thermodynamic forces $\underline{A}$, the real forces are not unlimited but restricted within the range of elasticity C (Already defined in Chap. 4) convex and containing the origin o. The dissipation D is expressed in terms of the rate $\dot{\underline{\alpha}}$ under the form:

$$D\left(\dot{\underline{\alpha}}\right) = Sup_{A^\star \in C}\underline{A^\star} : \dot{\underline{\alpha}} \geq 0. \tag{5.38}$$

The dissipation D is positively homogeneous function of degree 1 of the rate $\dot{\underline{\alpha}}$, e.g., $D\left(\lambda\dot{\underline{\alpha}}\right) = \lambda D\left(\dot{\underline{\alpha}}\right)$ *if* $\lambda \geq 0$.

The dissipation expression (5.38) is equivalent to the inequality:

$$\left(\underline{A} - \underline{A^\star}\right) : \dot{\underline{\alpha}} \geq 0 \; \forall \underline{A^\star} \in C \tag{5.39}$$

It is the principle of Hill maximum work (Hill 1948; Burlet et al. 1988). The definition (5.38) or inequality (5.39) translated as follows normality property: the rate $\dot{\underline{\alpha}}$ is a vector of the external normal cone in $\underline{A}$ of the elasticity domain C if $\underline{A}$ is a point interior to C, so $\dot{\underline{\alpha}} = \underline{0}$, the response is purely reversible ($D = 0$), is to say purely elastic. The parametres $\underline{\alpha}$ can only evolve if $\underline{A}$ reaches a threshold value that corresponds to the elasticity boundary domain C.

With reference to the uniaxial formulation, and in the case of perfect plasticity, it can be replaced $\underline{A}$ by $\underline{\sigma}$ and $\underline{\alpha}$ by $\underline{\varepsilon^p}$ to obtain a more conventional writing of the principle of maximum work:

$$\left(\underline{\sigma} - \underline{\sigma^\star}\right) : \dot{\underline{\varepsilon^p}} \geq 0 \; \forall \underline{\sigma^\star} \in C \tag{5.40}$$

The relationship $\left(\underline{\sigma} - \underline{\sigma^\star}\right) : \dot{\underline{\varepsilon^p}} \geq 0 \; \forall \underline{\sigma^\star} \in C$ gives, by taking $\underline{\sigma^\star} = \underline{\sigma}\,(t + \Delta t)$, then $\underline{\sigma^\star} = \underline{\sigma}\,(t - \Delta t)$, after going to the limit $\Delta t \to 0$:

$$\underline{\dot{\sigma} : \dot{\varepsilon^p}} = 0 \tag{5.41}$$

We can write the "normality rule":

$$\underline{\dot{\varepsilon^p}} = \dot{\lambda}_p \frac{\partial f}{\partial \underline{\sigma}},\ \lambda_p \geq 0 \ if\ f\left(\underline{\sigma}\right) = 0 \ and\ if\ \dot{f}\left(\underline{\sigma}\right) = 0 \tag{5.42}$$

where λ_p is called the "plastic multiplier" whose expression is:

$$\lambda_p = \frac{1}{\frac{\partial f}{\partial \underline{\sigma}} : \underline{\underline{L}} \frac{\partial f}{\partial \underline{\sigma}}} < \frac{\partial f}{\partial \underline{\sigma}} : \underline{\underline{L}} : \underline{\dot{\varepsilon}} > \tag{5.43}$$

where $< a >$ is the positive part of a: $< a >= 0$ if $a < 0$, $< a >= a$ if $a > 0$.

Figure 5.4 illustrates the principle of virtual work with $\underline{n} = {}^{\partial f}/_{\partial \underline{\sigma}}$.

5.4.3 Direction of Flow Associated with Conventional and Less Common Criteria

These flow directions are calculated initially for a material with perfect elastoplastic behavior.

5.4.3.1 Huber–Von Mises Criterion

The yield criterion $f\left(\underline{\sigma}\right) = \overline{\sigma} - \sigma_y$ is derived under the form:

$$\underline{n} = \frac{\partial f}{\partial \underline{\sigma}} = \frac{\partial \overline{\sigma}}{\partial \underline{\sigma}} = \frac{\partial \overline{\sigma}}{\partial \underline{s}} : \frac{\partial \underline{s}}{\partial \underline{\sigma}} = \frac{3}{2} \frac{\underline{\underline{s}}}{\overline{\sigma}} \tag{5.44}$$

Hence, it is $\frac{3s}{2} = \frac{3}{2}\mathrm{dev}\underline{\sigma}$ which gives the flow direction.

For simple tension:

$$\overline{\sigma} = \mid \sigma \mid ; \underline{n} = \underline{\begin{pmatrix} 1 & 0 & 0 \\ 0 & -{}^1/_2 & 0 \\ 0 & 0 & -{}^1/_2 \end{pmatrix}} sgn\sigma \tag{5.45}$$

For $f\left(\underline{\sigma}\right) = 0,\ and\ \dot{f}\left(\underline{\sigma}\right) = 0$:

$$\underline{\dot{\varepsilon^p}} = \dot{\lambda}_p \frac{\partial f}{\partial \underline{\sigma}} = \dot{\lambda}_p \underline{n}\ with\ \underline{n} : \underline{\dot{\sigma}} = 0 \tag{5.46}$$

5.4.3.2 Calculation of the Plastic Multiplier

The plastic multiplier is indetermined for a perfectly elastic-plastic material loaded at imposed stress rate.

It will be achieved by the consistency condition applied to the loading function $f\left(\underline{\sigma}\right)$ giving:

$$\dot{f}\left(\underline{\sigma}\right) = \underline{n} : \underline{\dot{\sigma}} = 0 \tag{5.47}$$

Knowing that the derivative of the stress tensor is derived from the derivative of the constitutive relation (5.46) such as:

$$\underline{\dot{\sigma}} = \underline{\underline{C}} : \underline{\dot{\varepsilon}}^{el} = \underline{\underline{C}} : \left(\underline{\dot{\varepsilon}} - \underline{\dot{\varepsilon}^{p}}\right) = \underline{\underline{C}} : \underline{\dot{\varepsilon}} - \dot{\lambda}_p \underline{\underline{C} : \underline{n}} \tag{5.48}$$

which leads to: $\underline{n} : \underline{\underline{C}} : \underline{\dot{\varepsilon}} - \dot{\lambda}_p \underline{n \underline{C} : \underline{n}} = 0$

$$\Longrightarrow \dot{\lambda}_p = \frac{\underline{n} : \underline{\underline{C}} : \underline{\dot{\varepsilon}}}{\underline{n} : \underline{\underline{C}} : \underline{n}} \tag{5.49}$$

5.4.3.3 Tresca Criterion

The flow rule is defined by sectors in the principal stress space.

For example:

$\sigma_1 > \sigma_2 > \sigma_3$, so $f\left(\underline{\sigma}\right) = \mid \sigma_1 - \sigma_3 \mid - \sigma_y$ and:

$$\underline{\dot{\varepsilon^{p}}} = \dot{\lambda}_p \begin{pmatrix} 1 & 0 & 0 \\ 0 & 0 & 0 \\ 0 & 0 & -1 \end{pmatrix} \tag{5.50}$$

Obviously the definition of the normal is a problem for stress states corresponding to the singular points.

In that case, Cailletaud et al. (2011) proposed:

$$if\ \sigma_1 > \sigma_2 = \sigma_3 = 0 :\ \underline{\dot{\varepsilon^{p}}} = \dot{\lambda}_p \begin{pmatrix} 1 & 0 & 0 \\ 0 & 0 & 0 \\ 0 & 0 & -1 \end{pmatrix} + \dot{\mu} \begin{pmatrix} 1 & 0 & 0 \\ 0 & -1 & 0 \\ 0 & 0 & 0 \end{pmatrix} \tag{5.51}$$

We will explicit the laws of elastic–plastic behavior according to the criteria and the chosen type of hardening.

5.4.3.4 Criterion Taking into Account the Asymmetry Between Tension and Compression (Lexcellent 2013a; Raniecki and Lexcellent 1998)

Let's recall that the equation of the surface of the initial domain is:

$$f\left(\underline{\sigma}\right) = \overline{\sigma} g\left(y\right) - \sigma_y \tag{5.52}$$

with $-1 \leq y \leq 1$

The choice of $g\,((y))$ is open but must ensure the convexity of the elastic domain (see Chap. 4).

Let's recall that it is the introduction of the third invariant that will allows us to take into account the asymmetry between tension and compression

with:

$$\underline{\dot{\varepsilon^p}} = \dot{\lambda}_p \frac{\partial f}{\partial \underline{\sigma}}$$

By $f\left(\underline{\sigma}\right)$ derivation versus $\underline{\sigma}$, one obtains:

$$\frac{\partial f}{\partial \underline{\sigma}} = \underline{n} = \underline{n_1} + \underline{n_2} \tag{5.53}$$

with:

$$\underline{n_1} = \sqrt{\frac{3}{2}} g(y) \underline{N} \tag{5.54}$$

$$\underline{n_2} = \sqrt{6} \frac{dg(y)}{dy} (\sqrt{6} (\underline{N^2} - 1/3) - y \underline{N}) \tag{5.55}$$

and: $\underline{N} = \frac{\underline{s}}{\left(\underline{s}:\underline{s}\right)^{\frac{1}{2}}} = \sqrt{\frac{3}{2}} \frac{\underline{S}}{\overline{\sigma}}$

We obtain the equivalent plastic strain:

$$\overline{\varepsilon^{pl}} = \sqrt{\frac{2}{3} \underline{\varepsilon^p} : \underline{\varepsilon^p}} = \overline{\gamma}(y) \tag{5.56}$$

with:

$$\overline{\gamma}(y) = \gamma \sqrt{(g(y))^2 + 9(1 - y^2)(\frac{dg(y)}{dy})^2} \tag{5.57}$$

Note that this formulation will also be used for shape memory alloys where we will replace the plastic deformation by pseudoelastic deformation associated with the phase transformation or the one described as pseudoplastic associated with the reorientation of martensite variants.

5.4.4 Prandtl–Reuss Law (Besson et al. 2001)

The criterion function is written as:

$$f\left(\underline{\sigma}, R\,(p)\right) = \overline{\sigma} - R\,(p) \tag{5.58}$$

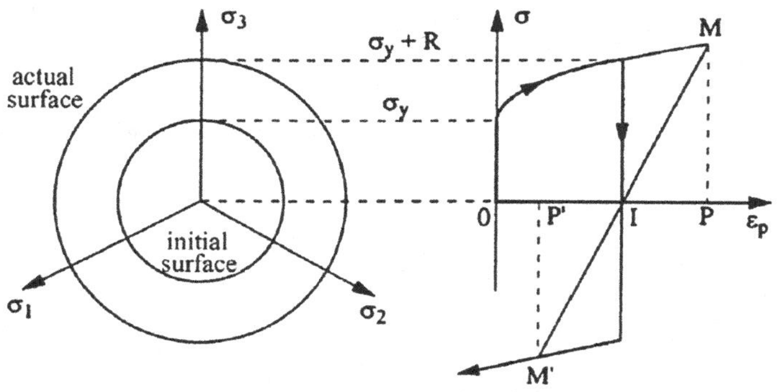

Fig. 5.5 Mapping of isotropic hardening; *Left* in the deviatoric plane; *right*, the stress versus plastic strain (Chaboche 2008)

We are in the case of a purely isotropic hardening as shown in Fig. 5.5.

With the Huber–von Mises equivalent stress $\overline{\sigma} = \left(\frac{3}{2}\underline{s} : \underline{s}\right)^{\frac{1}{2}}$
and $p(t) = \int_0^t \left(\frac{2}{3}\underline{\dot{\varepsilon^p}} : \underline{\dot{\varepsilon^p}}\right)^{\frac{1}{2}} d\tau$; $\underline{\dot{\varepsilon^p}} = \dot{\lambda}_p \underline{n}$,
the plastic hardening modulus $H(p)$ is defined by:

$$\frac{dR}{dp} = H(p)$$

$$\dot{f} = 0 \Longrightarrow \frac{\partial f}{\partial \underline{\sigma}} : \underline{\dot{\sigma}} + \frac{\partial f}{\partial R}\dot{R} = 0 \tag{5.59}$$

or: $\underline{n} = \frac{\partial f}{\partial \underline{\sigma}} = \frac{\partial \overline{\sigma}}{\partial \underline{\sigma}} = \frac{3}{2}\frac{\underline{s}}{\overline{\sigma}}$ *and* $\frac{\partial f}{\partial R} = -1$ it results:

$$\underline{n} : \underline{\dot{\sigma}} = \dot{R} \tag{5.60}$$

as: $\dot{p} = \left(\frac{2}{3}\underline{\dot{\varepsilon^p}} : \underline{\dot{\varepsilon^p}}\right)^{\frac{1}{2}} = \left(\frac{3}{2}\dot{\lambda}_p\underline{n} : \dot{\lambda}_p\underline{n}\right)^{\frac{1}{2}} = \dot{\lambda}_p$ *et* $\dot{R} = H\dot{\lambda}_p$ from where:

$$\dot{\lambda}_p = \frac{\underline{n} : \underline{\dot{\sigma}}}{H} = \dot{p} \tag{5.61}$$

and:

$$\underline{\dot{\varepsilon^p}} = \frac{\underline{n} : \underline{\dot{\sigma}}}{H}\underline{n}, \tag{5.62}$$

or by projection on the axis:

$$\left(\dot{\varepsilon_{ij}}\right)^p = \frac{9}{4}\frac{s_{kl}\dot{\sigma_{kl}}}{H\overline{\sigma}^2}s_{ij} \text{ or } d\varepsilon^p_{ij} = \frac{9}{4}\frac{s_{kl}d\sigma_{kl}}{H\overline{\sigma}^2}s_{ij} \tag{5.63}$$

The direction of plastic flow is defined by the tensor stress deviator. Only the projection on the normal $\underline{n}$ of the stress increment $d\underline{\sigma}$ is effective for the plastic increment $d\underline{\varepsilon^p}$

Exercise: Show that following expression in traction-compression:

$$\underline{\dot{\varepsilon^p}} = \frac{\dot{\sigma}}{H}\begin{pmatrix} 1 & 0 & 0 \\ 0 & -\frac{1}{2} & 0 \\ 0 & 0 & -\frac{1}{2} \end{pmatrix} \tag{5.64}$$

5.4.5 Proportional Loading Cases: Hencky–Mises Law (Besson et al. 2001)

Assumptions:

All external forces applied grows proportionally to a single parameter $\alpha\,(t)$.

The initial state is stress and strain free $\left(\underline{\sigma} = \underline{0}\ et\ \underline{\varepsilon} = \underline{0}\right)$.

The boundary conditions are given in terms of "forces."

The material follows the law of Prandtl–Reuss.

Let's $\underline{\sigma} = \alpha \underline{\Sigma}_0 \Longrightarrow \underline{\dot{\sigma}} = \dot{\alpha}\underline{\Sigma}_0$ $\underline{s} = \alpha \underline{S}_0 \Longrightarrow \overline{\sigma} = \alpha\, \overline{\Sigma_0}$

Thus: $\underline{n} = \frac{3}{2}\frac{\underline{\underline{s_0}}}{\overline{\Sigma_0}} = \underline{n}_0$ there is no normal rotation:

$$\underline{\dot{\varepsilon^p}} = \frac{\underline{n}_0 : \underline{\Sigma}_0}{H}\underline{n}_0\dot{\alpha}. \tag{5.65}$$

One can get the plastic strains tensor:

$$\underline{\varepsilon^p}\,(t) = \underline{n}_0 : \underline{\Sigma}_0\underline{n}_0\left(\int_0^t \frac{d\alpha}{H}\right) \tag{5.66}$$

If $H = cte$ (linear isotropic hardening), then:

$$\underline{\varepsilon^p} = \frac{9}{4}\frac{\underline{\Sigma_0} : \underline{S_0}}{\overline{\sigma_0}^2 H}\underline{S_0}\,(\alpha - \alpha_{seuil})\ \ with\ \overline{\sigma}\left(\alpha_{seuil}\underline{S_0}\right) = \sigma_y. \tag{5.67}$$

In simple tension:

$$\underline{\varepsilon^p} = \frac{\sigma - \sigma_y}{H}\begin{pmatrix} 1 & 0 & 0 \\ 0 & -\frac{1}{2} & 0 \\ 0 & 0 & -\frac{1}{2} \end{pmatrix}. \tag{5.68}$$

In simple torsion: $\overline{\sigma} = \sqrt{3}\tau$ $\underline{\sigma} = \begin{pmatrix} 0 & \tau & 0 \\ \tau & 0 & 0 \\ 0 & 0 & 0 \end{pmatrix}$ $\gamma^p = 2\varepsilon^p \Longrightarrow$

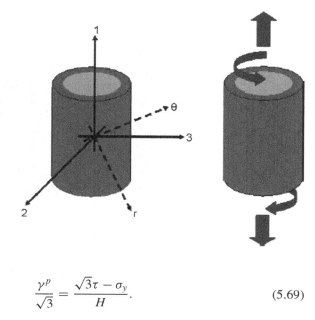

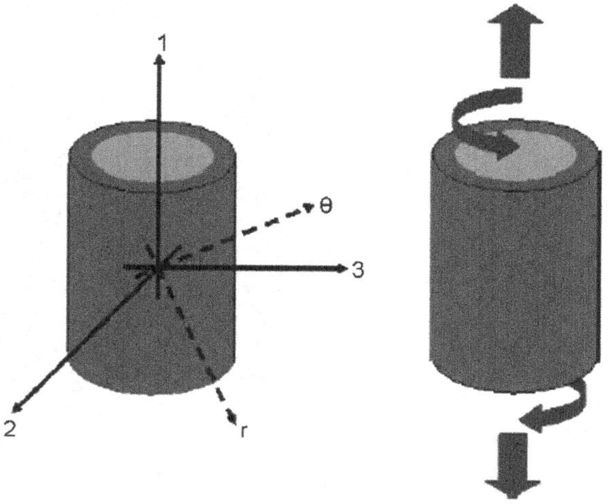

Fig. 5.6 Thin tube loaded in traction-torsion (Laverhne 2013)

$$\frac{\gamma^p}{\sqrt{3}} = \frac{\sqrt{3}\tau - \sigma_y}{H}. \tag{5.69}$$

5.4.6 *Exercise 5.2: Traction-Torsion of a Thin Tube for Isotropic Plasticity*

Burlet et al. (1988), Laverhne (2013)

Questions:

Let's consider a thin tube of average radius R and thickness e, loaded in tension–torsion by an axial force $F\overrightarrow{e_z}$ and a torque $C\overrightarrow{e_z}$ (Figs. 5.6 and 5.7).

The material behavior is assumed to be elastoplastic with linear isotropic hardening. Let's denote H the hardening modulus.

(1) stress state

(a) Determine the stress tensor in the cylindrical coordinate system $(\overrightarrow{e_r}, \overrightarrow{e_\theta}, \overrightarrow{e_z})$ associated to the tube.
(b) Calculate the equivalent stress Huber–von Mises.
(c) Give the shape of the elastic domain.
(d) Give the equations of 3D plasticity model.

(2) Response for proportional loadings
The force $\overrightarrow{F}$ and the torque $\overrightarrow{C}$ are assumed to vary proportionally.

(a) Show that one has in this case: $\tau = \sigma_{z\theta} = \alpha\sigma_{zz} = \alpha\sigma$ where α is a constant.
(b) Express the Huber–von Mises stress by introducing the coefficient α.

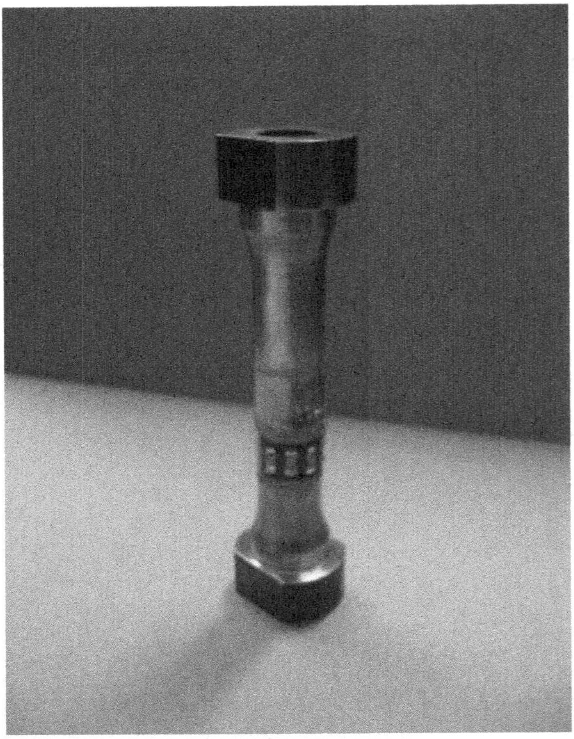

Fig. 5.7 Example of a tension–torsion test sample (Laverhne 2013)

(c) Calculate the normal $\vec{n}$ to the surface of plasticity and show that it is constant.
(d) Calculate the plastic strains tensor.
(e) Calculate the total strains tensor

(3) Response for nonproportional loading (a) tension then torsion (b) torsion then tension

Calculate the plastic strains in these two cases (a) and (b)

Calculate the total strains $\varepsilon_{rr}, \varepsilon_{\theta\theta}, \varepsilon_{zz}$ *and* $\varepsilon_{z\theta}$..

Correction:

Main steps of the resolution:

–Show incremental aspect of plasticity relations,

–Highlight dependence on the path followed (Fig. 5.8).

(1a) Calculation of the stress tensor.

Tensile stress:

$$\sigma_{zz} = \frac{F}{S} = \frac{F}{2\pi Re} = \sigma = cte \tag{5.70}$$

Torsional stress:

$$\sigma_{z\theta} = \frac{Cr}{I_0} = \frac{Cr}{\frac{\pi}{2}\left(R_1^4 - R_0^4\right)} \simeq \frac{C}{2\pi R^2 e} = \tau \tag{5.71}$$

Fig. 5.8 Three paths of mechanical loading (Cailletaud et al. 2011)

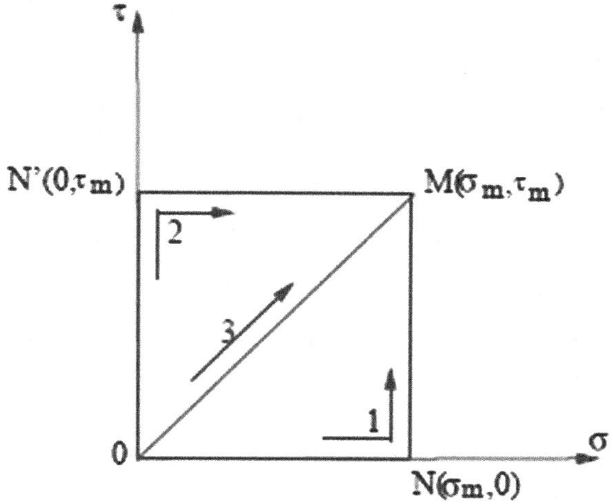

$$\underline{\sigma} M\,(r,\theta,z)=\begin{pmatrix}0&0&0\\0&0&\tau\\0&\tau&\sigma\end{pmatrix} \tag{5.72}$$

Deviatoric stress tensor: $\underline{s}M\,(r,\theta,z)=\begin{pmatrix}-\frac{\sigma}{3}&0&0\\0&-\frac{\sigma}{3}&\tau\\0&\tau&2\frac{\sigma}{3}\end{pmatrix}.$

$H-vM\;\overline{\sigma}=\sqrt{\sigma^2+3\tau^2}\;d\underline{\sigma}=\begin{pmatrix}0&0&0\\0&0&d\tau\\0&d\tau&d\sigma\end{pmatrix}.$

Path 3: proportional loading path OM

Let's consider $k=\frac{\tau\sqrt{3}}{\sigma}\Longrightarrow\overline{\sigma}=\sigma\sqrt{1+k^2}$ and $\underline{\sigma}=\begin{pmatrix}0&0&0\\0&0&\frac{k\sigma}{\sqrt{3}}\\0&\frac{k\sigma}{\sqrt{3}}&\sigma\end{pmatrix}$

Let's recall that: $d\varepsilon_{ij}^p=\frac{9}{4}\frac{s_{kl}d\sigma_{kl}}{H\overline{\sigma}^2}s_{ij}$ with $s_{kl}d\sigma_{kl}=\frac{2}{3}\sigma d\sigma+2\frac{k\sigma}{\sqrt{3}}\frac{kd\sigma}{\sqrt{3}}=\frac{2}{3}\sigma\left(1+k^2\right)d\sigma$
$\Longrightarrow d\varepsilon_{ij}^p=\frac{3}{2H}d\sigma\, s_{ij}\Longrightarrow d\varepsilon_{zz}^p=\frac{\sigma}{H}d\sigma\ et\ d\varepsilon_{z\theta}^p=\frac{3}{2H}d\tau$
That gives in M:

$$\varepsilon_{zz}^p\,(M)=\frac{\sigma_m-\sigma_y}{H};\ \varepsilon_{z\theta}^p\,(M)=\frac{3}{2H}\left(\tau_m-\frac{\sigma_y}{\sqrt{3}}\right) \tag{5.73}$$

with: $\varepsilon_{rr}^p\,(M)=\varepsilon_{\theta\theta}^p\,(M)=-\frac{\varepsilon_{zz}^p(M)}{2}\ et\ \varepsilon_{r\theta}^p\,(M)=\varepsilon_{rz}^p\,(M)=0$

Path 1: traction then torsion

–path ON: $d\varepsilon_{zz}^p=\frac{d\sigma}{H}\Longrightarrow\varepsilon_{zz}^p\,(N)=\frac{\sigma_m-\sigma_y}{H}$
–path NM: $d\varepsilon_{zz}^p=\frac{9}{4}\frac{2\tau d\tau}{H(\sigma_m^2+3\tau^2)}\frac{2}{3}\sigma_m;$

$d\varepsilon^p_{z\theta} = \frac{9}{4}\frac{2\tau d\tau}{H(\sigma_m^2+3\tau^2)}\tau = \frac{3}{2H}\left(1-\frac{\sigma_m^2}{\sigma_m^2+3\tau^2}\right)d\tau$
At the point M:

$$\varepsilon^p_{zz}(M) = \frac{\sigma_m-\sigma_y}{H} + \frac{\sigma_m}{2H}\ln\left(\frac{\sigma_m^2+3\tau_m^2}{\sigma_m^2}\right) \tag{5.74}$$

$$\varepsilon^p_{z\theta}(M) = \frac{3\tau_m}{2H} - \frac{\sqrt{3}\sigma_m}{2H}\text{arctg}\left(\frac{\sqrt{3}\tau_m}{\sigma_m}\right) \tag{5.75}$$

Path 2: torsion then traction

path ON': $d\varepsilon^p_{z\theta} = \frac{3\tau}{2H}d\tau \Longrightarrow \varepsilon^p_{z\theta}(N') = \frac{3}{2H}\left(\tau_m - \frac{\sigma_y}{\sqrt{3}}\right)$
path N'M: $d\varepsilon^p_{zz} = \frac{9}{4}\frac{\frac{2}{3}\sigma d\sigma}{H(\sigma^2+3\tau_m^2)}\frac{2}{3}\sigma = \frac{1}{H}\left(1-\frac{3\tau_m^2}{\sigma^2+3\tau_m^2}\right)$;
$d\varepsilon^p_{z\theta} = \frac{9}{4}\frac{\frac{2}{3}\sigma d\sigma}{H(\sigma^2+3\tau_m^2)}\tau_m = \frac{3\tau_m}{2H}\frac{\sigma d\sigma}{\sigma^2+3\tau_m^2}$
At the point M:

$$\varepsilon^p_{zz}(M) = \frac{\sigma_m}{H} - \frac{\sqrt{3}\tau_m}{H}\text{arctg}\left(\frac{\sigma_m}{\sqrt{3}\tau_m}\right) \tag{5.76}$$

$$\varepsilon^p_{z\theta}(M) = \frac{3}{2H}\left(\tau_m - \frac{\sigma_y}{\sqrt{3}}\right) + \frac{3\tau_m}{4H}\ln\left(\frac{\sigma_m^2+3\tau_m^2}{\tau_m^2}\right) \tag{5.77}$$

Don't forget that:

$$\varepsilon^p_{rr}(M) = \varepsilon^p_{\theta\theta}(M) = -\frac{\varepsilon^p_{zz}(M)}{2} \quad \textit{and} \ \varepsilon^p_{r\theta}(M) = \varepsilon^p_{rz}(M) = 0$$

For these tension–torsion problems, it is obvious that plastic strains are not the same depending on the load path used to reach the same stress point $M(\sigma_m, \tau_m)$.

Naturally, elastic strains are the same, regardless of the path followed:

$$\varepsilon^{el}_{zz}(M) = \frac{\sigma_m}{E};\ \varepsilon^{el}_{z\theta}(M) = (1+\nu)\frac{\tau_m}{E}; \tag{5.78}$$

$\varepsilon^{el}_{rr}(M) = \varepsilon^{el}_{\theta\theta}(M) = -\nu\varepsilon^{el}_{zz}(M);\ \varepsilon^{el}_{rz}(M) = \varepsilon^{el}_{r\theta}(M) = 0$
and the tensor of total strains is written:

$$\underline{\varepsilon}(M) = \underline{\varepsilon}^{el}(M) + \underline{\varepsilon}^{p}(M) \tag{5.79}$$

5.5 Isotropic and Kinematical Hardenings

As part of the generalized standard materials, loading function is used $f\left(\underline{\sigma}, A_i\right)$ to built the dissipation potential.

As established in Chap. 2, "the intrinsic dissipation" is written as:

$$\Phi = \underline{\sigma} : \underline{\dot{\varepsilon}} - \dot{\psi} = \underline{\sigma} : \underline{\dot{\varepsilon^p}} - A_I \dot{\alpha}_I \geq 0 \tag{5.80}$$

We introduce the multiplier $\dot{\lambda}_p$ and we built:

$$F\left(\underline{\sigma}, A_I\right) = \underline{\sigma} : \underline{\dot{\varepsilon^p}} - A_I \dot{\alpha}_I - \dot{\lambda}_p f\left(\underline{\sigma}, A_i\right) \tag{5.81}$$

The point where the partial derivatives with respect to $\underline{\sigma}$ and A_I are zero corresponds to an extreme value that will actually be a maximum if f is a convex function. It comes then:

$$\underline{\dot{\varepsilon^p}} = \dot{\lambda}_p \frac{\partial f}{\partial \underline{\sigma}} = \dot{\lambda}_p \underline{n} \; and \; \dot{\alpha}_I = -\dot{\lambda}_p \frac{\partial f}{\partial A_I} \tag{5.82}$$

We can talk about generalized normality.

Let's consider the isotropic hardening with the radius of the "variable domain" $R(p)$ such that $\dot{R} = b\left(R_s - R\right)\dot{p}$ with $R(0) = R_0$ and the kinematical hardening with the coordinates $\underline{X}$ of the «variable domain» center such that $\underline{\dot{X}} = \frac{2}{3} c \underline{\dot{\varepsilon^p}} - d\underline{X}\dot{p}$.

The criterion function f is written as:

$$f\left(\underline{\sigma}, \underline{X}, R\right) = \overline{\sigma - X} - R - \sigma_y = 0 \tag{5.83}$$

with: $\dot{f} = 0 \Longrightarrow$

$$\frac{\partial f}{\partial \underline{\sigma}} : \underline{\dot{\sigma}} + \frac{\partial f}{\partial \underline{X}} : \underline{\dot{X}} + \frac{\partial f}{\partial R}\dot{R} = 0$$

One obtains (Burlet et al. 1988):

$$\dot{\lambda}_p = \dot{p} = \frac{\underline{\dot{\sigma}} : \underline{n}}{H} \; avec \; H = c - d\underline{n} : \underline{X} + b\left(R_s - R\right) \tag{5.84}$$

for $b = d = 0 \Longrightarrow H = c$: the linear kinematical hardening also known as Prager (Fig. 5.9),

for $b = 0 \Longrightarrow H = c - d\underline{n} : \underline{X}$ the nonlinear kinematical hardening with a recall term also known as Armstrong–Frederick (Fig. 5.10),

for $c = d = 0 \Longrightarrow H = b\left(R_s - R\right)$ the nonlinear isotropic hardening

Remarks:

(1) Generally, the internal stress tensor is deviatoric: $\underline{X} = \mathrm{dev}\underline{X}$,

(2) the normal is written as t: $\underline{n} = \frac{\partial f}{\partial \underline{\sigma}} = \frac{3}{2}\frac{\underline{s} - \underline{X}}{\overline{\sigma - X}}$,

(3) for traction-compression:

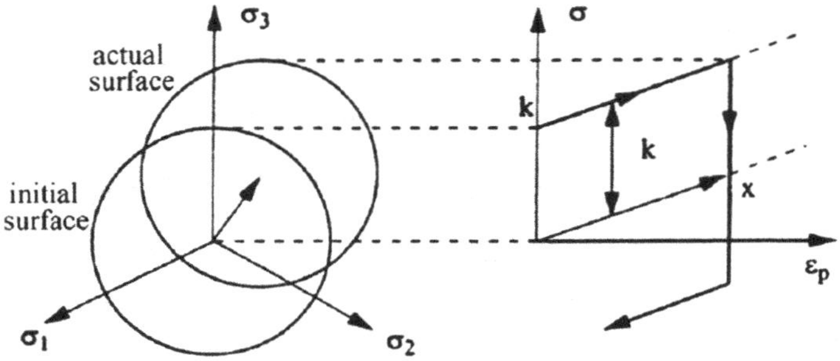

Fig. 5.9 Example of linear kinematical hardening; on *left* in the deviatoric plane; on *right*, the stress versus plastic strain (Chaboche 2008)

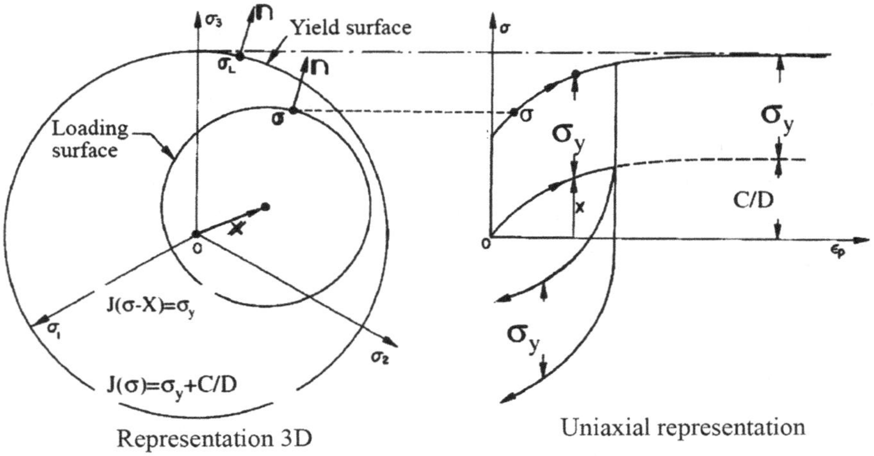

Fig. 5.10 Example of nonlinear kinematical hardening

$$\underline{X} = \begin{pmatrix} X & 0 & 0 \\ 0 & -\frac{X}{2} & 0 \\ 0 & 0 & -\frac{X}{2} \end{pmatrix} \Longrightarrow \underline{s} - \underline{X} = \begin{pmatrix} \frac{2}{3}(\sigma - X) & 0 & 0 \\ 0 & -\frac{1}{3}(\sigma - X) & 0 \\ 0 & 0 & -\frac{1}{3}(\sigma - X) \end{pmatrix}.$$

So:

$$\underline{n} = \frac{3}{2}\frac{\underline{s} - \underline{X}}{\underline{\underline{s - X}}} = \begin{pmatrix} 1 & 0 & 0 \\ 0 & -\frac{1}{2} & 0 \\ 0 & 0 & -\frac{1}{2} \end{pmatrix} \frac{\sigma - X}{\mid \sigma - X \mid} \tag{5.85}$$

$$\underline{\dot{\varepsilon}^p} = \underline{n}\dot{p} \ with \ \dot{p} = \frac{\dot{\sigma}}{H} sng\left(\sigma - X\right) \ and \ H = c - dXsng\left(\sigma - X\right) + b\left(R_s - R\right)$$

5.5.1 Case of Variable Temperature (Burlet et al. 1988)

It is necessary to take into account the evolution of the surface with the temperature:

$$R = R_0 + (R_s - R_0)\exp(-bp)$$

from where: $\frac{dR}{dT} = \frac{dR_0}{dT}(1 - \exp(-bp)) + \frac{dR_s}{dT}\exp(-bp)$

$\underline{X} = \frac{2}{3}c\underline{\alpha}$ with $\underline{\alpha}$ state variable independent of the temperature.

$\Longrightarrow \frac{d\underline{X}}{dT} = \frac{\underline{X}}{c}\frac{dc}{dT}$

Let's calculate $\frac{\partial f}{\partial T} = \frac{\partial \overline{\sigma - X}}{\partial T} - \frac{\partial R}{\partial T} = \frac{\partial \overline{\sigma - X}}{\partial \underline{X}}\frac{d\underline{X}}{dT} - \frac{\partial R}{\partial T}$

$\frac{\partial f}{\partial T} = -\frac{\underline{n}:\underline{X}}{c}\frac{dc}{dT} - \left(\frac{dR_0}{dT}(1 - \exp(-bp)) + \frac{dR_s}{dT}\exp(-bp)\right)$

$\dot{f}\left(\underline{\sigma}, \underline{X}, R, T\right) = 0$ is written now:

$$\frac{\partial f}{\partial \underline{\sigma}} : \underline{\dot{\sigma}} + \frac{\partial f}{\partial \underline{X}} : \underline{\dot{X}} + \frac{\partial f}{\partial R}\dot{R} + \frac{\partial f}{\partial T}\dot{T} = 0$$

with the same H as in the general case.

Now there is an additional "motor term." That means that the solid material can plastify without additional mechanical loading.

5.5.2 Summary of a Model Superimposing Kinematical and Isotropic Hardenings

Huber–von Mises criterion - isothermal conditions from the classic book by the LMT Cachan–Onera

Lemaitre et al. (2009)

$$\underline{\varepsilon} = \underline{\varepsilon}^{el} + \underline{\varepsilon}^{p}$$

$$\underline{\sigma} = \underline{\underline{C}}\left(\underline{\varepsilon} - \underline{\varepsilon}^{p}\right)$$

$$f\left(\underline{\sigma}, \underline{X}, R\right) = \overline{\sigma - X} - R - \sigma_y \le 0$$

$$\underline{\dot{\varepsilon}}^{P} = \frac{3}{2}\dot{\lambda}_p \frac{\underline{s} - \underline{X}}{\left(\overline{\sigma - X}\right)}$$

$$\underline{\dot{X}} = \frac{2}{3}c\underline{\dot{\varepsilon}}^{p} - \gamma(p)\underline{X}\dot{p}$$

$$\gamma(p) = \gamma_s + (\gamma_0 - \gamma_s)\exp\left(-K_y p\right)$$

$$\dot{R} = b\left(R_s - R\right)\dot{p}$$

$$\dot{\lambda_p} = \dot{p} = \frac{1}{h} H\left(f\right) < \frac{3}{2} \frac{\left(\underline{s} - \underline{X}\right) : \underline{\dot{\sigma}}}{\left(\sigma - X\right)} >$$

$$h = c - \frac{3}{2} \gamma\left(p\right) \frac{\left(\underline{s} - \underline{X}\right) : \dot{\underline{X}}}{\left(\sigma - X\right)} + b\left(R_s - R\right)$$

$$\begin{Bmatrix} f < 0 \; or \; \left(f = 0 \; and \, \dot{f} < 0\right) \rightarrow \dot{\lambda}_p = 0 \; (elasticity) \\ f = 0 \; and \, \dot{f} = 0 \rightarrow \dot{\lambda}_p > 0 \; (plasticity) \end{Bmatrix}$$

5.5.3 Inversion of Plasticity Relations

If one wishes to work on an increment of total strain, rather than a stress increment.

One can simply write:

$\underline{\dot{\sigma}} = \underline{\underline{C}} : \left(\dot{\underline{\varepsilon^{el}}}\right) = \underline{\underline{C}} : \left(\underline{\dot{\varepsilon}} - \underline{\dot{\varepsilon^p}}\right)$ from where: $H\dot{p} = \underline{n} : \underline{\dot{\sigma}} = \underline{n} : \underline{\underline{C}} : \left(\underline{\dot{\varepsilon}} - \underline{\dot{\varepsilon^p}}\right) \Longrightarrow$

$$\dot{p} = \frac{\underline{n} : \underline{\underline{C}} : \underline{\dot{\varepsilon}}}{H + \underline{n} : \underline{\underline{C}} : \underline{n}} \tag{5.86}$$

5.5.4 Exercise 5.3

Let's show that for isotropic elasticity with:

$C_{ijkl} = \lambda\delta_{ij}\delta_{kl} + \mu\left(\delta_{ik}\delta_{jl} + \delta_{il}\delta_{jk}\right)$ for material behavior such as "Huber–von Mises" one:

$$\dot{p} = \frac{2\mu\underline{n} : \underline{\dot{\varepsilon}}}{H + 3\mu} \tag{5.87}$$

5.5.5 Exercise 5.4 (Burlet et al. 1988)

Prediction with a nonlinear kinematic hardening model for uniaxial loading

Problem:

Simple tension: $\underline{\sigma} = \begin{pmatrix} \sigma & 0 & 0 \\ 0 & 0 & 0 \\ 0 & 0 & 0 \end{pmatrix}$ let $\varepsilon^p = \varepsilon^p_{11}$

$$\begin{Bmatrix} dX = c\left(ad\varepsilon^p - X \mid d\varepsilon^p \mid\right) \\ R = R_0 \end{Bmatrix}$$

Criterion function:

$$f = \mid \sigma - X \mid - R \Longrightarrow \left\{ \begin{array}{c} traction : \sigma = X + R \\ compression : \sigma = X - R \end{array} for\ plastic\ flow \right\}$$

Let's consider: $\dot{\varepsilon}^p = constante$

(1) Find the function: $X = X(\varepsilon^p)$ for monotonous traction.

(2)Integration on a "rising" monotonous branch between:

$\left(\varepsilon^p_{min}, X_{min}\right) \Longrightarrow \left(\varepsilon^p_{max}, X_{max}\right)$.

(3)Integration on a "down" monotonous branch between:

$\left(\varepsilon^p_{max}, X_{max}\right) \Longrightarrow \left(\varepsilon^p_{min}, X_{min}\right)$.

(4) In the case of symmetric cycles, find the relationship between ΔX *and* $\Delta\varepsilon^p$.

(5) In the case of asymmetric cycles, calculate the ratcheting $\delta\varepsilon^p$ per cycle (Fig. 5.11).

Correction:

This is a problem of integration between "variable" bounds.

(1) Monotonous tension

$:dX = c(a - X)\, d\varepsilon^p\ avec\ X = 0\ when\ \varepsilon^p = 0 \Longrightarrow X = a\left(1 - e^{-c\varepsilon^p}\right)$.

(2) Integration on a "rising" monotonous branch

$X = a + (X_{min} - a)\exp\left(-c\left(\varepsilon^p - \varepsilon^p_{min}\right)\right)\ X_{max}\ obtained\ when\ \varepsilon^p = \varepsilon^p_{max}$.

(3) Integration on a "down" monotonous branch

$X = -a + (X_{max} + a)\exp\left(+c\left(\varepsilon^p - \varepsilon^p_{max}\right)\right)\ X_{min}\ obtained\ when\ \varepsilon^p = \varepsilon^p_{min}$.

(4) Symmetric cycles

$\Delta X = 2a\,\text{th}\left(c\frac{\Delta\varepsilon^p}{2}\right)\ with\ \Delta X = X_{max} - X_{min}\ and\ \Delta\varepsilon^p = \varepsilon^p_{max} - \varepsilon^p_{min}$.

(5) asymmetric cycles, the ratcheting $\delta\varepsilon^p$ per cycle:

plastic strain increment is observed for example during cyclic tests between two fixed levels of stress that we call σ_{max} *and* σ_{min}. This progressive deformation can lead to the failure of the structure.

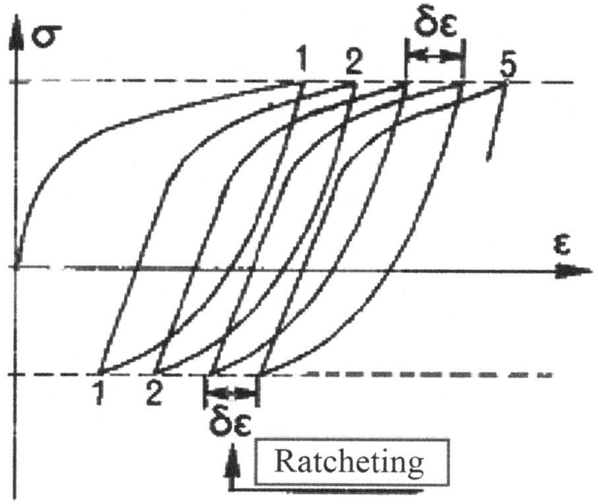

Fig. 5.11 Ratcheting effect linked to the cycling between two imposed stress (Lemaitre et al. 2009)

It is calculated on the ascending branch X_{max} corresponding to σ_{max} and then on the descending branch X_{min} corresponding to σ_{min} and one obtains:

$$\delta\varepsilon^p = \frac{1}{c}\ln\left(\frac{a^2 - (\sigma_{min} + R_0)^2}{a^2 - (\sigma_{max} - R_0)^2}\right).$$

5.6 Conclusion

We adapt a "classic" approach in "generalized standard materials" which incorporates the teachings of «the Ecole des Mines de Paris», «the Ecole Nationale Supérieure de Cachan» and finally «the Ecole Polytechnique».

Sometimes the "classic" models are not sufficient to take into account the unusual experimental observations.

To this end, we must adopt more "complex" approaches by considering:
–the loading surface,
–the description of the internal variables.

A simple solution is the addition of several kinematical hardenings.

For nonsymmetric flows in tension-compression, the loading surface with «correction function» $g(y)$ multiplied by the Huber–von Mises invariant may be considered.

The shape of the kinematic hardening can therefore be adjusted:

$$\underline{X_i} = C\frac{\tanh\left(D\alpha_{eq}\right)}{D\alpha_{eq}}\underline{\alpha_i}\ \alpha_{eq} = \left(3/2\underline{\alpha} : \underline{\alpha}\right)^{\frac{1}{2}}\ \underline{\alpha} = \underline{\varepsilon^p}$$

We write multi-models mechanisms like: two mechanisms and criteria (or one criterion).

The models that have been described so far use the same function as the limit of the elasticity range, for determining the direction of flow and the evolution of hardening variables

Yield function $: f$ *flow* $\underline{\dot{\varepsilon}^p} = \dot{\lambda}_p \frac{\partial f}{\partial \underline{\sigma}}$ *hardening* $\dot{\alpha}_I = -\dot{\lambda}_p \frac{\partial f}{\partial A_I}$.

Chapter 6
Viscoelasticity

Abstract «Soft solids» such as polymers, organic matrix composites, and bitumen have dissipative phenomena associated with elasticity also known as «viscosity». The viscoelastic characteristics of a body are the dependence of its response on the strain rate and a long recovery time of the initial state after the removal of mechanical loading. Different models can be constructed from the rheological description: springs and dampers in series or in parallel.

6.1 Introduction

Polymers, organic matrix composites, bitumen that Lemaitre et al. (2009) have called "soft solids", have dissipative phenomena associated with elasticity also known as "viscosity". This viscosity is a parameter linking stress and strain rate. First, the viscoelastic characteristics of a body are the dependence of its response to the strain rate and a long time to recover the non-deformed initial state after removal of the mechanical loading. One can speak of a "Reversible" behavior in terms of deformation, but thermodynamically irreversible due to the viscous dissipation. The recovery to the free stress state comes along with the absence of permanent deformation as this is not the case in plasticity or viscoplasticity.

The aging of polymers will not be included in this chapter.

In dynamics, the very low damping for metals $\left(10^{-4}\right)$, at room temperature, can reach 10^{-2} *to* 1 for polymers and can be modeled by the theory of viscoelasticity.

Finally, the viscosity generates creep under imposed constant stress and relaxation under deformation kept constant.

6.2 Viscoelasticity 1D (Pattofatto 2009)

Different models can be constructed from rheological descriptions: springs and dampers in series or in parallel with behavior laws:

$$for\ the\ spring\ : \sigma^e = E_r \varepsilon^{el}$$

C. Lexcellent, *Linear and Non-linear Mechanical Behavior of Solid Materials*, DOI 10.1007/978-3-319-55609-3_6

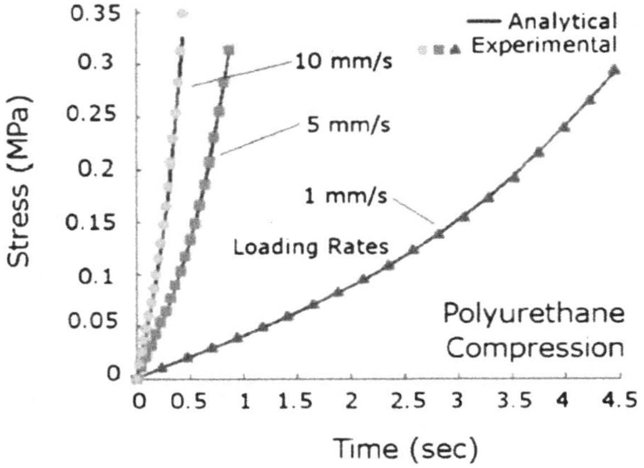

Fig. 6.1 Compression tests on a silicon PU

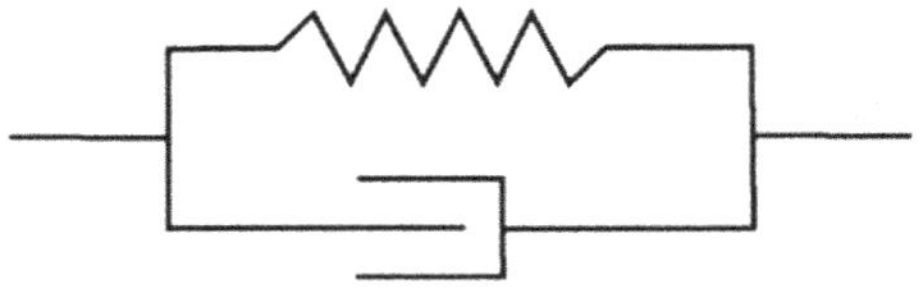

Fig. 6.2 Kelvin–Voigt rheological model

$$for\ the\ damper\ : \sigma^{an} = \eta\dot{\varepsilon}^{an}.$$

The parameter identification E_r *Relaxed Young modulus and* η *viscosity* can be done through « holding time » basic tests (creep or relaxation) or conventional tests (tensile, compression, shear) (Fig. 6.1) or frequency tests (DMA).

6.2.1 Kelvin–Voigt Model

The Kelvin–Voigt model (Fig. 6.2) is represented by a shock absorber "in parallel" with a spring.

In isotropic three-dimensional (3D) cases, the free energy (the same as in elasticity) and the dissipation potential are written as follows:

$$\rho_0\psi\left(\underline{\varepsilon}^{el}\right) = \tfrac{1}{2}\left[\lambda\left(\mathrm{tr}\underline{\varepsilon}^{el}\right)^2 + 2\mu\underline{\varepsilon}^{el} : \underline{\varepsilon}^{el}\right]$$

$$\phi\left(\underline{\dot{\varepsilon}}\right) = \tfrac{1}{2}\left[\Lambda\left(\mathrm{tr}\underline{\dot{\varepsilon}}^{el}\right)^2 + 2M\underline{\dot{\varepsilon}}^{el} : \underline{\dot{\varepsilon}}^{el}\right]$$

with the behavior law

$$\underline{\sigma} = \rho_0\frac{\partial\psi}{\partial\underline{\varepsilon}^{el}} + \frac{\partial\phi}{\partial\underline{\dot{\varepsilon}}^{el}}$$

that gives

$$\underline{\sigma} = \lambda \left(\mathrm{tr}\underline{\varepsilon^{el}} + \tau \mathrm{tr}\underline{\dot{\varepsilon^{el}}}\right) \underline{1} + 2\mu \left(\underline{\varepsilon^{el}} + \tau \underline{\dot{\varepsilon^{el}}}\right) \; with\ \tau = \frac{\Lambda}{\lambda} = \frac{M}{\mu}. \tag{6.1}$$

In 1D case:

$$\rho_0 \psi \left(\varepsilon^{el}\right) = \frac{1}{2} E \varepsilon^{el2}$$

$$\phi \left(\dot{\varepsilon}\right) = \frac{1}{2} \eta \dot{\varepsilon^{el}}^2.$$

This leads to

$$\sigma = E_r \varepsilon^{el} + \eta \dot{\varepsilon}^{el} = E_r \left(\varepsilon^{el} + \tau \dot{\varepsilon^{el}}\right) \; with\ \tau = \frac{\eta}{E_r}. \tag{6.2}$$

Response to a creep test/recovery 1D (Fig. 6.3):

$$\sigma(t) = \begin{Bmatrix} \sigma_0\ if\ 0 \le t \le T \\ 0\ if\ t > T \end{Bmatrix} \Longrightarrow \varepsilon(T) \begin{Bmatrix} \frac{\sigma_0}{E}\left[1 - \exp\left(-\frac{t}{\tau}\right)\right]\ if\ 0 \le t \le T \\ \varepsilon(T) \exp\left(-\frac{t-T}{\tau}\right)\ if\ t > T \end{Bmatrix}.$$

The prediction of a relaxation test is not correct.

Indeed,

$$\varepsilon^{el}(t) = \begin{Bmatrix} \varepsilon_0\ if\ 0 \le t \le T \\ 0\ if\ t > T \end{Bmatrix} \Longrightarrow \sigma(t) = \begin{Bmatrix} \sigma_0 = E\varepsilon_0\ if\ 0 \le t \le T \\ 0\ if\ t > T \end{Bmatrix}.$$

The stress remains constant, which is contrary to observations.

Response to a harmonic load (Fig. 6.4):

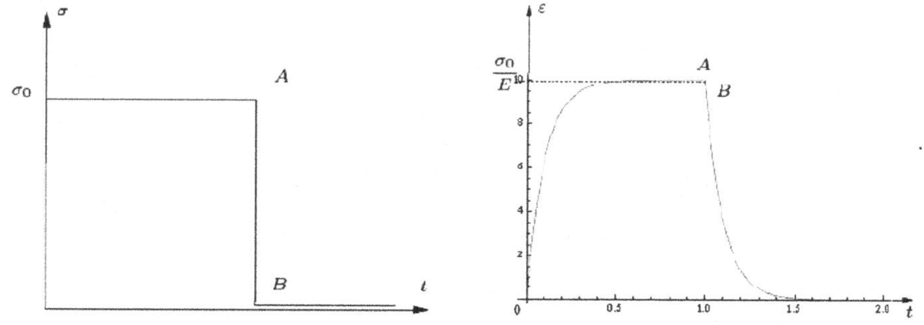

(a) Loading (b) Response

Fig. 6.3 Response to a creep test/recovery 1D: Kelvin–Voigt model (Boukamel 2009)

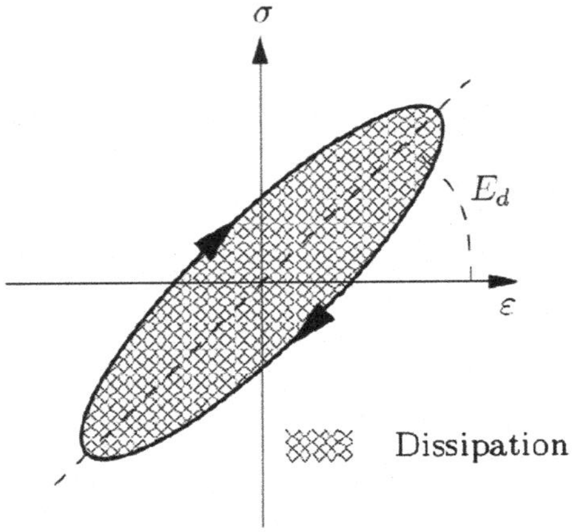

Fig. 6.4 Hysteresis loop predicted by the Kelvin–Voigt model (Boukamel 2009)

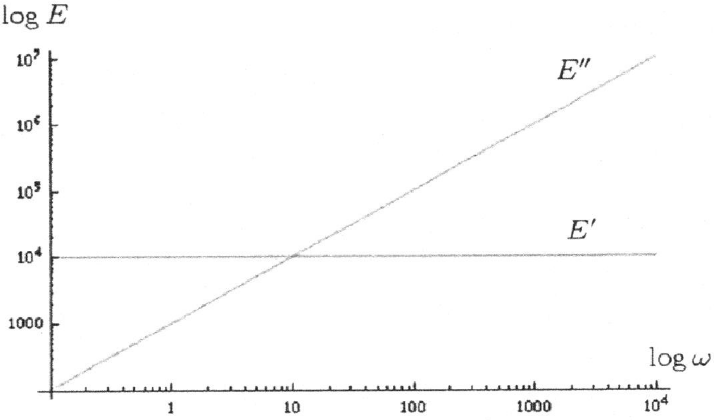

Fig. 6.5 Modulus dependence with the pulsation ω (Boukamel 2009)

either

$$\varepsilon^{el}(t) = \varepsilon_0 R_e\left[\exp\left(j\omega t\right)\right] \Longrightarrow \sigma(t) = E_d \varepsilon_0 R_e\left[\exp j\left(\omega t + \delta\right)\right]$$

with $E_d = \sqrt{E^2 + \eta^2\omega^2}$ dynamic modulus and $\delta = \arctan\left(\frac{\eta\omega}{E}\right)$.

The plot of the stress σ depending on the deformation ε gives us the hysteresis loop whose area represents the dissipation.

The complex modulus $\overline{E}$ is defined by

$$\overline{E} = \frac{\sigma_0}{\varepsilon_0}\exp\left(j\delta\right) = E' + jE'' \left\{\begin{array}{l} E' = E_d\cos\delta = E_r \\ E'' = E_d\sin\delta = \eta\omega \end{array}\right\}$$

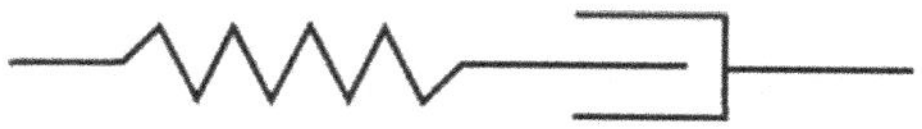

Fig. 6.6 Maxwell rheological model

with E' called "storage modulus" and E'' "loss modulus" (Fig. 6.5).

6.2.2 *Maxwell Model*

The Maxwell model (Fig. 6.6) is defined by a shock absorber "in series" with a spring;

In one-dimensional case:

$$\rho_0\psi\left(\varepsilon,\varepsilon^v\right)=\frac{1}{2}E(\varepsilon-\varepsilon^v)^2,\ \phi\left(\dot{\varepsilon}^v\right)=\frac{1}{2}\eta(\dot{\varepsilon}^v)^2\Longrightarrow\sigma=E\left(\varepsilon-\varepsilon^v\right)=\eta\dot{\varepsilon}^v$$

$$\begin{aligned}\sigma^1&=E\varepsilon^e\\ \sigma^2&=\eta\dot{\varepsilon^v}\end{aligned}$$

$$\begin{gathered}\sigma^1=\sigma^2=\sigma\\ \varepsilon^e+\varepsilon^v=\varepsilon\Longrightarrow\dot{\varepsilon^e}+\dot{\varepsilon^v}=\dot{\varepsilon}\Longrightarrow\end{gathered}$$

$$\dot{\varepsilon}=\frac{\dot{\sigma}}{E}+\frac{\sigma}{\eta}. \tag{6.3}$$

solution: $\sigma\left(t\right)=\sigma_0 exp\left(\frac{-t}{\tau}\right)+E\varepsilon_0$.

In 3D isotropic situation:

$$\left\{\begin{array}{c}\rho_0\psi\left(\underline{\varepsilon}\right)=\frac{1}{2}\left[\lambda\left(\mathrm{tr}(\underline{\varepsilon}-\underline{\varepsilon}^v\right)^2+2\mu(\underline{\varepsilon}-\underline{\varepsilon}^v):(\underline{\varepsilon}-\underline{\varepsilon}^v)\right]\\ \phi\left(\underline{\dot{\varepsilon}}\right)=\frac{1}{2}\left[\Lambda\left(\mathrm{tr}\underline{\dot{\varepsilon}^v}\right)^2+2M\underline{\dot{\varepsilon}}^v:\underline{\dot{\varepsilon}}^v\right]\end{array}\right\},$$

that causes

$$\underline{\sigma}=\lambda\mathrm{tr}\left(\underline{\varepsilon}-\underline{\varepsilon}^v\right)\underline{1}+2\mu\left(\underline{\varepsilon}-\underline{\varepsilon}^v\right)=\Lambda\mathrm{tr}\underline{\dot{\varepsilon}}^v\underline{1}+2M\underline{\dot{\varepsilon}}^v. \tag{6.4}$$

Response to a creep test/recovery 1D (Fig. 6.7):

$$\sigma\left(t\right)=\left\{\begin{array}{c}\sigma_0\ if\ 0\leq t\leq T\\ 0\ if\ t>T\end{array}\right\}\Longrightarrow\varepsilon\left(T\right)\left\{\begin{array}{c}\frac{\sigma_0}{E}\left[1+\frac{t}{\tau}\right]\ if\ 0\leq t\leq T\\ \varepsilon\left(T\right)\ if\ t>T\end{array}\right\}.$$

Response for a relaxation test/reset (Fig. 6.8):

$$\varepsilon\left(t\right)=\left\{\begin{array}{c}\varepsilon_0\ if\ 0\leq t\leq T\\ 0\ if\ t>T\end{array}\right\}\Longrightarrow\sigma\left(t\right)=\left\{\begin{array}{c}\sigma_0=E\varepsilon_0\exp\left(-\frac{t}{\tau}\right)\ if\ 0\leq t\leq T\\ E\varepsilon_0\left[\exp\left(-\frac{t}{\tau}\right)-1\right]\exp\left(-\frac{t-T}{\tau}\right)\ if\ t>T\end{array}\right\}.$$

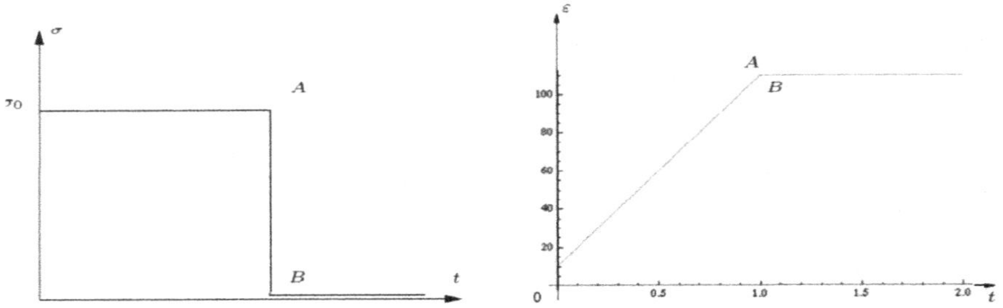

(a) Loading (b) Response

Fig. 6.7 Prediction for creep tests with the Maxwell model (Boukamel 2009)

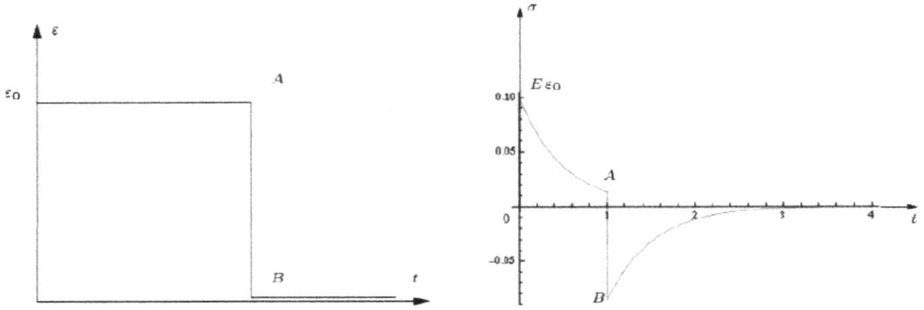

(a) Loading (b) Response

Fig. 6.8 Prediction for relaxation tests with the Maxwell model (Boukamel 2009)

Response to harmonic loads.
either:

$$\varepsilon\left(t\right)=\varepsilon_0 R_e\left[\exp\left(j\omega t\right)\right]\Longrightarrow\sigma\left(t\right)=E_d\varepsilon_0 R_e\left[\exp j\left(\omega t+\delta\right)\right],$$

with $E_d=\frac{E\eta\omega}{\sqrt{E^2+\eta^2\omega^2}}$ and the dynamic modulus $\delta=\arctan\left(\frac{E}{\eta\omega}\right)$.
Complex modulus:
The complex modulus $\overline{E}$ is defined by

$$\overline{E}=\frac{\sigma_0}{\varepsilon_0}\exp\left(j\delta\right)=E^{'}+jE^{''}\left\{\begin{array}{l}E^{\prime}=\frac{E\eta^2\omega^2}{E^2+\eta^2\omega^2}\\E^{''}=\frac{E^2\eta\omega}{E^2+\eta^2\omega^2}\end{array}\right\}$$

with $E^{'}$ called "storage modulus" and $E^{''}$ "loss modulus" (Fig. 6.9).

The Kelvin–Voigt model is well suited to the description of creep but not to relaxation, while the Maxwell model fits well the relaxation but not creep.

It is for this reason that a mixed rheological model was introduced. This is the standard linear model or Zener one (Fig. 6.10).

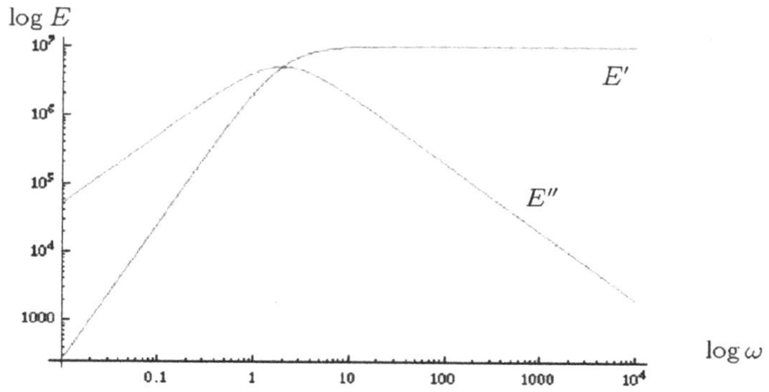

Fig. 6.9 Modulus dependence with the pulsation ω (Boukamel 2009)

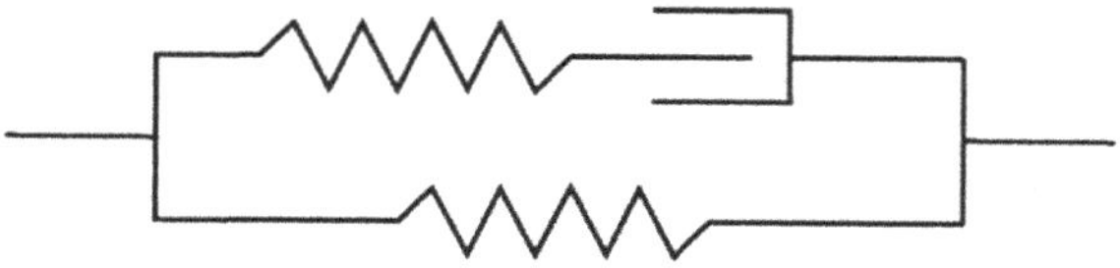

Fig. 6.10 Standard linear rheological model

6.2.3 Zener Model:

Complex models, with many branches, further illustrate the actual viscoelastic behavior. The standard linear model consists of a parallel association of a spring E_e and a rheological Maxwell model E_1, η.

Exercise: Show the differential behavior equation:

$$\sigma + \tau\dot{\sigma} = E_e\varepsilon + (E_1 + E_e)\,\tau\dot{\varepsilon}, \tag{6.5}$$

with $\tau = {}^{\eta}/_{E_1}$ the time characteristic time associated to Maxwell model.

Response:

$$\dot{\varepsilon_1} = \frac{\dot{\sigma_1}}{E_1} + \frac{\sigma_1}{\eta}$$
$$\varepsilon_2 = \frac{\sigma_2}{E_2}$$

with $\varepsilon = \varepsilon_1 = \varepsilon_2 \Longrightarrow \dot{\varepsilon} = \dot{\varepsilon_1} = \dot{\varepsilon_2}$ et $\sigma = \sigma_1 + \sigma_2 \Longrightarrow \dot{\sigma} = \dot{\sigma_1} + \dot{\sigma_2}$

$$\dot{\sigma}_1 = E_1\dot{\varepsilon} - \frac{(\sigma - E_e\varepsilon)}{\tau}$$

$$\dot{\sigma}_2 = E_e\dot{\varepsilon}$$

$\Longrightarrow \sigma + \tau\dot{\sigma} = E_e\varepsilon + (E_1 + E_e)\,\tau\dot{\varepsilon}.$

For convenience we may transform this equation with $E_1 = e$ in Boukamel (2009):

$$(E+e)\left(\dot{\sigma}+\frac{\varepsilon}{\tau_F}\right)=\dot{\sigma}+\frac{\sigma}{\tau_R}\ with\ \tau_F=\eta\frac{e+E}{Ee}\ and\ \tau_R=\tau=\frac{\eta}{e}. \tag{6.6}$$

Response to a creep test/recovery (Fig. 6.11):

$$\sigma(t)=\left\{\begin{matrix}\sigma_0\ if\ 0\le t\le T\\ 0\ if\ t>T\end{matrix}\right\}\Longrightarrow\varepsilon(T)\left\{\begin{matrix}\frac{\sigma_0}{E}\left[1-\frac{e}{E+e}\exp\left(-\frac{t}{\tau_F}\right)\right]\ if\ 0\le t\le T\\ \varepsilon(T)\exp\left(-\frac{t-T}{\tau_F}\right)\ if\ t>T\end{matrix}\right\}$$

Response for a relaxation test/reset (Fig. 6.12):

$$\varepsilon(t)=\left\{\begin{matrix}\varepsilon_0\ if\ 0\le t\le T\\ 0\ if\ t>T\end{matrix}\right\}\Longrightarrow\sigma(t)=\left\{\begin{matrix}\sigma_0=\varepsilon_0(E+e\exp\left(-\frac{t}{\tau_R}\right))\ if\ 0\le t\le T\\ e\varepsilon_0\left[\exp\left(-\frac{T}{\tau_R}\right)-1\right]\exp\left(-\frac{t-T}{\tau_R}\right)\ if\ t>T\end{matrix}\right\}$$

Complex modulus $\overline{E}$: (Fig. 6.13):

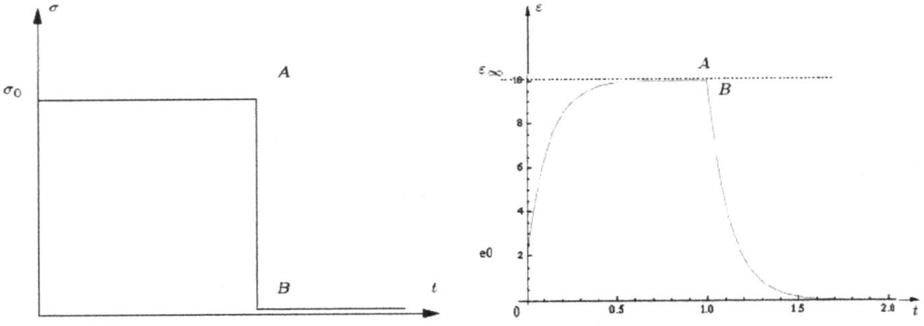

(a) Loading (b) Response

Fig. 6.11 Prediction for creep with the Zener model (Boukamel 2009)

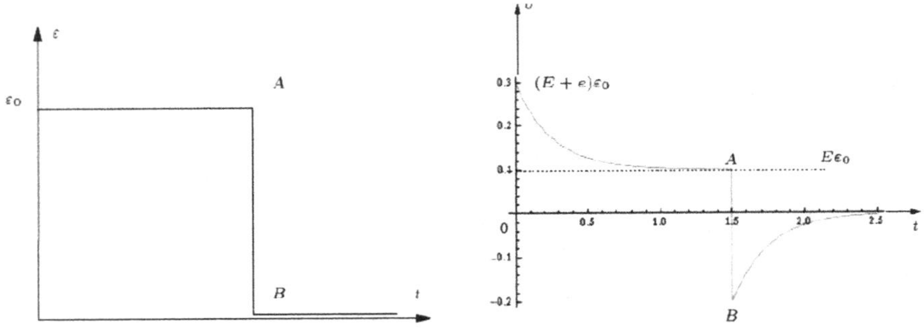

(a) Loading (b) Response

Fig. 6.12 Prediction for relaxation with the Zener model (Boukamel 2009)

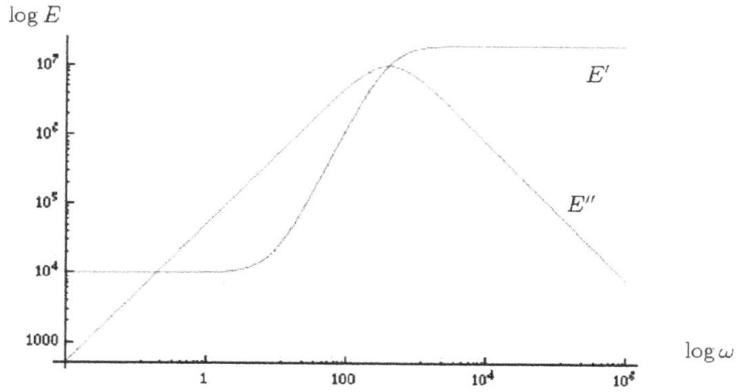

Fig. 6.13 Modulus dependence with the pulsation ω (Boukamel 2009)

$$\overline{E} = \frac{\sigma_0}{\varepsilon_0}\exp(j\delta) = E' + jE'' \left\{ \begin{array}{c} E' = E + \frac{e\eta\omega^2}{e^2+\eta^2\omega^2} \\ E'' = \frac{e^2\eta\omega}{e^2+\eta^2\omega^2} \end{array} \right\}$$

6.2.4 *The Generalized Maxwell Model*

The generalized Maxwell model accounts for more complex behaviors by adding other simple items like Maxwell body in parallel (Fig. 6.14).

Fig. 6.14 Generalized Maxwell model (Pattofatto 2009)

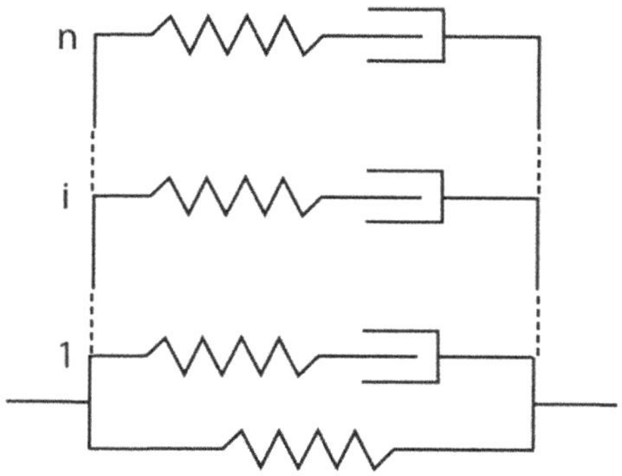

For 1D case, the free energy is written as

$$\begin{aligned}\rho_0\psi\left(\varepsilon,\varepsilon^1,\ldots,\varepsilon^n\right) &= \rho_0\psi_0\left(\varepsilon\right)+\textstyle\sum_{i=1}^n\rho_0\psi_i\left(\varepsilon,\varepsilon_i\right)\\ &= \tfrac{1}{2}E\varepsilon^2+\textstyle\sum_{i=1}^n\tfrac{1}{2}e_i\left(\varepsilon-\varepsilon^i\right)^2,\end{aligned}$$

and the dissipation potential as

$$\phi\left(\dot{\varepsilon}^1,\ldots,\dot{\varepsilon}^n\right)=\sum_{i=1}^n\phi_i\left(\dot{\varepsilon^i}\right)$$

that gives

$$\begin{aligned}\sigma &= \sigma_0+\textstyle\sum_{i=1}^n\sigma_i\\ &= E\varepsilon+\textstyle\sum_{i=1}^n e_i\left(\varepsilon-\varepsilon^i\right)\\ 0 &= \eta_i\dot{\varepsilon^i}-e_i\left(\varepsilon-\varepsilon^i\right).\end{aligned} \tag{6.7}$$

Harmonic loadings (Fig. 6.15):

$$\varepsilon\left(t\right)=\varepsilon_0R_e\left[\exp\left(j\omega t\right)\right]\Longrightarrow\sigma\left(t\right)=\sigma_0R_e\left[\exp j\left(\omega t+\delta\right)\right]$$

with dynamic modulus $E_d=\sqrt{E'^2+E''^2}$ and loss angle $\delta=\arctan\left(\frac{E''}{E'}\right)$. The complex modulus is written as follows (Fig. 6.16):

$$\overline{E}=\frac{\sigma_0}{\varepsilon_0}\exp\left(j\delta\right)=E'+jE''\left\{\begin{array}{l}E'=\frac{\sigma_0}{\varepsilon_0}\cos\delta\\ E''=\frac{\sigma_0}{\varepsilon_0}\sin\delta\end{array}\right\}.$$

Fig. 6.15 Hysteresis loop predicted by generalized Maxwell model

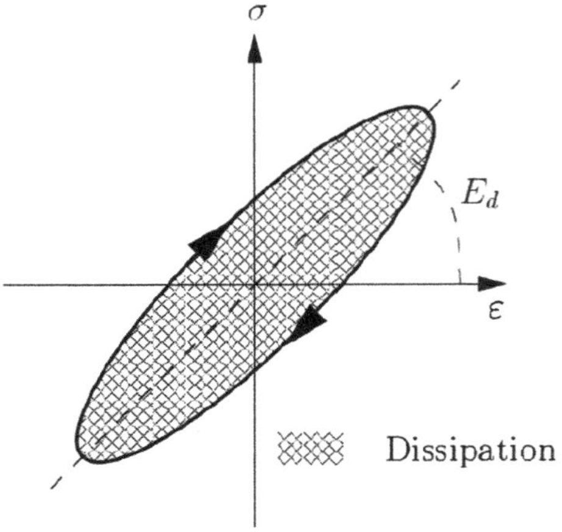

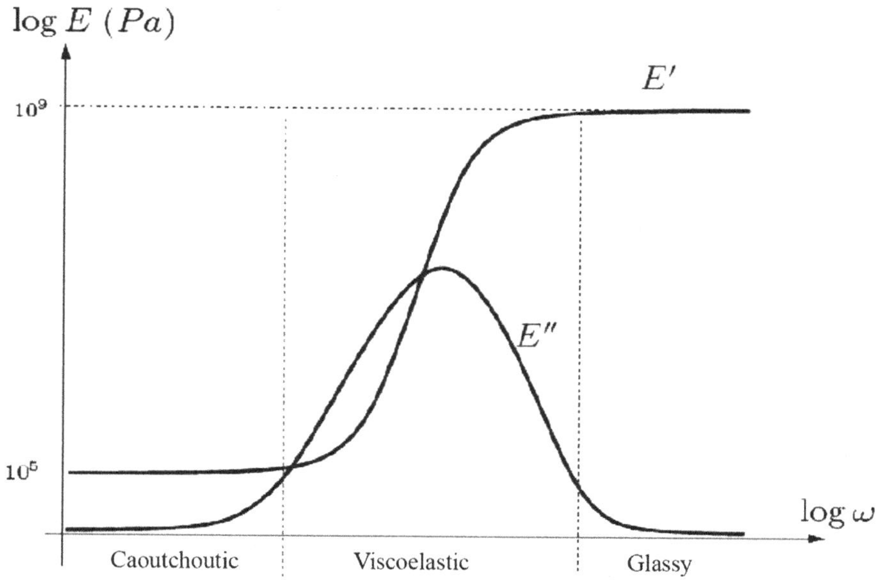

Fig. 6.16 Modulus dependence with the pulsation ω (Boukamel 2009)

We can calculate the energy dissipated per cycle $\triangle W = \pi E^{''} \varepsilon_0^2$ and $W = \frac{1}{2} E^{'} \varepsilon_0^2$ and deduce

$$E^{'} = \frac{2W}{\varepsilon_0^2}, \; E^{''} = \frac{\triangle W}{\pi \varepsilon_0^2} \; and \; \frac{E^{''}}{E^{'}} = \tan\delta = \frac{\triangle W}{2\pi W}.$$

Chapter 7
Viscoplasticity

Abstract The viscoplasticity is defined as a «time-dependent» behavior. The creep mechanisms (and by duality relaxation) are the first manifestations of material viscoplasticity. The general form of the viscoplastic-unified constitutive model comes from Chaboche in 2008. A comparison between «plasticity» and «viscoplasticity» ended the chapter.

"Any solid is a fluid that is not known" (Lemaitre et al. 2009).

7.1 Creep Mechanisms

7.1.1 Ashby Deformation Maps

As a preamble, we will examine the creep mechanisms who are the first manifestation of materials viscoplasticity. Let us recall that the flow corresponds to the deformation, over time, of a body subjected to external mechanical stresses. The Ashby maps itemize in a diagram (normalized stress σ/μ and reduced temperature T/T_f) representative of different mechanisms of deformation for a great number of metals (Ashby 1972) (Fig. 7.1).

7.1.2 Stress and Temperature Effects

1. High stress levels induce the slip dislocations, that is to say a plastic behavior.
2. At intermediate stresses the recovery creep occurs. It is a thermally activated creep.

There are two antagonists mechanisms: hardening which is a function of the deformation and the restoration due to time (the diffusion).

The measurement of the internal stress σ_i in 1D viscoplasticity is carried out by means of "dip test" technique (technique of stress decrement) (Figs. 7.2 and 7.3).

C. Lexcellent, *Linear and Non-linear Mechanical Behavior of Solid Materials*, DOI 10.1007/978-3-319-55609-3_7

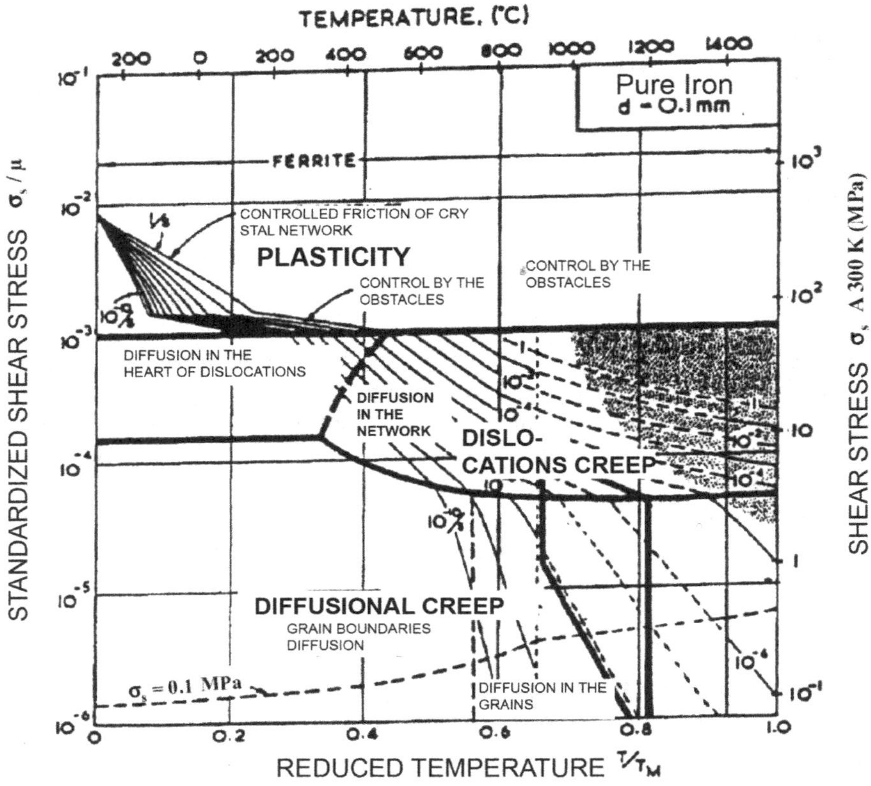

Fig. 7.1 Deformation mechanisms maps. The iso-rate lines of deformation are given from decade to decade. (Frost and Ashby (1982) and Gourgues-Lorenzo (2008))

"dip tests" were performed on Fe–Co–V tubes of in order to obtain the flow direction for biaxial creep loads in tension–torsion $(\sigma_{zz}, \tau_{z\theta})$ (Delobelle et al. 1982) (Figs. 7.4 and 7.5).

In the one-dimensional situation, we defined the internal stress 1D, $\sigma_i = \sigma_i\,(\varepsilon^{vp}, t)$ whose kinetics equation is written as:

$$\left\{ \begin{array}{c} \dot{\sigma_i} = \left(\frac{\partial \sigma_i}{\partial \varepsilon^{vp}}\right) \dot{\varepsilon}^{vp} + \left(\frac{\partial \sigma_i}{\partial t}\right) \\ \dot{\varepsilon}^{vp} = k\,(\sigma - \sigma_i)^n \end{array} \right\} \tag{7.1}$$

There is a stationnary rate when restoring balance hardening called stationary creep rate:

$$\dot{\varepsilon_s}^{vp} = \frac{-\left(\frac{\partial \sigma_i}{\partial t}\right)}{\frac{\partial \sigma_i}{\partial \varepsilon^{vp}}}$$

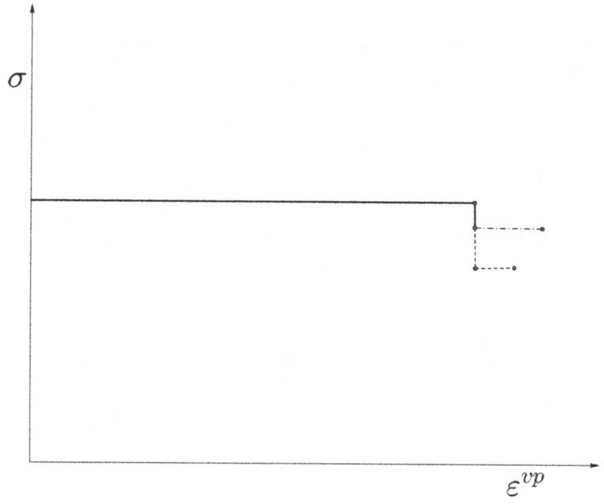

Fig. 7.2 Low stress decrement: creep hesitation $\dot{\varepsilon}^{vp} \geq 0$

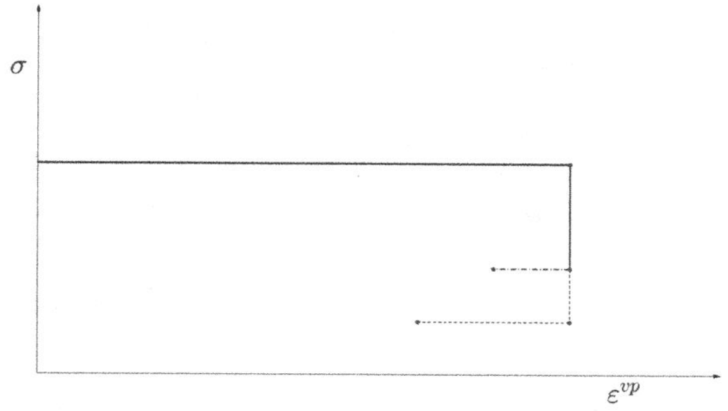

Fig. 7.3 Strong stress decrement strain rate back: $\dot{\varepsilon}^{vp} \leq 0$

For example, Friedel proposed for pure metals a stationary creep law:

$$\dot{\varepsilon_s}^{vp} = K\sigma^3 exp\left(-\frac{Q_D}{kT}\right)$$

where Q_D constitutes the metal self-diffusion energy.
3. Low stress levels correspond to diffusion creep $\left(T \geq 0.7T_f\right)$.
a. Intergranular diffusion (Herring-Nabarro creep):

$$\dot{\varepsilon}^{vp} = \frac{K\sigma}{d^2kT}$$

with d average grain size and K function of the self-diffusion, to the atomic volume and grain shape.
b. Diffusion at grain boundaries: preponderant creep at low temperatures (Coble):

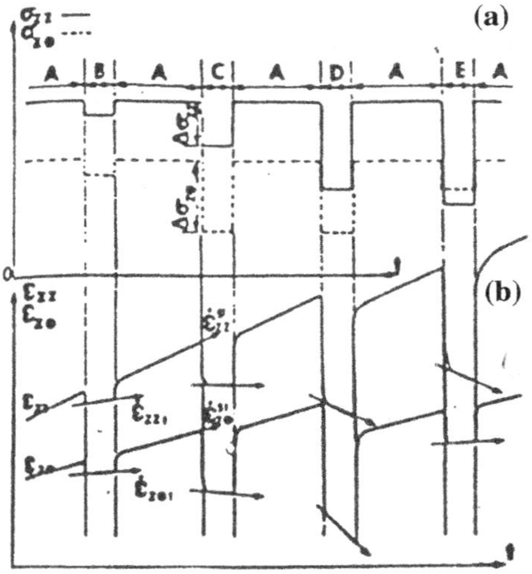

Fig. 7.4 Dip test for biaxial traction–torsion creep tests. Alloy Fe–Co–V at 680 °C (Delobelle et al. 1982)

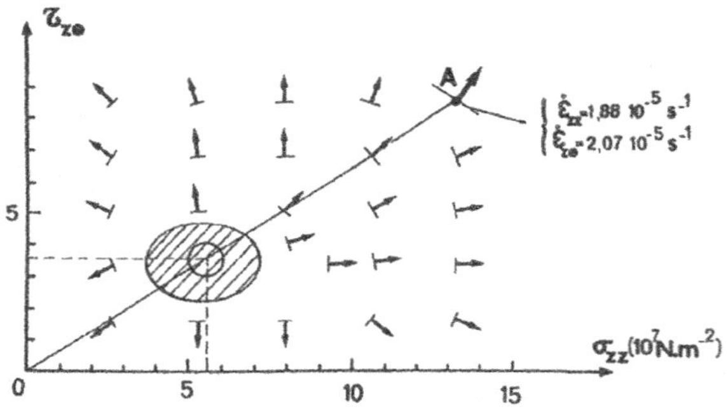

Fig. 7.5 Flow directions for creep unloading $\sigma_{zz} - \tau_{z\theta}$ after stationnary creep in point A. Alloy Fe–Co–V at 680 °C (Delobelle et al. 1982)

$$\dot{\varepsilon}^{vp} = \frac{K\sigma}{d^3 kT}$$

with K function of the self-diffusion, of the atomic volume and "effective" grain boundary thickness.

A creep rate being an affine function of σ is identified as "viscous-Newtonian" type (see fluid mechanics).

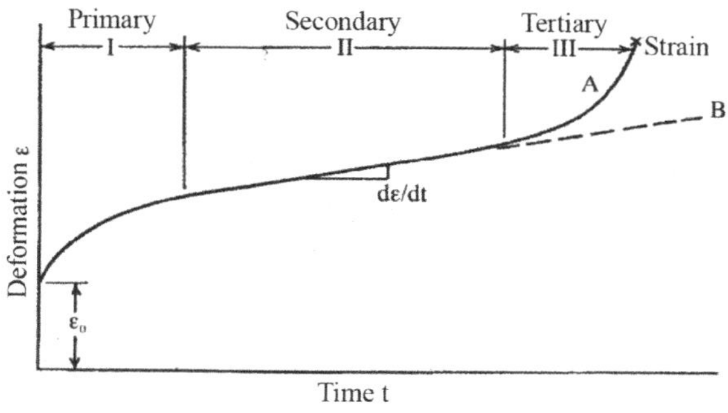

Fig. 7.6 Schematic representation of a creep–restoration curve

7.1.3 *Creep Description (Phenomenological Models)*

At low temperatures and at fixed stress, there may be a logarithmic creep which gives:

$$\varepsilon_{11}^{vp}(t) \simeq \ln(1+\alpha t).$$

At high temperatures, creep is reported as primary, secondary, and tertiary until failure (Fig. 7.6). The secondary creep rate can be predicted by the Norton 1D law:

$$\dot{\varepsilon}_{11s}{}^{vp} = \left(\frac{|\,\sigma_{11}\,|}{K}\right)^n sgn\,(\sigma_{11})$$

n may optionally be very large (10–20). The primary creep corresponds to the hardening induced by the deformation:

$$\dot{\varepsilon}_{11}{}^{vp}(t) = \left(\frac{|\,\sigma_{11}\,|}{K\left(\varepsilon_{11}^{vp}\right)}\right)^n sgn\,(\sigma_{11})$$

Classically: $K\left(\varepsilon_{11}^{vp}\right) = K_0\left(|\,\varepsilon_{11}^{vp}\,|^{1/m}\right)$

What gives in the one-dimensional situation:

$$|\,\sigma_{11}\,| = K_0\,|\,\dot{\varepsilon}_{11}{}^{vp}\,|^{1/n}|\,\varepsilon_{11}^{vp}\,|^{1/m}$$

Under constant stress primary creep law fits: $\sigma = \sigma_0$ such as:

$$\varepsilon_{11}^{vp} = \left\{\frac{n+m}{m}\left(\frac{\sigma_0}{K_0}\right)^n\right\}^{\frac{m}{n+m}} t^{\frac{m}{n+m}}$$

In the particular case corresponding to great values of n ($50 < n < 200$), $K\left(\varepsilon_{11}^{vp}\right)$ describes the limit curve of the material; plasticity is obtained as the limit of viscoplasticity at high rates.

It is worth noticing that when simply assuming $\dot{\varepsilon}^{vp} = \left(\frac{\sigma}{K}\right)^n$, stress evolution can be obtained from initial σ_0 under the strain maintained constant:

$$\dot{\varepsilon} = 0 = \frac{\dot{\sigma}}{E} + \left(\frac{\sigma}{K}\right)^n \Longrightarrow \frac{1}{\sigma_0^{n-1}} - \frac{1}{\sigma^{n-1}} = -\frac{(n-1)\,E}{K}t$$

This expression is often used but does not correctly fit the experimental datas.

There are other equations for steady flow.

"hyperbolic sine" law for high stress: $\dot{\varepsilon_s}^{vp} = A\,(\mathrm{sh}\alpha\sigma)^n$

and the «adimensional» Dorn equation for dislocation creep:

$$\frac{\dot{\varepsilon_s}^{vp} kT}{D\mu b} = A\left(\frac{\sigma}{\mu}\right)^n$$

with D the effective diffusion coefficient (to choose for the highly concentrated solid solutions (Lexcellent 1989)), with b Burgers vector and k Boltzman constant μ the shear elastic modulus.

7.2 Exercise 7.1: Steady Creep of an Fe–Cr–Al Alloy for Metal Catalytic Converters

(Gourgues-Lorenzo 2008).

Figure 7.7 shows the results of creep tests of this type of alloy in terms of $\ln\dot{\varepsilon_s}^{vp}$ *versus* $\ln\sigma$.

The experimental data are shown in Table 7.1.

What are the forms of laws given by experimental results?

Determine the corresponding exponents.

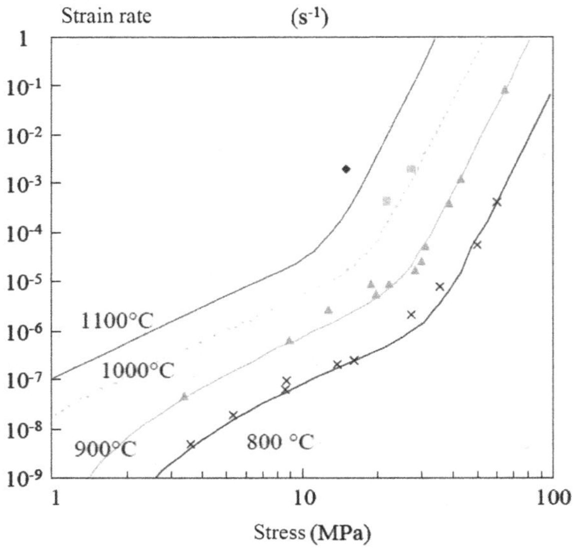

Fig. 7.7 Variation of the creep stationnary rate as function of applied stress for Fe–Cr–Al alloy (Gourgues-Lorenzo 2008)

Table 7.1 Creep experimental data for Fe–Cr–Al alloy (Gourgues-Lorenzo 2008)

T = 800 °C σ (MPa) eps pt (s^{-1})		T = 900 °C σ (MPa) eps pt (s^{-1})		T = 1000 °C σ (MPa) eps pt (s^{-1})		T = 1100 °C σ (MPa) eps pt (s^{-1})	
60,1	4,26 10^{-4}	64,2	8,32 10^{-2}	27,3	1,99 10^{-3}	15,1	1,98 10^{-3}
49,8	5,48 10^{-5}	43,1	1,20 10^{-3}	21,8	4,37 10^{-4}		
35,2	8.06 10^{-6}	38,5	3,84 10^{-4}				
27,3	2,17.48 10^{-6}	31,0	5,28 10^{-4}				
16,1	2,52 10^{-7}	30,1	2,52 10^{-5}				
13,9	2,06 10^{-7}	28,3	1,74 10^{-5}				
8,7	9,47 10^{-8}	22,2	8,89 10^{-5}				
8,5	6,33 10^{-8}	19,9	5,55 10^{-6}				
5,3	1,95 10^{-8}	18,9	8,88 10^{-6}				
3,6	5,06 10^{-9}	12,7	2,64 10^{-6}				
		8,9	6,43 10^{-7}				
		3,4	4,80 10^{-8}				

eps pt:vitesse de déformation stationnaire en fluage

Let us calculate the value of the activation energy for creep different domains. If one increases the temperature by 100 K, of how much should be increased the strain rate?

Correction.

Stress effect:

Figure 7.7 shows that there are two stress regimes: the domains of high stress and that of low stress levels.

Thus, at constant temperature, there must be a power law:

$$\dot{\varepsilon_s}^{vp} = B_0\sigma^n.$$

If one takes two representative points 1 and 2 on each of the two curves at high and low stress; it comes:

$$n = \frac{\ln\left(\frac{\dot{\varepsilon_2}}{\dot{\varepsilon_1}}\right)}{\ln\left(\frac{\sigma_2}{\sigma_1}\right)}$$

The results are given in Table 7.2. Standard stresses are the shear stress divided by the shear modulus at 900 °C; normalized deformation are shear deformations.

– Effect of temperature:

The steady creep behavior of law is as follows:

$$\dot{\varepsilon_s}^{vp} = B\sigma^n \exp\left(-\frac{\Delta H}{kT}\right) \tag{7.2}$$

with:

$$\Delta H = k\frac{\ln\left(\frac{\dot{\varepsilon_2}}{\dot{\varepsilon_1}}\right) - n\ln\left(\frac{\sigma_2}{\sigma_1}\right)}{\left(\frac{1}{T_1} - \frac{1}{T_2}\right)}$$

Table 7.2 Extraction of creep tests (Gourgues-Lorenzo 2008)

T = 900 °C				
Domain of high stresses				
σ (MPa)	eps pt (s^{-1})	σ standardized	eps pt standardized (s^{-1})	n
31.0	5.28 10^{-5}	2.98 10^{-4}	9.15 10^{-5}	10.1
64.2	8.32 10^{-2}	6.18 10^{-4}	1.44 10^{-1}	
Domain of low stresses				
σ (MPa)	eps pt (s^{-1})	σ standardized	eps pt standardized (s^{-1})	n
3.40	4.80 10^{-8}	3.27 10^{-5}	8.32 10^{-8}	2.7
8.90	6.43 10^{-7}	8.56 10^{-5}	1.11 10^{-6}	

The corresponding values are given in Table 7.3.

A comparison with the Ashby maps shows that the creep rates are overestimated by the map by a factor of 10.

This simply means that the Ashby map for pure iron cannot be used for quantitative data for the Fe–Cr–Al alloys (Gourgues-Lorenzo 2008).

7.3 General Form of the Viscoplastic Constitutive Equation (Chaboche 2008)

In the framework of the thermodynamic of irreversible process (see Chap. 2 and Halphen and N'Guyen (1975), Germain (1973)), the viscoplastic constitutive equations can be established:

for extended plasticity, by using a viscoplastic potential $\Omega(f)$. The stress state goes beyond the elastic range, with a positive value $\sigma_v = f > 0$ which can be called the viscous stress.

In this case, the normality rule is written:

Table 7.3 Activation energies determination (Gourgues-Lorenzo 2008)

Domain of high stresses				
σ (MPa)	eps pt (s^{-1})	T (°C)	n	ΔH(kJ/mol)
28,3	1,74 10^{-5}	900	10,1	633
27,3	1,99 10^{-3}	1000		
Domain of low stresses				
σ (MPa)	eps pt (s^{-1})	T (°C)	n	ΔH(kJ/mol)
8,5	6,33 10^{-8}	800	2,7	230
8,9	6,43 10^{-7}	900		

eps pt: stationary strain rate

$$\underline{\dot{\varepsilon}^{vp}} = \frac{\partial \Omega (f)}{\partial \underline{\sigma}} = \frac{\partial \Omega (f)}{\partial f} \frac{\partial f}{\partial \underline{\sigma}} = \dot{p}^{vp} \underline{n} \tag{7.3}$$

with:

$$\underline{n} = \frac{3}{2} \frac{\underline{s} - \underline{X}}{\overline{\sigma - X}} \quad \dot{p}^{vp} = \left(\frac{2}{3} \underline{\dot{\varepsilon}^{vp}} : \underline{\dot{\varepsilon}^{vp}} \right)^{\frac{1}{2}}$$

and:

$$f = \left(\overline{\sigma - X}\right) - R - \sigma_y \leq 0. \tag{7.4}$$

Chaboche (2008) decomposed the tensor of applied stresses as follows:

$$\underline{\sigma} = \underline{X} + \left(R + \sigma_y + \sigma_v (\dot{p}) \right) \underline{n} \tag{7.5}$$

where $\underline{X}$ is the kinematical internal variable; R the isotropic variable, σ_y the initial elastic limit and σ_v the viscous stress.

We can write for uniaxial tensile tests:

$$\sigma = X + R + \sigma_y + \sigma_v \tag{7.6}$$

Clearly, it is the existence of the viscous stress σ_v which makes the difference between viscoplasticity and plasticity. It remains to make the choice of the viscoplastic potential $\Omega (f)$ and also the choice of the hardening functions for all internal variables.

They are called a_j $(j = 1, 2, \ldots, N)$ for the moment and can be scalar or tensorial. Their general form includes a term of deformation hardening by and a term of dynamic recovery and also a static one.

$$\dot{a}_j = h_j (\ldots) \underline{\dot{\varepsilon}^{vp}} - r_j^D (\ldots) a_j \underline{\dot{\varepsilon}^{vp}} - r_j^S (\ldots) a_j \tag{7.7}$$

The first term is related to a_j whose values increase with viscoplastic deformation, the second term is a recall or evanescent memory and the third; a term of time restoration. The role of temperature is taken into account in the term r_j^S.

This term is used to take into account the effect of thermal agitation, including the rise of dislocations mechanisms of and their corresponding annihilation. We may note a strong analogy with the physical equations concerning the dislocation density ρ given by Estrin (1996) for uniaxial loadings:

$$d\rho = M \left(k_0 + k_1 \sqrt{\rho} - k_2 \rho \right) d\varepsilon^{vp} - r^S \left(\sqrt{\rho}, T \right) dt \tag{7.8}$$

7.3.1 Viscosity Function Choice

Let us consider:

$$\dot{p}^{vp} = < \frac{\overline{\sigma}^{vp}}{D} >^n \tag{7.9}$$

with:

$$\overline{\sigma}^{vp} = \left(\tfrac{3}{2}\underline{\sigma}^{vp} : \underline{\sigma}^{vp}\right)^{\frac{1}{2}}$$
$$\dot{p}^{vp} = \left(\tfrac{2}{3}\underline{\dot{\varepsilon}}^{vp} : \underline{\dot{\varepsilon}}^{vp}\right)^{\frac{1}{2}}$$

where $<.>$ are the Mc Cauley hooks $< x > = max\,(x, 0)$, n depends on the material, the considered strain rate range and also the temperature (see Exercise 7.7), without forgetting the "diffusion creep" when n = 1. It is observed that $3 \leq n \leq 30$ for usual materials.

The advantage of the expression (7.9) is that it can be easily derived from a viscoplastic potential:

$$\Omega = \frac{D}{n+1} < \frac{\overline{\sigma}^{vp}}{D} >^{n+1} \tag{7.10}$$

7.3.2 Isotropic Hardening Equations

If we consider the expression of the strain rate norm

$$\dot{p}^{vp} = < \frac{\overline{(\sigma - X)} - R - \sigma_y}{D} >^n \tag{7.11}$$

Three choices can be considered to introduce an isotropic hardening:

(i) with the variable R by increasing the size of the elastic range,
(ii) by increasing the constraint D,
(iii) by coupling with the law of evolution of the kinematic hardening variable $\underline{X}$.

One can choose:

$$R = R\left(p^{vp}\right),\ D = D\left(p^{vp}\right)$$

or:

$$D\left(p^{vp}\right) = K + \xi R\left(p^{vp}\right)$$

finally, by decomposing the Huber–von Mises stress:

$$\overline{\sigma} = \sigma_y + R\left(p^{vp}\right) + \overline{\sigma}_v\left(\dot{p}^{vp}, p^{vp}\right) = \sigma_y + R\left(p^{vp}\right) + D\left(p^{vp}\right)\left(\dot{p}^{vp}\right).^{\frac{1}{n}}$$

If one neglects the initial elastic domain $\left(\sigma_y = 0\right)$ and taking a power law for D, we can find the three-dimensional version of the multiplicative law (Rabotnov 1969):

$$\overline{\sigma} = K\left(p^{vp}\right)^{\frac{1}{m}}\left(\dot{p}^{vp}\right)^{\frac{1}{n}} \tag{7.12}$$

As we have seen previously in the (1D) case, the multiplicative form of hardening is very convenient to be determined (Lemaitre 1971), and gives good results in a large range; at least for near-proportional monotonic loadings.

7.3.3 Kinematic Hardening Equations

As in plasticity, the linear kinematic hardening of Prager is given by (Prager 1949):

$$\dot{\underline{X}} = \frac{2}{3}c\dot{\underline{\varepsilon}}^{vp} \; and \; \underline{X} = \frac{2}{3}c\underline{\varepsilon}^{vp} \tag{7.13}$$

An evolution of the model was initially proposed by Armstrong and Frederick with dynamic recovery:

$\dot{\underline{X}} = \frac{2}{3}c\dot{\underline{\varepsilon}}^{vp} - \gamma\underline{X}\dot{p}^{vp}.$

The integration provides for a uniaxial loading:

$$X = \nu\frac{c}{\gamma} + \left(X_0 - \nu\frac{c}{\gamma}\right)\exp\left(-\nu\gamma\left(\varepsilon^{vp} - \varepsilon_0^{vp}\right)\right) \tag{7.14}$$

in which $\nu = \pm 1$ gives the flow direction and X_0, ε_0^{vp} are the values of X *et* ε^{vp} at the beginning of the considered branch.

7.3.4 Hardening equations in viscoplasticity

It is used in the same relationship as in plasticity, that is to say independent of time for $\dot{R}$ and $\dot{\underline{X}}$, and a recovery term is added:

$$\dot{\underline{X}} = \frac{2}{3}c\dot{\underline{\varepsilon}}^{vp} - \gamma\underline{X}\dot{p}^{vp} - \underline{X}\left(\frac{\overline{X}}{M}\right)^{m-1} \tag{7.15}$$

with $\overline{X} = \left(\frac{3}{2}\underline{X} : \underline{X}\right)^{\frac{1}{2}}$.

For the 1D case, stationary rate when $\dot{X} = 0 \Longrightarrow$

$$\left\{\begin{array}{c}\frac{2}{3}\dot{\varepsilon_s}^{vp} - \gamma X \mid \dot{\varepsilon_s}^{vp} \mid - \left(\frac{X}{M}\right)^m = 0 \\ \sigma_0 = X_s + R_s + K\,(\dot{\varepsilon_s}^{vp})^{\frac{1}{2}}\end{array}\right. .$$

The internal stress predicted by the model decreases with the applied stress σ_0.

7.4 Exercise: Viscoplastic Behavior Predicted by a Rheological Model

(Burlet et al. 1988)

Statement:

(A) Let us consider the rheological model consisting of (Fig. 7.8):
a spring (proportional strain deformation of the branch 1 (c)
a pad (slip if $\mid \sigma \mid > \sigma_y$)
a damper (stress proportional to the strain rate (μ))
a spring 2 (stress proportional to the strain$(E\ Young\ modulus)$

(1) Model equations in terms σ, ε^{vp}, $\dot{\varepsilon}^{vp}$?

Response:

(a) if $\dot{\varepsilon}^{vp} = 0, \Longrightarrow \sigma_3 = 0, \ \mid \sigma_2 \mid = \mid \sigma - c\varepsilon^{vp} \mid \leq \sigma_y$
Summary:

$$\left\{\begin{array}{c}\sigma_1 = c\varepsilon^{vp} \\ \mid \sigma_2 \mid = \sigma_y \\ \sigma_3 = \mu\dot{\varepsilon}^{vp}\end{array}\right\}$$

(b) If $\dot{\varepsilon}^{vp} > 0\ so\ \sigma = \sigma_y + c\varepsilon^{vp} + \mu\dot{\varepsilon}^{vp}$
(c) If $\dot{\varepsilon}^{vp} < 0\ so\ \sigma = -\sigma_y + c\varepsilon^{vp} + \mu\dot{\varepsilon}^{vp}$

(2) infinitely fast loading ($\dot{\varepsilon}^{vp} = 0$) until $\sigma_0 > \sigma_y$. Let us find $\varepsilon^{vp}\ (t)$.
Response:

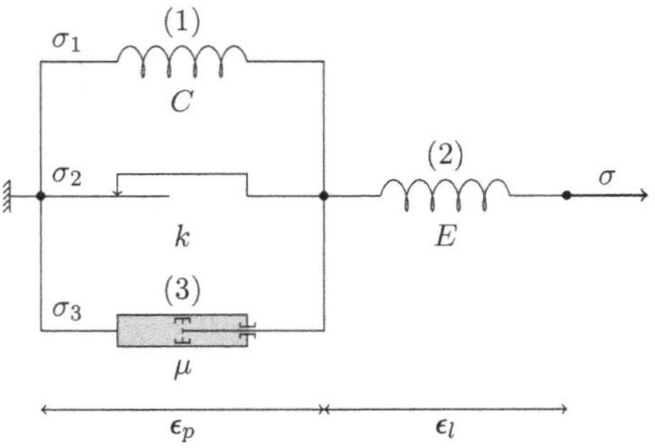

Fig. 7.8 First rheological model

Initial conditions at $t = 0$: σ_0, $\varepsilon^{vp} = 0$; $\dot{\varepsilon}^{vp} = \frac{\sigma_0 - \sigma_y}{\mu}$
Differential equation verified: $\mu\dot{\varepsilon}^{vp} + c\varepsilon^{vp} = \sigma_0 - \sigma_y$
Solution: $\varepsilon^{vp}(t) = \frac{\sigma_0 - \sigma_y}{c}\left(1 - \exp\left(-\frac{c}{\mu}t\right)\right)$
creep deformation stops when $t \Longrightarrow +\infty$ at $\varepsilon^{vp} = \frac{\sigma_0 - \sigma_y}{c}$

(3) We reduce the stress $\sigma = 0$. What is going on ?
You have to cross the elastic range. If one returns to the other side:
$\sigma = 0 = -\sigma_y + c\varepsilon^{vp} + \mu\dot{\varepsilon}^{vp} \Longrightarrow c\left(\varepsilon_{p0} - x\right) = \sigma_0 - 2\sigma_y \Longrightarrow x = \frac{\sigma_y}{c}$
If the creep time is as $\varepsilon^{vp} < \frac{k}{c}$, nothing happens
if $\sigma_0 < 2\sigma_y$, nothing happens $\forall t$
if $\sigma_0 > 2\sigma_y$ and if the time is sufficient, a recovery was observed until $\varepsilon^{vp} = \frac{k}{c}$

(B) We make the model more complicated by adding a second damper of characteristic η in branch 1 (Fig. 7.9).
We call X the stress in the branch 1.

(1) Model equations in terms σ, X, $\dot{\varepsilon}^{vp}$.
In the branch 1: $\dot{\varepsilon}^{vp} = \dot{\varepsilon}^{c} + \dot{\varepsilon}^{c}$
with:

$$\begin{cases} \eta\dot{\varepsilon}^{a} = X \\ c\varepsilon^{c} = X \end{cases} \Longrightarrow \dot{\varepsilon}^{vp} = \frac{\dot{X}}{c} + \frac{X}{\eta} \; or \; \dot{X} = c\dot{\varepsilon}^{vp} - \frac{c}{\eta}X$$

(a) if $\dot{\varepsilon}^{vp} = 0 \mid \sigma - X \mid \leq \sigma_y$
(b) if $\dot{\varepsilon}^{vp} > 0$ $\sigma = \sigma_y + X + \mu\dot{\varepsilon}^{vp} \Longrightarrow \dot{\varepsilon}^{vp} = \left(\frac{|\sigma - X| - \sigma_y}{\mu}\right) sgn\,(\sigma - X)$.
(c) if $\dot{\varepsilon}^{vp} < 0$ $\sigma = -\sigma_y + X + \mu\dot{\varepsilon}^{vp}$.

(2) Behavior of the creep model $\sigma = \sigma_0$
One finds:

$$X = \eta\frac{\sigma_0 - \sigma_y}{\eta + \mu}\left(1 - \exp\left(-c\frac{\eta + \mu}{\eta\mu}t\right)\right)$$

$$\dot{\varepsilon}^{vp} = \frac{\sigma_0 - \sigma_y}{\eta + \mu}\left(1 + \frac{\eta}{\mu}\exp\left(-c\frac{\eta + \mu}{\eta\mu}t\right)\right)$$

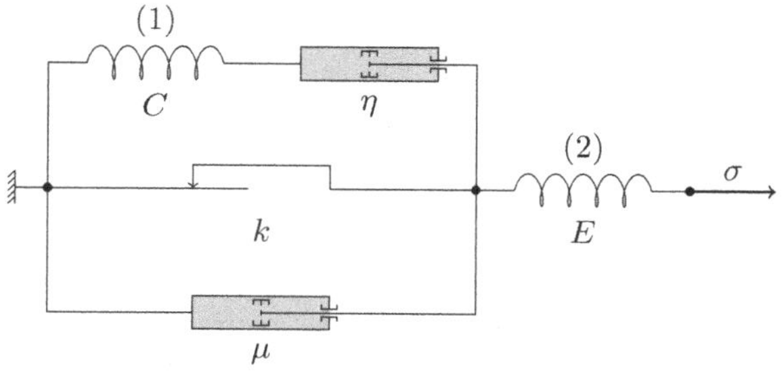

Fig. 7.9 Second rheological model

$$\varepsilon^{vp} = \frac{\sigma_0 - \sigma_y}{\eta + \mu}\left(t + \frac{\eta^2}{c\,(\eta + \mu)}(1 - \exp\left(-c\frac{\eta + \mu}{\eta\mu}t\right)\right)$$

We have primary and secondary creeps: $\dot{\varepsilon_s}^{vp} = \frac{\sigma_0 - \sigma_y}{\eta + \mu}$.

Saturation of X that dépend on applied load: $X_s = \eta\frac{\sigma_0 - \sigma_y}{\eta + \mu}$.

(3) Behavior when returning to zero stress.

The behavior is qualitatively identical to the predictions given by the previous model, but there are less recovery because there is a restoration of X: $\dot{X} = c\dot{\varepsilon}^{vp} - \frac{c}{\eta}X$. Finally $X = \sigma_y$.

7.5 Exercise 7.3: Viscoplastic Flow for Plane Strains (in the Annals pp 195–199 of Cailletaud et al. 2011)

Statesment:

(1) Consider a cube of side 1 whose edges are parallel to the axes $\overrightarrow{x_1}$, $\overrightarrow{x_2}$, $\overrightarrow{x_3}$ of an orthonormal frame. This cube is uniformly loaded in the direction $\overrightarrow{x_1}$; the faces with normal $\overrightarrow{x_2}$ are free, and the faces with normal $\overrightarrow{x_3}$ are blocked. Give the form of the stress tensor and then the strain one.

Elastic resolution.

No shearing action $\Longrightarrow \sigma_{12} = \sigma_{13} = \sigma_{23} = 0$

Plane strain hypothesis $\Longrightarrow \varepsilon_{13} = \varepsilon_{23} = \varepsilon_{33} = 0$

$$\underline{\sigma} = \begin{pmatrix} \sigma_{11} & 0 & 0 \\ 0 & 0 & 0 \\ 0 & 0 & \sigma_{33} \end{pmatrix}, \quad \underline{\varepsilon} = \begin{pmatrix} \varepsilon_{11} & 0 & 0 \\ 0 & \varepsilon_{22} & 0 \\ 0 & 0 & 0 \end{pmatrix}$$

Isotropic and elastic behavior law:

$$\underline{\varepsilon}^{el} = \frac{1+\nu}{E}\underline{E} - \frac{\nu}{E}\mathrm{tr}\left(\underline{\sigma}\right)\underline{1}$$

$$\begin{cases} E\varepsilon_{11}^{el} = \sigma_{11} - \nu\sigma_{33} \\ E\varepsilon_{22}^{el} = -\nu\sigma_{11} - \nu\sigma_{33} \end{cases} \qquad E\varepsilon_{33}^{el} = 0 = -\nu\sigma_{11} + \sigma_{33} \Longrightarrow \sigma_{33} = \nu\sigma_{11}$$

one obtains finally:

$$\begin{cases} \varepsilon_{11}^{el} = \frac{1-\nu^2}{E}\sigma_{11} \\ \varepsilon_{22}^{el} = \frac{-\nu(1+\nu)}{E}\sigma_{11} \end{cases} \qquad \varepsilon_{33}^{el} = 0$$

(2) Set the value of σ_{11} for which the material will reach the limit of elasticity range for the Huber–Von Mises and Tresca criteria.

State: Huber–Von Mises:

$f\left(\underline{\sigma}\right) = \overline{\sigma} - \sigma_y = 0$

$$\underline{\sigma} = \begin{pmatrix} \sigma_{11} & 0 & 0 \\ 0 & 0 & 0 \\ 0 & 0 & \nu\sigma_{11} \end{pmatrix}, \ \underline{s} = \text{dev}\left(\underline{\sigma}\right) = \frac{\sigma_{11}}{3}\begin{pmatrix} 2-\nu & 0 & 0 \\ 0 & -1-\nu & 0 \\ 0 & 0 & 2\nu-1 \end{pmatrix}$$

$$\Longrightarrow \overline{\sigma} = \left(\frac{3}{2}\text{dev}\underline{\sigma} : \text{dev}\underline{\sigma}\right)^{\frac{1}{2}} = \mid \sigma_{11} \mid \sqrt{1-\nu+\nu^2} \Longrightarrow$$

$$\sigma_{11} = \pm\frac{\sigma_y}{\sqrt{1-\nu+\nu^2}}$$

Tresca criterion:

$f\left(\underline{\sigma}\right) = \mid \sigma_{11} \mid - \sigma_y = 0 \Longrightarrow \sigma_{11} = \pm\sigma_y$

(3) It is assumed that the material follows a viscoplasticity law with a threshold σ_y which is written in simple traction with <x>= max (x, 0)

$$\dot{\varepsilon}_{11}^{vp} = < \frac{\sigma_{11} - \sigma_y}{K} >^n$$

Huber–von Mises criterion:

We generalize this law to the case of three-dimensional loadings by using the Huber–von Mises criterion.

The existence of the viscoplastic potential provides the equation:

$$\underline{\dot{\varepsilon}^{vp}} = \frac{\partial\Omega}{\partial\underline{\sigma}} = \frac{\partial\Omega}{\partial f}\frac{\partial f}{\partial\underline{\sigma}}$$

with $f\left(\underline{\sigma}\right) = \overline{\sigma} - \sigma_y$ $\underline{n} = \frac{\partial f}{\partial\underline{\sigma}}$ one verify $\underline{n} : \underline{n} = \frac{3}{2}$
it comes:

$$\underline{\dot{\varepsilon}^{vp}} = \frac{3}{2}\frac{\text{dev}\left(\underline{\sigma}\right)}{\overline{\sigma}}\left(\frac{\overline{\sigma} - \sigma_y}{K}\right)^n \quad (7.16)$$

$$\underline{\dot{\varepsilon}^{vp}} = \pm\frac{1}{2\sqrt{1-\nu+\nu^2}}\begin{pmatrix} 2-\nu & 0 & 0 \\ 0 & -1-\nu & 0 \\ 0 & 0 & 2\nu-1 \end{pmatrix}\left(\frac{\overline{\sigma} - \sigma_y}{K}\right)^n \quad (7.17)$$

(4) Calculate then the system evolution (stress, total strain, viscoplastic strain) in the following two cases:

– We block the stress σ_{11} at the maximal value σ_m

– We block the total strain ε_{11} at the maximal value ε_m.

State:

$$\dot{\varepsilon}_{11} = \dot{\varepsilon}_{11}^{el} + \dot{\varepsilon}_{11}^{vp}$$

$$\dot{\varepsilon}_{11} = \frac{1-\nu^2}{E}\dot{\sigma}_{11} \pm \frac{1}{2\sqrt{1-\nu+\nu^2}}\left(\frac{\sigma_{11}-\sigma_y}{K}\right)^n$$

It is assumed that the loading is instantaneous, so that at t = 0:

$$\varepsilon_s = \frac{1-\nu}{E}\sigma_s$$

If the stress value at σ_m is imposed, you have $\sigma_s = \sigma_m$ and $\dot{\sigma}_{11} = \dot{0}$ and:

$$\varepsilon_{11}(t) = \frac{1-\nu}{E}\sigma_s \pm \frac{1}{2\sqrt{1-\nu+\nu^2}}\left(\frac{\sigma_{11}-\sigma_y}{K}\right)^n t \tag{7.18}$$

We note that the deformation evolution is linear with time.
– If the strain value at ε_m is imposed, we have $\varepsilon_s = \varepsilon_m$ and $\dot{\varepsilon}_{11} = 0$
What gives:

$$\dot{\varepsilon}_{11} = 0 = \frac{1-\nu^2}{E}\dot{\sigma}_{11} \pm \frac{1}{2\sqrt{1-\nu+\nu^2}}\left(\frac{\sigma_{11}-\sigma_y}{K}\right)^n$$

Therefore:

$$\frac{1}{-n+1}\left(\left(\frac{\sigma_{11}-\sigma_y}{K}\right)^{-n+1}\right) = \frac{E}{1-\nu^2}\frac{1}{2\sqrt{1-\nu+\nu^2}}t + cte \tag{7.19}$$

Constant at $t = 0$ is:

$$\sigma_s = \frac{E}{1-\nu}\varepsilon_m$$

It can be shown that with the Tresca criterion, the problem is simpler and gives solutions also in relaxation (annals pp 195–199 (Cailletaud et al. 2011)).

7.5.1 Tresca Criterion

With $f\left(\underline{\sigma}\right) = max_{i,j} \mid \sigma_i - \sigma_j \mid$
one has:

$$\underline{\dot{\varepsilon}^{vp}} = \left(\frac{f\left(\underline{\sigma}\right)}{K}\right)^n \frac{\partial f}{\partial \underline{\sigma}}.$$

Let choose $\sigma_1 > \sigma_2 > \sigma_3$, the criterion will take the form $f\left(\underline{\sigma}\right) = \mid \sigma_1 - \sigma_3 \mid$, hence:

$$\underline{\dot{\varepsilon}^{vp}} = \left(\frac{f\left(\underline{\sigma}\right)}{K}\right)^n \begin{pmatrix} 1 & 0 & 0 \\ 0 & 0 & 0 \\ 0 & 0 & -1 \end{pmatrix}$$

Then calculate the system evolution (stress, strain, viscoplastic deformation) in the following two cases:

– We impose the stress σ_{11} at the maximum reached value σ_m.
– We impose the total strain ε_{11}at the maximum value ε_m.

$$\dot{\varepsilon}_{11} = \dot{\varepsilon}_{11}^{el} + \dot{\varepsilon}_{11}^{vp}$$

$$0 = \dot{\varepsilon}_{33}^{el} + \dot{\varepsilon}_{33}^{vp}$$

What gives:

$$\begin{cases} \dot{\varepsilon}_{11} = \frac{1}{E}\dot{\sigma}_{11} - \frac{\nu}{E}\dot{\sigma}_{33} + \left(\frac{\sigma_{11}-\sigma_y}{K}\right)^n \\ 0 = \frac{1}{E}\dot{\sigma}_{33} - \frac{\nu}{E}\dot{\sigma}_{11} - \left(\frac{\sigma_{11}-\sigma_y}{K}\right)^n \end{cases}$$

$\Longrightarrow$

$$\dot{\varepsilon}_{11} = \frac{1-\nu^2}{E}\dot{\sigma}_{11} + (1-\nu)\left(\frac{\sigma_{11}-\sigma_y}{K}\right)^n$$

It is assumed that the loading is instantaneous, so that at t = 0:

$$\varepsilon_s = \frac{1-\nu}{E}\sigma_s$$

If the stress is imposed at σ_m, we have $\sigma_s = \sigma_m$ and $\dot{\sigma}_{11} = \dot{0}$ and:

$$\varepsilon_{11}(t) = \frac{1-\nu}{E}\sigma_m + (1-\nu)\left(\frac{\sigma_m-\sigma_y}{K}\right)^n t \tag{7.20}$$

We note that the deformation evolution is linear with time.
– If the strain is imposed at ε_m, we have $\varepsilon_s = \varepsilon_m$ and $\dot{\varepsilon}_{11} = 0$

What gives:

$$\dot{\sigma}_{11} + \frac{1}{1+\nu}\left(\frac{\sigma_{11} - \sigma_y}{K}\right)^n = 0$$

The exponent n is generally greater than 1, the evolution of σ_{11} is described as a power function:

$$\sigma_{11}(t) = \sigma_y + K\left(\left(\frac{\sigma_m - \sigma_y}{K}\right)^{1-n} + \frac{E(n-1)}{(1+\nu)K}t\right)^{\frac{1}{1-n}} \tag{7.21}$$

with:

$$\sigma_m = \frac{E}{1-\nu}\varepsilon_m$$

7.6 Viscoplasticity Unified Constitutive Models

7.6.1 Viscoplasticity Flow Laws:

In these models, there is always an equation of state or so-called "viscosity" one.

For Lemaitre et al. (2009):

$$\left\{\begin{array}{c} \underline{\varepsilon} = \underline{\varepsilon}^{el} + \underline{\varepsilon}^{vp} \; \underline{\sigma} = \underline{\underline{C}}\left(\underline{\varepsilon} - \underline{\varepsilon}^{vp}\right) \\ \dot{\underline{\varepsilon}}^{vp} = \frac{3}{2}\frac{\underline{\underline{s}} - \underline{\underline{X}}}{\overline{\sigma - X}}\dot{p}^{vp} \\ \dot{p}^{vp} = < \frac{\overline{\sigma - X} - R - \sigma_y}{K(p^{vp})} >^n \end{array}\right\} \tag{7.22}$$

For Delobelle (1988):

$$\left\{\begin{array}{c} \dot{\underline{\varepsilon}}^{vp} = \frac{3}{2}\frac{\underline{\underline{s}} - \underline{\underline{X}}}{\overline{\sigma - X}}\dot{p}^{vp} \\ \dot{p}^{vp} = p_0(T)\sinh\left(\frac{\sigma_v}{D(p^{vp}, T)}\right)^n \; with \; \sigma_v = \overline{\sigma - X} \end{array}\right\} \tag{7.23}$$

In his viscoplastic models' review, Chaboche (2008) has proposed the models of Bodner, Robinson, Krempl, Miller, Johnson, Kocks, Walker that have the same formalism with the viscosity equation and the evolution of the internal variables laws.

Their predictions of the viscous stress σ_v as function of the viscoplastic strain rate $\dot{\varepsilon}^{vp}$ are gathered in Fig. 7.10.

Three models give identical results, with the same exponent $n_0 = 20$ in the field of low rates: those in hyperbolic sine Delobelle version and two versions of Onera.

In addition, Chaboche overlooks changes in the coefficient of Norton's power law with the viscous stress (Fig. 7.11).

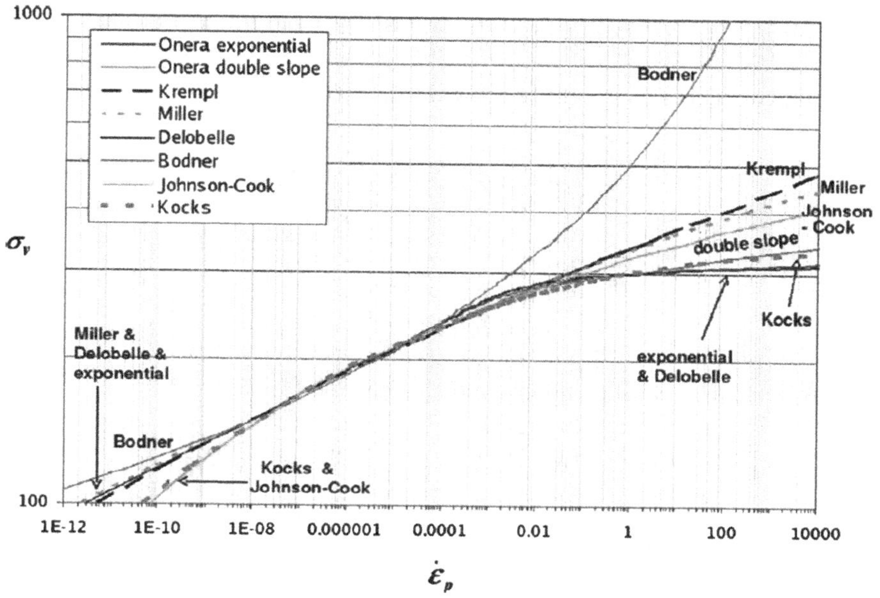

Fig. 7.10 Predictions of the viscous stress σ_v as function of the viscoplastic strain rate $\dot{\varepsilon}^{vp}$. Norton exponent $n_0 = 20$ for intermediate regime $10^{-8} \leq \dot{\varepsilon}^{vp} \leq 10^{-4}s^{-1}$. (Chaboche 2008)

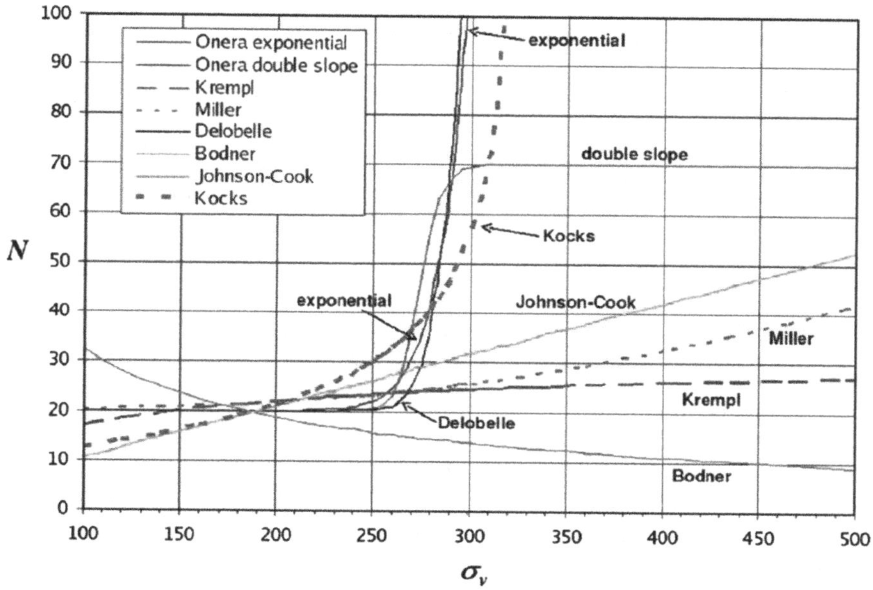

Fig. 7.11 Power law coefficients as a function of the viscous stress for different viscosity equations

7.6.2 Internal Variables Kinetic Equations

In a "light" version of their model Lemaitre et al. (2009) have defined:

$$\underline{\dot{X}} = \frac{2}{3}c\underline{\dot{\varepsilon}}^{vp} - \gamma \underline{X}\dot{p}^{vp} - \underline{X}\left(\frac{\overline{X}}{M}\right)^{m-1} \tag{7.24}$$

and:

$$R = R_\infty \left(1 - e^{-bp}\right) \tag{7.25}$$

whereas Delobelle (1988) proposed a more complicated version:

$$\begin{aligned}
\underline{\dot{X}} &= C\left[\tfrac{2}{3}\gamma\left(p^{vp}\right)\dot{\varepsilon}^{vp} - \left(\underline{X} - \underline{X}_1\right)\dot{p}^{vp}\right] - r_m\left(T\right)\sinh\left(\tfrac{||X||}{X_0(T)}\right)^m \tfrac{\underline{X}}{||X||} \\
\underline{X}_1 &= C_1\left[\tfrac{2}{3}\gamma\left(p^{vp}\right)\dot{\varepsilon}^{vp} - \left(\underline{X}_1 - \underline{X}_2\right)\dot{p}^{vp}\right] \\
\underline{X}_2 &= C_2\left[\tfrac{2}{3}\gamma\left(p^{vp}\right)\dot{\varepsilon}^{vp} - \underline{X}_2\dot{p}^{vp}\right]
\end{aligned} \tag{7.26}$$

7.6.3 Other Types of Modeling

7.6.3.1 Partition Between Creep and Plasticity

The "initial" way to describe the creep and plasticity consisted of adding two anelastic strains:

$$\underline{\varepsilon} = \underline{\varepsilon}^{el} + \underline{\varepsilon}^{p} + \underline{\varepsilon}^{c}$$

with $\underline{\varepsilon}^{p}$ plasticity and $\underline{\varepsilon}^{c}$ creep

On one hand:

$$\dot{a^{p}}_j = h_j^p\left(\ldots\right)\underline{\dot{\varepsilon}^{p}} - r_j^p\left(\ldots\right)a_j^p\underline{\dot{\varepsilon}^{p}}$$

and on the another hand:

$$\overline{\varepsilon}^c\left(\overline{\sigma}, t\right) = A_1\left(\overline{\sigma}\right)t^{\frac{1}{p}} + A_2\left(\overline{\sigma}\right)t$$

We can find unified approaches by coupling hardening plasticity and creep (e.g., Contesti and Cailletaud 1989):

$$\dot{a^{p}}_j = h_j^p\left(\ldots\right)\underline{\dot{\varepsilon}^{p}} + h^{pc}\underline{\dot{\varepsilon}^{c}} - r_j^p\left(...\right)a_j^p$$

$$\dot{a^c}_j = h_j^{cp}(\ldots)\,\underline{\dot{\varepsilon}^p} + h^c\underline{\dot{\varepsilon}^c} - r_j^c(\ldots)\,a_j^c$$

with all possible variations.

7.6.3.2 Other Approaches

- Multiple mechanisms and multi-criteria
- Micro–macro transition scales:

It is used of the basic equations of the crystal plasticity, by writing into the model the slip systems that can be activated, for different orientations of grains constituting the polycrystal (representative volume element). It is an homogenization work that will not be discussed here.

7.7 Comparison Between "Plasticity" and "Viscoplasticity"

7.7.1 The Threshold Representing the Elastic Range f

Plasticity:

Behavior: threshold f (Fig. 7.12)

$$\left\{\begin{array}{c} f < 0:\ elasticity \\ f = 0,\ \dot{f} < 0:\ elastic\ unloading \\ f = 0,\ \dot{f} = 0:\ \underline{\dot{\varepsilon}^p} = \dot{\lambda}\frac{\partial f}{\partial \underline{\sigma}} \end{array}\right\}$$

Remark: f cannot be positive.

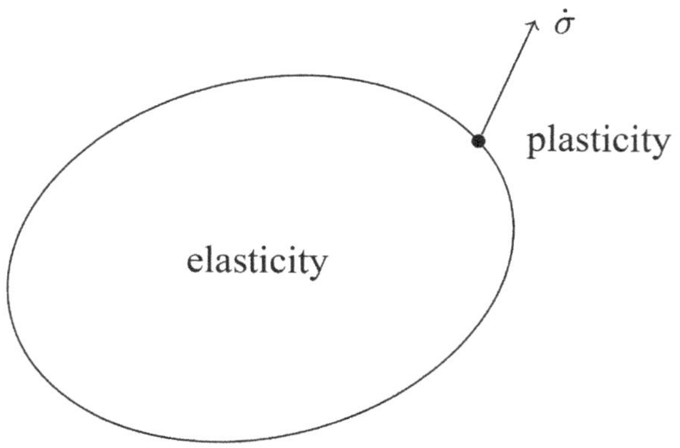

Fig. 7.12 Plasticity representation

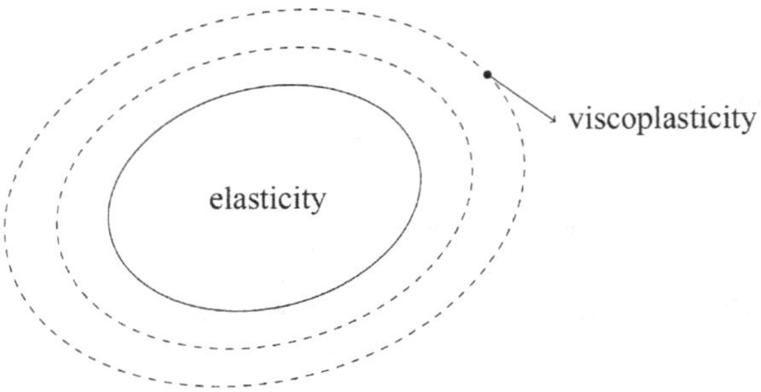

Fig. 7.13 Viscoplasticity representation

Viscoplasticity:

Behavior: threshold f (Fig. 7.13)

$$\left\{ \begin{array}{c} f \leq 0 : \ elasticity \\ f > 0 : \ \underline{\dot{\varepsilon^{vp}}} = \dot{p}^{vp} \frac{\partial f}{\partial \underline{\sigma}} \end{array} \right\}$$

Remark: f can be positive.

7.7.2 Behavior Models

Plasticity:

The consistency condition $\dot{f} = 0$ allows us to determine the plastic multiplier $\dot{\lambda_p}$.

The strain rate depends on the stress rate $\underline{\dot{\sigma}}$ and on the values of the hardening variables.

In 1D: $\sigma = X + R$ (Fig. 7.14).

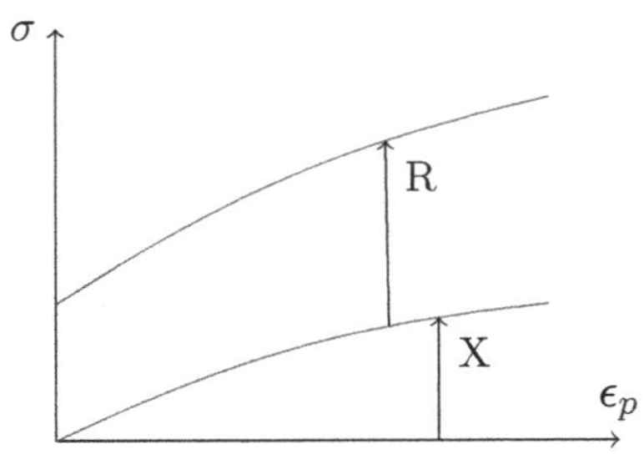

Fig. 7.14 Plasticity behavior: in 1D $\sigma = X + R$

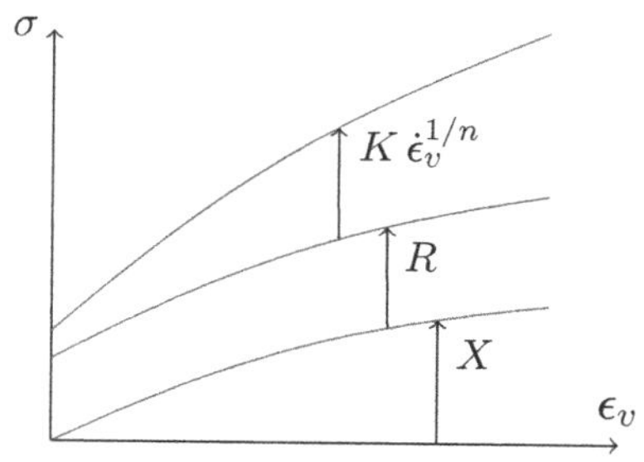

Fig. 7.15 Viscoplastic behavior : : in 1D: $\sigma = X + R + \sigma_v$

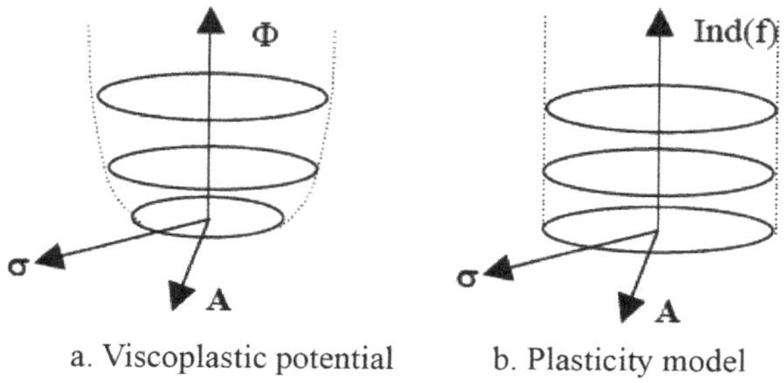

Fig. 7.16 Comparison between plasticity and viscoplasticity theories

For viscoplasticity:

we choose:

$$\dot{p}^{vp} = \left(\frac{\sigma - X - R}{K} \right)^n$$

Strain rate depends on the current values of the stresses $\underline{\sigma}$ and on the values of the hardening variables.

In 1D: $\sigma = X + R + \sigma_v$ (Fig. 7.15).

with the viscous stress:

$$\sigma_v = K \left(\dot{\varepsilon}^{vp} \right)^{\frac{1}{n}}$$

Figure 7.16 also shows the comparison between viscoplasticity and plasticity.

Note that plasticity occurs at almost zero rate or "infinite" and finite rate for viscoplasticity (Fig. 7.17).

Most models do not separate the plastic strain rate $\underline{\dot{\varepsilon}^p}$ and the viscoplastic strain rate $\underline{\dot{\varepsilon}^{vp}}$ (Chaboche 2008; Delobelle 1988).

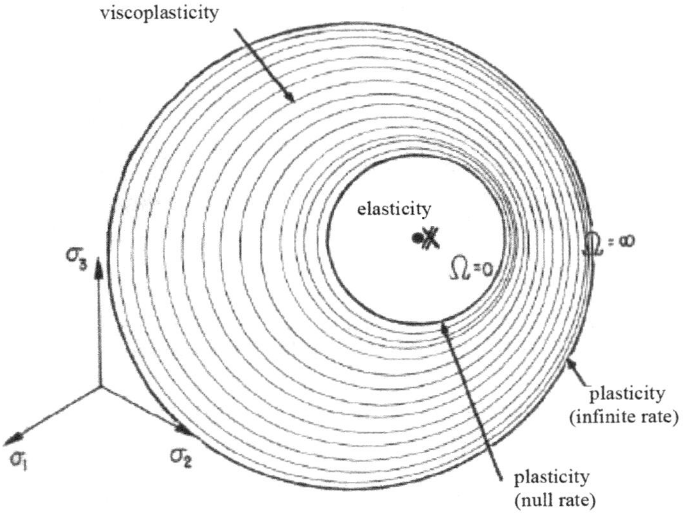

Fig. 7.17 Equipotential flow surfaces

Chapter 8
The Shape-Memory Alloys

Abstract One focuses on «shape memory alloys» (SMAs), where a phase transformation (martensite) can be induced by stress and (or) temperature variations. A process of martensite variants reorientation can also take place. The SMAs functional properties such as pseudoelasticity, one-way shape memory effect, recovery stress, double-shape memory effect (training) are described. In the framework of thermodynamics of irreversible process applied to the generalized standard materials, a macroscopic model with internal variables is built. At last, a design of SMAs elements is done.

8.1 Some General Points About SMAs

8.1.1 Introduction

What are these alloys that we call « shape-memory alloys »?

To begin with, they are metallic alloys with two, three or even four components, with very special compositions.

There are two primary families of SMAs:

- « copper-based » materials Cu–Al (Zn, Ni, Be, etc.);
- « nickel titanium-X » (where X is an element present in small proportion) Ni-Ti (Fe, Cu, Co, etc.).

These materials are called « shape - memory » materials; meaning that they have the property of « remembering » thermomechanical treatments to which they have been subjected (tension, torsion, flexion, etc.).

More specifically, the geometric shapes they have, at high and low temperatures, constitute two states which they « remember ». This memory is developed by way of training, e.g., often by the repetition of the same thermodynamical loading: this is in terms of imposed stress or strain and/or in terms of temperature.

The physical key to « shape memory » is based on a phase transformation between a parent phase called austenite (A) and a produced phase called martensite (M). For SMAs, this phase transformation is described as thermoelastic. It involves a change

C. Lexcellent, *Linear and Non-linear Mechanical Behavior of Solid Materials*, DOI 10.1007/978-3-319-55609-3_8

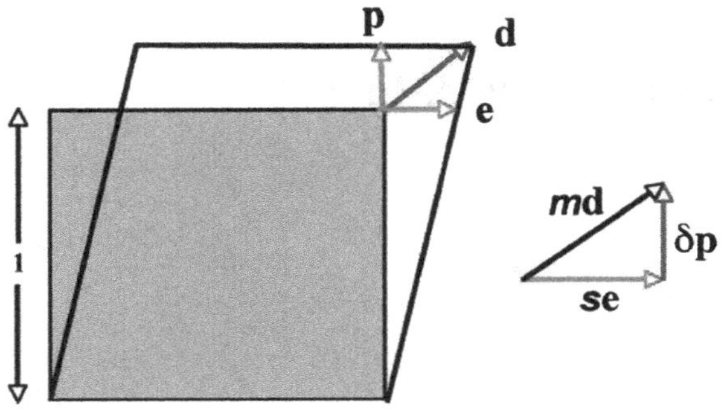

Fig. 8.1 Geometrical martensitic transformation (austenite in *gray*) (Kundu and Bhadeshia 2007)

of the crystalline lattices between the phase A, also known as the « high temperature » phase and a phase M, also known as the « low temperature » phase.

This change is called a « martensitic transformation ». The austenite is transformed into « martensite variants »

A martensite variant is classically defined by an habit plane and a displacement vector $\overrightarrow{d}$ in this plane (Fig. 8.1) or by a « Bain » matrix or a stretch one $\underline{U_\nu}$.

For example, for a transformation austenite cubic $(lattice\ a_0) \Longrightarrow martensite$ quadratic $(lattices\ a, c)$

$$\underline{U_1} = \begin{bmatrix} \beta & 0 & 0 \\ 0 & \alpha & 0 \\ 0 & 0 & \alpha \end{bmatrix} \quad with\ \alpha = {}^{a}/_{a_0},\ \beta = {}^{c}/_{a_0} \tag{8.1}$$

As shown by the photographs taken by Chu et James on the « copper-based SMAs » (Chu 1993) (Fig. 8.2), the microstructure may prove to be very complex, which makes it difficult to analyze them (Fig. 8.2).

8.1.2 Why Are SMAs of Interest for Industry?

SMAs belong to the category of so-called « adaptive » materials. Not only, they are useful as structural elements, but also appreciable for their mechanical properties such as toughness; they are also capable of fulfilling functions such as sensor or actuator.

They are widely used in domains with high technical added value. For example:

- biomedical industry: used for implants, protheses, or stents which are « latticed » tubes that are inserted into a conduit, e.g., a bronchus;
- aeronautics: filtering out of harmful frequencies, noise reduction (Boeing);
- aerospace: deployment of antennas;

Fig. 8.2 Optical micrograph of a microstructure of a Cu–Al–Ni alloy: a « corner- » type microstructure, horizontal extension 0.75 mm: reproduced with kind permission of C. Chu et R.D. James (Bhattacharya 2003)

- watch manufacturing: insertion of a SMA spring into a watch mechanism;
- nuclear industry: pipes.

8.1.3 Crystallographic Theory of Martensitic Transformation

Let us first examine the crystallographic aspect of martensitic transformation. At a low temperature, the phase M (hereafter called M_T self- accommodating martensite) obtained from A, by simple cooling of the alloy and therefore isotropic redistribution of the variants, may, under external stresses, produces major deformations associated with the reorientation of martensite variants. This behavior is qualified as « pseudoplastic » (Fig. 8.3).

« Shape memory » implies a particular manifestation of crystalline phase transformation known as « martensitic phase transformation ». This a solid-to-solid phase

Fig. 8.3 Reorientation of martensite variants under stress

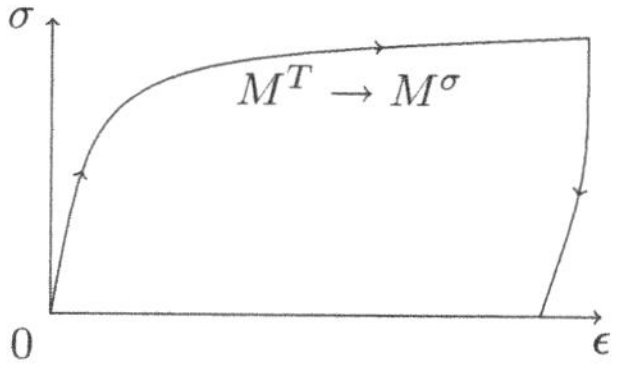

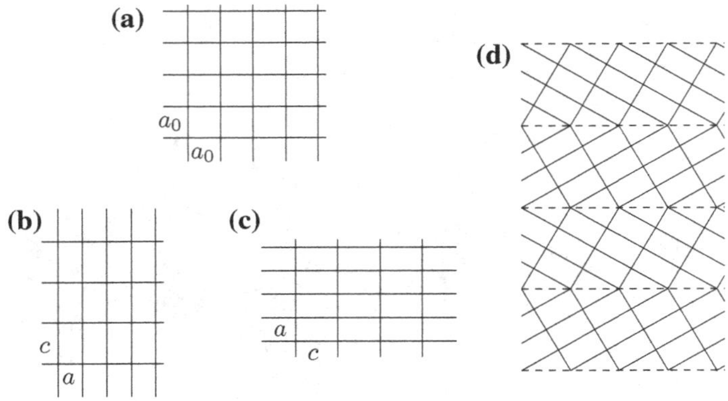

Fig. 8.4 An illustration of martensitic transformation: **a** austenite **b**, **c** martensite variants, **d** a coherent arrangement between martensite variants (Bhattacharya 2003)

transformation, where the parameters of the crystalline lattice change suddently (e.g., A $\longrightarrow M_t$ when cooled) at a specific temperature depending of the alloy.

Although the change is abrupt and the distorsion of the lattice is very significant, there is no diffusion and no alteration in the relative positions of the atoms during the transformation. The transformation is said to be « displacive », of the first order (sudden change of crystalline parameters) (Fig. 8.4).

If the alloy is heated, it undergoes a thermal expansion until the reverse transformation (M $\longrightarrow$ A) occurs at another critical temperature. The difference between the two critical temperatures (A $\longrightarrow$ M) and (M $\longrightarrow$ A) show that the SMA behavior is hysteretic. The conventional SMA exhibits a small amount of hysteresis. Figure 8.5 illustrates the so-called « pseudoelastic » curve representing a phase transformation under tensile stress.

Fig. 8.5 The pseudoelastic stress/strain curve

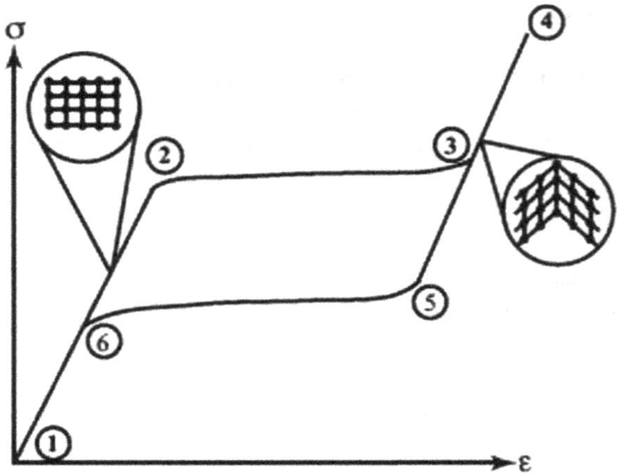

One observable characteristic of martensitic transformation is the resulting microstructure. In a typical transformation, the austenite which is often « cubic » has a greater degree of symmetry than the produced phase. This is shown schematically in two dimensions in Fig. 8.4, where the austenite is a square (a) and the martensite a rectangle (b and c). Consequently, we have multiple martensite variants, in this case two: (b and c). The number ν of variants obtained depends on the change of symmetry during the transformation.

More specifically:

$$\nu = \frac{number\ of\ rotations\ in\ P_a}{number\ of\ rotations\ in\ P_m} \tag{8.2}$$

where Pa (Pm) is the symmetry group of A(M).

Indeed, there is no reason the austenite crystal should transform into only one martensite variant. However, the microstructure must be consistent and may be presented in the architecture shown in Fig. 8.4d, which is corroborated by the transmission electron microscope (TEM) observations on a nickel–aluminum alloy (see the images of microstructures in (Bhattacharya, 2003)).

The need of the crystals to form mixtures of variants, while the whole must remain consistent, gives rise to complex structures which we refer to as the microstructure of the martensite.

8.2 The World of Shape-Memory Alloys

8.2.1 Introduction and General Points

The term martensite was first put forward in honor of Adolf Martens, a german metallurgist who was the first to observe this microstructure in quenched steel. Regarding SMAs, Chang and Read (1951) were the first to detect a phase transformation in a Au–Cd alloy by metallographic observations and measurements of electrical resistance. Pseudoelasticity was born.

Two decades later, the shape-memory effect was observed on a bent bar of that same alloy. In 1963, Buelher et al. (1963) discovered the same properties on an equiatomic Ni–Ti alloy. In the wake of this discovery, the world witnessed, in 1969, the first industrial application with the one-way shape-memory effect used on a sleeve of hydraulic lines of a fighter plane. This created a particular interest in these new materials.

Yet the process of gaining true knowledge of the material and its thermomechanical behavior was slow and would cause difficulties in piratical perennial use in the 1970–1980s. In fact for scientific researchers, this topic truly began in the 1980s associated with the first experimental investigations and the earliest modeling attempts. At the first mechanical tests were, naturally, uniaxial (tension-compression) as were the models. As in the study of any solid material, multiaxial tests proportional or otherwise, such tension–torsion-internal pressure tests, were performed, which required the development of more complex models (Orgeas and Favier 1998; Gall et al. 1998; Bouvet et al. 2002).

Finally, in the 2000s, SMAs came to include the category of « smart materials », but it is more accurate to speak of functional materials or adaptive materials.

To begin with, we may ask a simple question, what is the advantage of SMAs over conventional materials?

Classical metallic alloys exhibit a restricted domain of elasticity. For this reason, an elastic limit of 0, 2% deformation has been defined for simple traction tests.

SMAs are able to accommodate extremely significant reversible deformations of about 6–7% with a few memory Ni–Ti alloys (Tobushi et al. 1998). However, it is rare for SMAs to have the memory property from the moment of its elaboration. In order to obtain this memory effect, a heat treatment must be performed.

8.2.2 *Basic Metallurgy of SMAs, by Michel Morin*

« An overview of the basic metallurgy of commonly used SMAs » has been written by Michel Morin of the Materials and Engineering Sciences MATEIS lab. at INSA Lyon (France).

The interested reader can find information in the book of Christian Lexcellent in French (Lexcellent 2013a) or in English (Lexcellent 2013b).

8.2.3 *Measurement of phase transformation temperatures*

There are two classic techniques:
(1) « *Differential Scanning Calorimetry* » (DSC). This test enables to measure the four characteristic temperatures associated with phase transformation:

- M_s^0: appearance of the first martensite platelet in the austenite during cooling,
- M_f^0: appearance of the last martensite platelet in the austenite during cooling i.e., when austenite has completely disappeared,
- A_s^0: disappearance of the first martensite platelet in the austenite during heating (for energy reasons, this is the platelet which appeared the last),
- A_f^0: disappearance of the last martensite platelet in the austenite during heating (for energy reasons, this is the platelet which appeared the first).

In addition, the amount of heat given off during the forward transformation A $\longrightarrow$ M and absorbed during the reverse transformation M $\longrightarrow$ A can be measured (see Fig. 8.6).

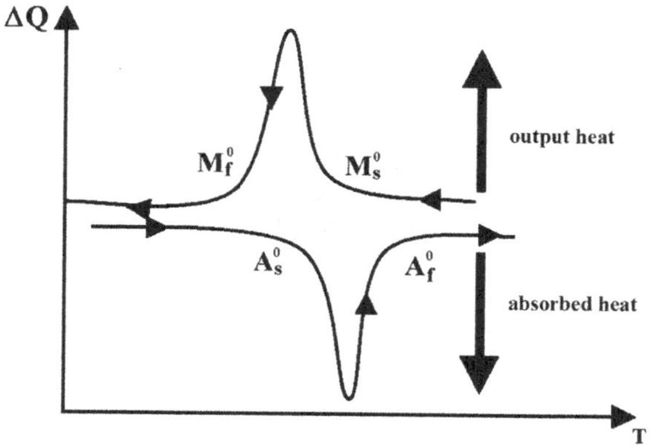

Fig. 8.6 Measurement of the characteristic temperatures of phase transformation by DSC

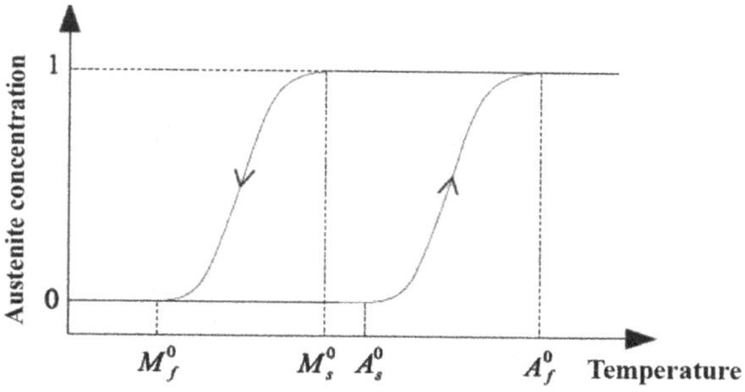

Fig. 8.7 Volume fraction of austenite depending on the temperature: characterization of the four phase transformation temperatures in the stress-free state

(2) Measurements of electrical resistance can be used, as the resistivity of austenite is different from that of martensite. Thus, the measurement of the electrical resistance becomes an indicator of the phase transformation advance.

The curve on Fig. 8.7 shows that the response (martensite volume fraction versus temperature) is hysteretic.

Finally, DSC or electrical resistance measurements can be used to validate the heat treatment performed to « force β phase formation ».

8.2.4 Self-accommodating Martensite and Stress-Induced Martensite

However, there is a distinction to be made between the martensite obtained by simply cooling the alloy, when no particular direction is favored during the phase transformation, and the martensite induced by stress, when the stress direction favors the appearance of certain variants over others.

This distinction between thermally induced martensite M_T and stress-induced martensite M_σ was drawn by Brinson (1993).

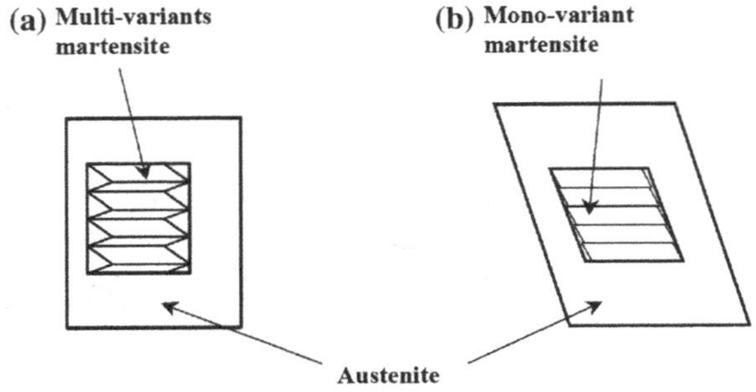

Fig. 8.8 Schematic representation: **a** multivariants martensite **b** monovariant martensite

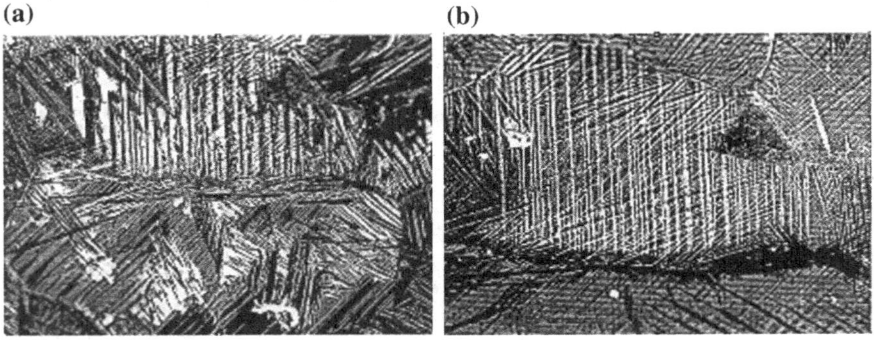

Fig. 8.9 Micrograph: **a** self-accommodating martensite **b** oriented martensite

Figure 8.8 gives a schematic representation of the difference between the two martensite microstructures.

Figure 8.9, in the case of a CuAlNiMn polycrystal, presents images of self-accommodating martensite on the one hand, « oriented » martensite on the other hand.

Leclercq and Lexcellent (1996). We can clearly distinguish the martensite platelets in both cases.

As we shall see, this dichotomy is very useful, in terms of macroscopic simulation, particularly in the case of anisothermal loading.

8.2.5 *Functional Properties of SMAs*

Martensitic transformation and the process of reorientation of martensite variants provide SMAs specific functional properties.

8.2.5.1 The Pseudoelastic Effect

Let us consider an alloy in the austenitic state, at stress-free state; e.g., at $T \geq A_f^0$, the stress–strain curve (Fig. 8.10) can be decomposed as follows:

- elastic austenitic strain at beginning of the loading,
- an $A \longrightarrow M_\sigma$ phase transformation for $\sigma = \sigma_s^{AM}$ up to σ_f^{AM}. Even for Ni–Ti alloys, the phase transformation is rarely total. Plastification of the sample often, occurs at the end of loading;
- if the phase transformation is total, when the stress is released, the martensite exhibits elastic behavior;
- then the reverse transformation $M_\sigma \longrightarrow A$,
- and finally an elastic behavior on the austenite until the stress goes back to zero.

The physical phenomenon of pseudoelasticity is clearly the forward and reverse phases under mechanical loading.

The behavior tends to be hysteretic because the direct and reverse paths are not the same.

With regard to monocrystalline alloys, the detection of the transformation threshold stress for polycrystal is more random, as shown on Fig. 8.10.

Application: significant pseudoelastic strains (up to several per cent) can be recovered.

Among the main medical applications, one might cite:

- An arch for dental brace: a Ni–Ti wire exerts a small but near-constant force (see $M_\sigma \Rightarrow A$ plateau) on badly aligned teeths,
- a pseudoelastic nerve lag,
- glasses' arms,
- pseudoelastic stents (Fig. 8.11).

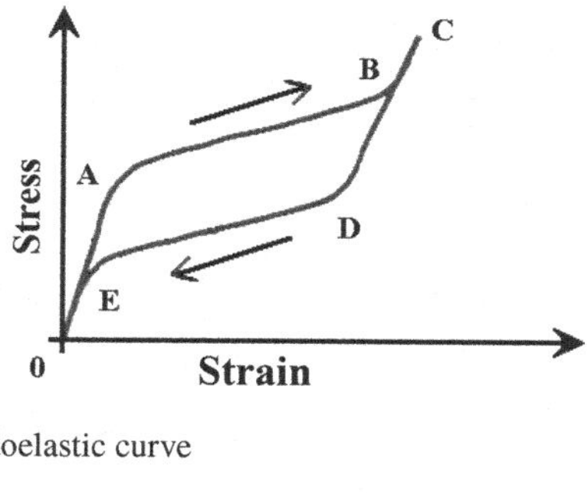

Fig. 8.10 Schematic pseudoelastic curve

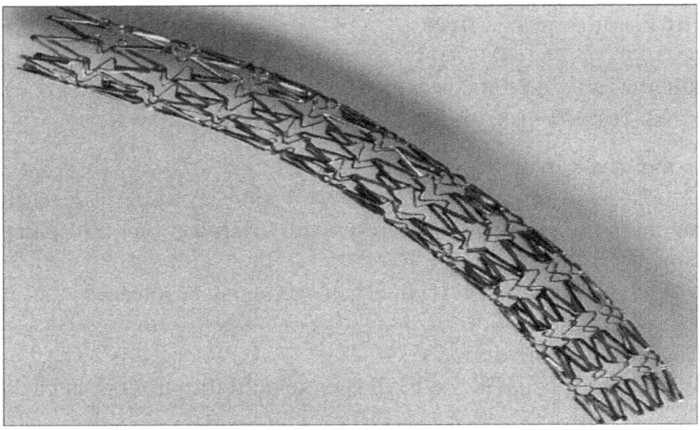

Fig. 8.11 Pseudoelastic stents

8.2.5.2 One-Way Shape-Memory Effect (Fig. 8.12)

Let us consider a SMA at stress-free state, in the austenitic state (at $T_0 > A_f^0$).

First stage: cooling to $T_1 < M_f^0$: $A \longrightarrow M_T$ (M_T thermal martensite).

Second stage: with T_1 kept constant, a stress is applied σ: $M_T \longrightarrow M_\sigma$ (reorientation of martensite platelets).

Third stage: at T_1, the stress on the sample is entirely released. We obtain a point B ($T = T_1$, $\varepsilon = \varepsilon_R$ some %).

We only need to provide heat to the material from T_1 to T_0 in order to recover the initial state: point A $\varepsilon = 0$. $(T_1 - T_0)$ is approximately 30 à 60° which is small. This constitutes an excellent example of actuation.

Applications: the main applications are based on cut-off switches and sensors:

$\longrightarrow$ smoke detector

$\longrightarrow$ « smart » fryer (Patoor and Berveiller 1990)

$\longrightarrow$ thermomechanical actuator,

$\longrightarrow$ connection of two pipes.

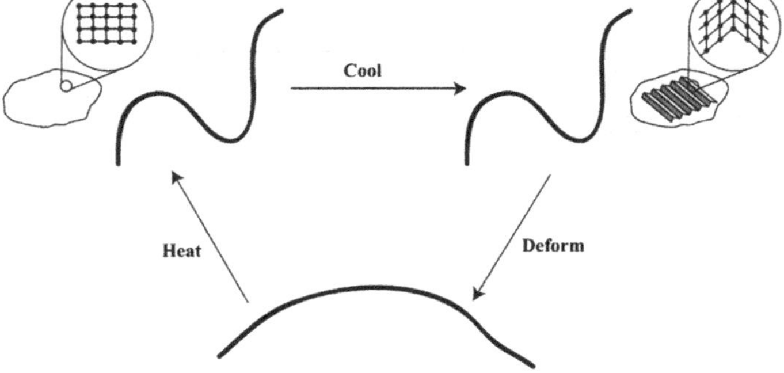

Fig. 8.12 Schematic representation of the one-way shape memory effect

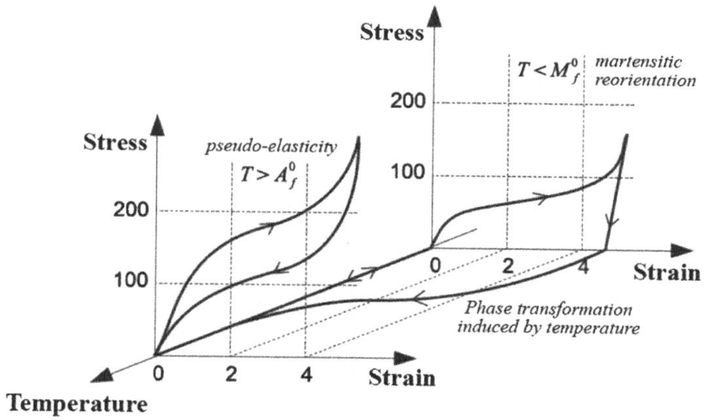

Fig. 8.13 Pseudoelasticity and one-way shape-memory effect in a SMA

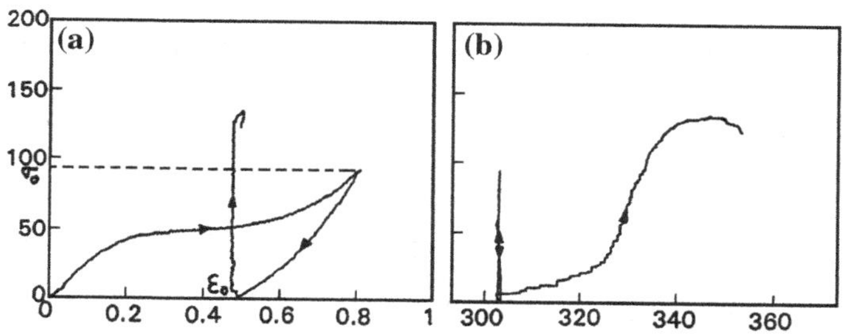

Fig. 8.14 Measurement of a recovery stress for a Ni–Ti alloy

Pseudoelasticity and the one-way shape-memory effect are shown on Fig. 8.13 in a ternary diagram $\{\sigma, \varepsilon, T\}$.

8.2.5.3 Recovery Stress

In reference to the one-way shape-memory effect, the procedure is strictly the same up to point B ($\sigma = 0$, $T = T_1$, and $\varepsilon = \varepsilon_R$). Then, the material is heated, while the deformation ε_R is kept constant.This is a « countered phase return » between two antagonistic processes. Heating naturally tends to transform M_σ into A but maintaining a constant deformation prevents the intrinsic deformation associated with M_σ from reducing, causing a high degree of stress called « recovery stress » (see Fig. 8.14).

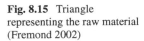

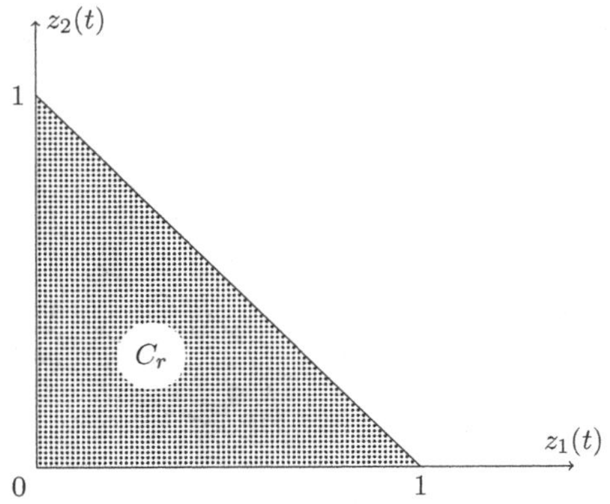

Fig. 8.15 Triangle representing the raw material (Fremond 2002)

8.2.5.4 Double Shape-Memory Effect: Training

If we apply a specific thermomechanical treatment called « training treatment » to a SMA, it can memorize two geometric shapes: one in the austenitic phase and one in its martensitic phase, under simple thermal loading. As training is often repeated, such treatments often constitute cycling loadings e.g. thermal cycles under constant stress. The shape memorized in the martensitic phase « results from the preferred formation » of variants oriented by the internal stress field generated during the training process (Patoor et al. 1987).

This can be represented schematically by the approach proposed by Frémond (Fremond 1998; Fremond 2002); with a model with two martensite variants M_1 of volume fraction z_1 and M_2 of volume fraction z_2. Training a SMA is tantamount to « cutting a corner off the triangle ».

For the raw material, the domain of martensite existence covers entirely the triangle C_r (Fig. 8.15).

For the trained material, the domaine of existence of trained martensite is located in the triangle C_{re} (Fig. 8.16). This means that any and every point in the triangle C_r can be reached, « a priori », by the raw material. With trained material, only the triangle C_{re} remains accessible.

8.3 Macroscopic Models with Internal Variables

8.3.1 Introduction

Most of these models are constructed in the framework of thermodynamics of irreversible process (TIP) applied to the generalized standard materials made popular by Bernard Halphen et N'Guyen Quoc Son (Halphen and N'Guyen 1974).

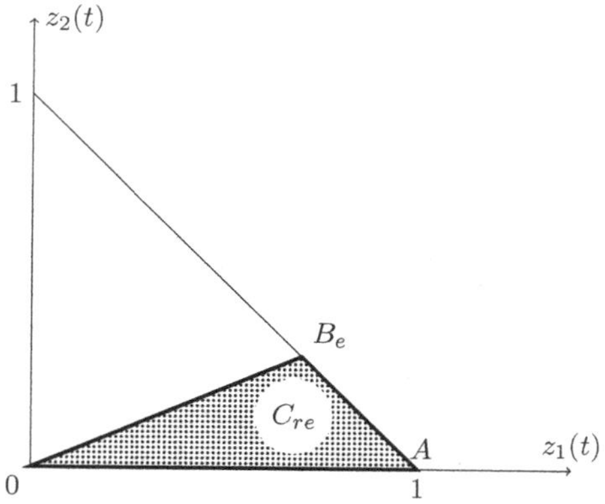

Fig. 8.16 Triangle representing the trained material (Fremond 2002)

The main difference between these models is their description of the hysteresis phenomenon. « Generalized standard materials » type models use approach similar to plasticity and are therefore based on the definition of threshold functions for the transformations $A \leftrightarrows M$. Models with non-convex free energies describe hysteresis in the form of instability phenomena. The hereditary approach considers rheological models based on springs and frictional dampers, putting forward a thermomechanical conception of hysteresis.

For the other phenomenological investigations, it is primarily the choice of internal variables associated with the martensitic transformation that distinguishes these models from one another. The behavior of the solid material is regulated by the choice of a specific free energy and two potentials of dissipation, e.g., two criteria functions for phase transformation (forward $A \Rightarrow M$ or reverse $M \Rightarrow A$).

Historically, the first "empirical" (so to speak) models were those advanced by Tanaka and Nagaki (1982),Tanaka et al. (1986) which use the volume fraction of martensite z as an internal variable and the total strain ε and temperature T as observable variables.

- σ is the traction stress;

– ε^{tr}, ε^{th}, ε^{el} the uniaxial traction deformations, respectively associated with phase transformation, thermal elongation, and elastic contribution note that $\varepsilon^{tr} = \gamma z$ and $\varepsilon^{th} = \alpha\,(T - T_0)$
–E Young's modulus.

They give a one-dimensional (1D) constitutive law:

$$\sigma = E(\varepsilon - \varepsilon^{tr} - \varepsilon^{th}) = E(\varepsilon - \gamma z - \alpha(T - T_0)) = E(\varepsilon^{el}) \tag{8.3}$$

with phase transformation kinetics in exponential laws.

For $A \Rightarrow M$:

$$z = 1 - \exp(a_1 < \sigma - b_1(T - M_s^0) >) \tag{8.4}$$

For $M \Rightarrow A$:

$$z = \exp(-a_2 < b_2(T - A_S^0) - \sigma >) \tag{8.5}$$

with $< x >= x$ if $x \geq 0$ and 0 if $x \leq 0$ and $a_1,\ a_2,\ b_1,\ b_2$ constants.

This theory was extended in 1990 by Liang and Rogers, who gave a "sinusoidal form" to the phase transformation kinetics (Liang and Rogers 1990).

In 1993, Brinson (1993) introduced a distinction between martensite of purely thermal origin M_T (created simply by cooling of austenite, i.e. the stress-free state) and a martensite obtained under the influence of an applied stress called M_σ.

In 1992, Raniecki et al. (1992) improved the model initially delivered by Tanaka and Nagaki (1982); Tanaka et al. (1986) and placed them in the context of thermodynamics of irreversible processes.

In this chapter, it is no question of giving a complete inventory of all the existing macroscopic models relating to SMAs; there are far too many of them in the literature. We shall only highlight a few of the salient points of some of these models.

However, we shall shift from a one-dimensional (1D) formulation, where all the variables are scalar, to a three-dimensional formulation where the strains and stresses become second-order tensors. For instance, $\varepsilon \Longrightarrow \underline{\varepsilon}$ ou $\sigma \Longrightarrow \underline{\sigma}$ etc. This results from the desire to take account of the multiaxial mechanical loads such as tension/torsion on a tube.

8.4 R_L Model

This model was published by Raniecki et al. (1992) and extended to the anisothermal case by Leclercq and Lexcellent (1996).

We define the representative elementary volume (REV) by martensite platelets (whose overall volume fraction is z) disseminated in the austenitic parent phase (with overall volume fraction $(1 - z)$).

The free energy of the two-phase mixture $(A + M)$ can be defined by

$$\psi(\underline{\varepsilon},\ T,\ z) = (1 - z)\psi_A(\underline{\varepsilon}_A,\ T) + z\psi_M(\underline{\varepsilon}_M,\ T) + \psi_{int}(z,\ T) \tag{8.6}$$

where ψ_A (ψ_M) represents the free energy of the austenite A (or of the martensite M) and ψ_{int} represents the interaction energy between the austenite and martensite. The choice of this expression proved to be both delicate and crucial in terms of modeling. We shall therefore come back to it later.

Let us examine the simplest choice: $\psi_{int}(z, T) = 0$, which will give a reversible model referred to as R.

8.4.1 Reversible R Model

In this case, when $\psi_{int}(z, T) = 0$, the free energy becomes

$$\psi(\underline{\varepsilon}, T, z) = (1 - z)\psi_A(\underline{\varepsilon}_A, T) + z\psi_M(\underline{\varepsilon}_M, T) \tag{8.7}$$

Let us consider the free energy of the α phase:

$$\rho\psi^{\alpha}(\underline{\varepsilon}_{\alpha}, T) = u_{\alpha}^{0} - Ts_{\alpha}^{0} + \frac{1}{2}(\underline{\varepsilon}_{\alpha} - \underline{\varepsilon}_{\alpha}^{\text{tr}} - \underline{\varepsilon}_{\alpha}^{\text{th}})\underline{\underline{L}}(\underline{\varepsilon}_{\alpha} - \underline{\varepsilon}_{\alpha}^{\text{tr}} - \underline{\varepsilon}_{\alpha}^{\text{th}}) + C_v\left(T - T_0 - Ln\left(\frac{T}{T_0}\right)\right) \tag{8.8}$$

$\alpha = 1$ for austenite and $\alpha = 2$ for martensite.
In the expression of the free energy $\psi_{\alpha}(\underline{\varepsilon}_{\alpha}, T)$, the first term on the right represents the chemical contribution, the second the elastic energy and the third the thermal contribution.
First hypothesis

We assume the same values for the elastic constants $\underline{\underline{L}}$, density ρ, specific heat C_v, and thermal expansion coefficient α for both austenite and martensite. It is possible not to do so, but the expressions become more complicated, for a gain of only 5–8% in terms of precision,

Thus,

$$\underline{\varepsilon}_{\alpha}^{\text{th}} = \underline{\varepsilon}_{1}^{\text{th}} = \underline{\varepsilon}_{2}^{\text{th}} = \underline{\varepsilon}^{\text{th}} = \alpha(T - T_0)\underline{\mathbf{1}} \tag{8.9}$$

By definition,

$$\underline{\sigma}_{\alpha} = \rho\frac{\partial\psi_{\alpha}}{\partial\underline{\varepsilon}_{\alpha}} = \underline{\underline{L}}(\underline{\varepsilon}_{\alpha} - \underline{\varepsilon}_{\alpha}^{\text{tr}} - \underline{\varepsilon}_{\alpha}^{\text{th}}) = \underline{\underline{L}}(\underline{\varepsilon}_{\alpha}^{el}) \tag{8.10}$$

where

$\underline{\varepsilon}_{\alpha}$ represents the total deformation of the α phase;

$\underline{\varepsilon}_{\alpha}^{\text{tr}}$ represents the transformation deformation associated with the α phase;

$\underline{\varepsilon}_{\alpha}^{\text{th}}$ represents the thermal deformation of the α phase;

$\underline{\varepsilon}_{\alpha}^{\text{el}}$ represents the elastic deformation of the α phase.

Second hypothesis

$\underline{\varepsilon}_{1}^{\text{tr}} = 0, \underline{\varepsilon}_{2}^{\text{tr}} = \underline{\kappa}$

Third hypothesis

We apply the mixture law to the total deformation:

$$\underline{\varepsilon} = (1 - z)\underline{\varepsilon}_1 + z\underline{\varepsilon}_2 \tag{8.11}$$

with

$\underline{\varepsilon}_{\alpha} = (\underline{\varepsilon}_{\alpha}^{\text{el}} + \underline{\varepsilon}_{\alpha}^{\text{tr}} + \underline{\varepsilon}_{\alpha}^{\text{th}})$.

Fourth hypothesis

The stress state is indifferent from the phase state - in other words, it comes

$$\underline{\sigma}_{\alpha} = \underline{\sigma}_1 = \underline{\sigma}_2 = \underline{\sigma} \Longrightarrow \underline{\varepsilon}_{\alpha}^{el} = \underline{\varepsilon}_1^{el} = \underline{\varepsilon}_2^{el} = \underline{\varepsilon}^{el} \tag{8.12}$$

and:

$$\underline{\varepsilon} = \underline{\varepsilon}^{\mathrm{el}} + \underline{\kappa} z + \alpha(T - T_0)\underline{1} \tag{8.13}$$

The free energy of the $(A + M)$ biphase is written:

$$\rho\psi(\underline{\varepsilon}, z, T) = u_1^0 - Ts_1^0 - z\Pi_0^f(T) + \tfrac{1}{2}(\underline{\varepsilon} - \underline{\kappa} z - \alpha(T - T_0)\underline{1})\underline{\underline{L}}(\underline{\varepsilon} - \underline{\kappa} z - \alpha(T - T_0)\underline{1}) \\ + C_v\left(T - T_0 - \ln\left(\frac{T}{T_0}\right)\right) \tag{8.14}$$

with:

$$\underline{\sigma} = \rho\frac{\partial\psi}{\partial\underline{\boldsymbol{\varepsilon}}} = \underline{\underline{L}} : (\underline{\varepsilon} - \underline{\kappa} z - \alpha(T - T_0)\underline{1});\ s = -\frac{\partial\psi}{\partial T} \tag{8.15}$$

where s constitutes the specific entropy and $\Pi_0^f(T) = \Delta u - T\Delta s$ with $\Delta u = u_1^0\text{-}u_2^0$; $\Delta s = s_1^0\text{-}s_2^0$

The increment of mechanical dissipation (in reference to Chap. 2) is written as:

$$dD = \Pi^f dz \geq 0 \tag{8.16}$$

with:

$$\Pi^f(\underline{\sigma}, z, T) = -\frac{\partial\psi}{\partial z} = \Pi_0^f(T) + \frac{\underline{\kappa} : \underline{\sigma}}{\rho} \tag{8.17}$$

or:

$$\Pi^f(\underline{\varepsilon}, z, T) = -\frac{\partial\psi}{\partial z} = \Pi_0^f(T) + \frac{\underline{\kappa} : \underline{\underline{L}}(\underline{\varepsilon} - \underline{\kappa} z - \alpha(T - T_0)\underline{1})}{\rho} \tag{8.18}$$

A (thermodynamically) reversible model such as R means that the dissipation is null, and consequently, during the forward or reverse phase transformation $\forall dz$; $z\epsilon\,[0, 1]$:

$$\Pi^f(\underline{\sigma}, z, T) = 0;\ \Pi^f(\underline{\varepsilon}, z, T) = 0 \tag{8.19}$$

1st application: uniaxial tensile stress.

In this case, we can reduce $\underline{\boldsymbol{\kappa}}$ to a scalar γ and the threshold stress is written:

$$\sigma^{am} = -\rho\frac{\Pi_0^f(T)}{\gamma} \tag{8.20}$$

In the Clausius–Clapeyron diagram (σ, T), we can deduce the slope of domain separation between the austenitic and martensitic ones:

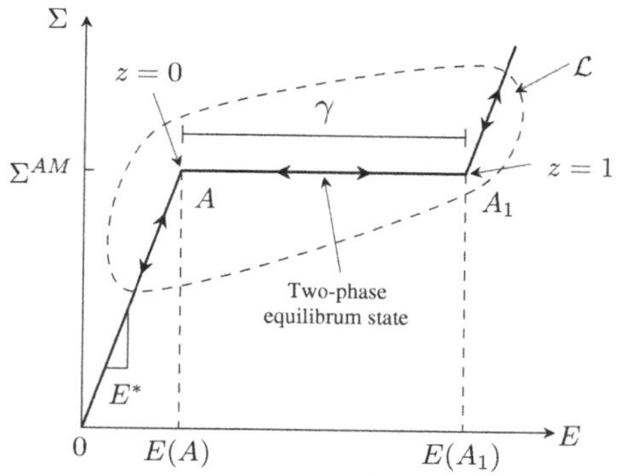

Fig. 8.17 Tensile curve corresponding to model R (Raniecki et al. 1992)

$$\frac{d\sigma^{am}}{dT} = \rho\frac{\Delta s}{\gamma} \tag{8.21}$$

As shows Fig. 8.17, the deformation at the onset of phase transformation $\varepsilon^{am}(z=0)$ is governed by $\varepsilon^{am}(z=0) = {}^{\sigma^{am}}/_{E}$ and at the end of the transformation $\varepsilon^{am}(z=1) = {}^{\sigma^{am}}/_{E} + \gamma$ and an intermediary deformation $\varepsilon^{am}(z) = {}^{\sigma^{am}}/_{E} + \gamma z$.

This simple model is particularly well adapted to describe the behavior of monocrystals with reduced hysteresis such as Cu–Zn–Al (Patoor et al. 1987).

8.4.2 R_L Model with a Hysteresis Loop (Muller 1989)

In this case, the free energy includes the mixture law for the two phases in the R model (Eq. (8.14)) and the interaction term $\psi_{int}(z, T)$:

$$\rho\psi(\underline{\varepsilon}, z, T) = u_1^0 - Ts_1^0 - z\Pi_0^f(T) + \frac{1}{2}(\underline{\varepsilon} - \underline{\kappa}z - \alpha(T - T_0)\underline{1}) : \underline{\underline{L}} : (\underline{\varepsilon} - \underline{\kappa}z - \alpha(T - T_0)\underline{1}) + C_v\left(T - T_0 - \ln\left(\frac{T}{T_0}\right)\right) + \psi_{int}(z, T) \tag{8.22}$$

with:

$$\underline{\sigma} = \rho\frac{\partial\psi}{\partial\underline{\boldsymbol{\varepsilon}}} = \underline{\underline{L}} : (\underline{\varepsilon} - \underline{\kappa}z - \alpha(T - T_0)\underline{1});\ s = -\frac{\partial\psi}{\partial T} \tag{8.23}$$

which represents the same expression as Eq. (8.15) in the R model.

In terms of the modeling of multiaxial experiments (taking into account the asymmetry between tension and compression, for example) the choice of tensor $\underline{\boldsymbol{\kappa}}$ is very important. The choice of $\psi_{int}(z, T)$ turns out to be crucial with regards to dissipation and hysteresis.

A logical reasoning would be to hold that there is no interaction when only one of the two phases is present (although this assumption means somewhat to overlook the fact that the martensitic phase, "when transforming" into the parent phase, may have traces left!). We write:

$$\psi_{int}(z=0) = \psi_{int}(z=1) = 0 \tag{8.24}$$

Hence, a natural choice was made first by Muller (1989) then by MULLER and XU (1991) in their one-dimensional theory regarding the formation of hysteresis loops due to phase transformation:

$$\psi_{int}(z, T) = Az(1-z) \tag{8.25}$$

Raniecki et al. (1992) replaced the constant A by $\psi_{it}(T) = u_0 - Ts_0$:

This may remind the thoughts of Licht, published in *"z(1 – z), j'aime assez"* ("z(1 – z): I quite like that") (Licht 1998).

We can also identify the mathematical expression of $\psi_{int}(z, T)$ by determining parameters on the experimental hysteretic traction curves. For instance, as outlined by Bouvet et al. (2002); Boubakar and Lexcellent (2007). This can be done in the context of finite transformations (Vieille et al. 2003).

In comparison to the R model, dissipation exists, and is written, in reference to Chap. 2:

$$dD = \Pi^f dz \geq 0 \tag{8.26}$$

with:

$$\Pi^f(\underline{\sigma}, z, T) = \Pi_0^f(T) + \frac{\underline{\kappa} : \underline{\sigma}}{\rho} - (1-2z)\psi_{it}(T) \tag{8.27}$$

The threshold stress tensor $\underline{\sigma}^{am}$ $A \Rightarrow M$ is driven by:

$$\Pi^f(\underline{\sigma}^{am}, z=0, T) = \Pi_0^f(T) + \frac{\underline{\kappa} : \underline{\sigma}^{am}}{\rho} - \psi_{it}(T) = 0 \tag{8.28}$$

and the stress tensor $\underline{\sigma^{ma}}$ for the onset of the forward $(A \Rightarrow M)$ transformation is governed by:

$$\Pi^f(\underline{\sigma}^{ma}, z=1, T) = \Pi_0^f(T) + \frac{\underline{\kappa} : \underline{\sigma}^{ma}}{\rho} + \psi_{it}(T) = 0 \tag{8.29}$$

The forward transformation $A \Rightarrow M$ occurs when $dz > 0 \Rightarrow \Pi^f \geq 0$ and the reverse transformation $M \Rightarrow A$ when $dz < 0 \Rightarrow \Pi^f \leq 0$.

In order to specify the equations for the phase transformation kinetics, we assume that there are two functions $\phi_\alpha(\Pi^f, z)$ $(\alpha = 1, 2)$ such that an active process of parent phase decomposition $(dz > 0\ A \Rightarrow M)$ can take place only if $\phi_1 = const$ $(d\phi_1 = 0)$. Similarly, an active process of martensite decomposition $(dz < 0\ M \Rightarrow A)$ can only progress if $\phi_2 = cont$ $(d\phi_2 = 0)$:

$$\phi_1 = \Pi^f - k_1(z),\ \phi_2 = -\Pi^f + k_2(z) \tag{8.30}$$

In order to be in agreement with metallurgists such as Koistinen and Marburger

Koistinen and Marburger (1959) regarding the expression of the phase transformation kinetics, we choose:

$$k_1(z) = 2\psi_{it}\left(M_S^0\right) z + \frac{s_0 - \Delta s - 2s_0 z}{a_1}\ln(1-z) \tag{8.31}$$

$$k_2(z) = 2\psi_{it}\left(A_S^0\right)(1-z)\,\frac{s_0 + \Delta s - 2s_0(1-z)}{a_1}\ln(z) \tag{8.32}$$

Thus, the phase transformation kinetics in both directions can be calculated:

$$d\phi_1 = 0 \Rightarrow dz^{a\to m} \tag{8.33}$$

$$d\phi_2 = 0 \Rightarrow dz^{m\to a} \tag{8.34}$$

The hysteresis size is defined in one-dimension (e.g. simple tension) by:

$$\Delta\sigma = \sigma^{am} - \sigma^{ma} \tag{8.35}$$

Note that, in this case, $\underline{\kappa} : \underline{\sigma} = \gamma\sigma$ where σ is the traction stress and γ is the complete transformation deformation,

with σ^{am} defined by: $\Pi^f(\sigma^{am}, z = 0, T) = 0$ therefore giving:

$$\sigma^{am} = \frac{\rho}{\gamma}\left(-\Pi_0^f(T) + \psi_{it}(T)\right) \tag{8.36}$$

Similarly σ^{ma} defined by $\Pi^f(\sigma^{ma}, z = 1, T) = 0$ therefore giving:

$$\sigma^{ma} = \frac{\rho}{\gamma}\left(-\Pi_0^f(T) - \psi_{it}(T)\right) \tag{8.37}$$

We get the hysteresis size:

$$\Delta\sigma = \sigma^{am} - \sigma^{ma} = \frac{\rho}{\gamma}2\psi_{it}(T) \tag{8.38}$$

If we wish to analyze the thermomechanical coupling, we need to introduce the heat equation. More specifically that the phase transformations are highly exothermic for $A \Rightarrow M$ and endothermic for $M \Rightarrow A$.

8.4.2.1 Heat Equation

Considering that performing an "isothermal test" is a convenient but incorrect use of the language. We can simply impose a constant temperature at the boundary of the sample, which constitutes a structure whose internal temperature is not controlled. Studies on thermomechanical coupling, with infrared imagery and estimations of the energy stored in the material, can be found in Chrysochoos and Louche (2000), Chrysochoos et al. (2009).

This equation is crucial in order to establish the contribution of thermal effects of a SMA to the "slow" dynamic response (in comparison to piezoelectric materials).

As stated in Chap. 2, the first principle of thermodynamics applied to the "Representative Elementary Volume" is written as follows:

$$\rho \dot{e} = \underline{\sigma} : \underline{\dot{\varepsilon}} + \mathrm{r} - \operatorname{div} \vec{q} \tag{8.39}$$

with $e = (\psi + Ts)$
and:

$$\rho\psi(\underline{\varepsilon}, z, T) = u_1^0 - Ts_1^0 - z\Pi_0^f(T) + \frac{1}{2}(\underline{\varepsilon} - \underline{\kappa}z - \alpha(T - T_0)\underline{1})\underline{\underline{L}}(\underline{\varepsilon} - \underline{\kappa}z - \alpha(T - T_0)\underline{1}) + C_v\left(T - T_0 - \ln\left(\frac{T}{T_0}\right)\right) + \psi_{int}(z, T) \tag{8.40}$$

and:

$$\begin{aligned} \rho\dot{\psi} &= \underline{\sigma} : \underline{\dot{\varepsilon}} + r - \operatorname{div}\vec{q} - \rho s\dot{T} - \rho\dot{s}T \\ \rho\dot{\psi} &= \rho\frac{\partial\psi}{\partial\underline{\varepsilon}} : \underline{\dot{\varepsilon}} + \rho\frac{\partial\psi}{\partial z}\dot{} + \rho\frac{\partial\psi}{\partial \dot{T}}\dot{T} \\ \rho\dot{\psi} &= \underline{\sigma} : \underline{\dot{\varepsilon}} - \rho\Pi^f\dot{z} - \rho s\dot{T} \end{aligned} \tag{8.41}$$

and also:

$$\rho\dot{s}T = -\rho T\frac{\partial^2\psi}{\partial\underline{\varepsilon}\partial T} : \dot{\varepsilon} - \rho T\frac{\partial^2\psi}{\partial T^2}\dot{T} - \rho\frac{\partial^2\psi}{\partial z\partial T}\dot{z} \tag{8.42}$$

Finally, we find:

$$\rho C_v\dot{T} - \underline{\kappa} : \underline{\sigma}\dot{z} - \rho(\Delta u - (1 - 2z)u_0)\dot{z} - r + \operatorname{div}\vec{q} + 3\alpha T K_0 tr\underline{\dot{\varepsilon}} = 0 \tag{8.43}$$

In practice, not all the terms have the same importance, and often, the heat equation is reduced to:

$$\rho C_v\dot{T} - \underline{\kappa} : \underline{\sigma}\dot{z} - \rho(\Delta u)\dot{z} - r + \operatorname{div}\vec{q} = 0 \tag{8.44}$$

Let us examine the case of a SMA wire with diameter d and length l (Benzaoui et al. 1997). Its exterior surface and its volume are, $S_{amf} = \pi dl$ and $V_{amf} = \frac{\pi d^2 l}{4}$ respectively.

For a wire with electrical resistance R heated by the Joule effect due to an electrical current i, the internal heat source r is written:

$$r = \frac{Ri^2}{V_{amf}} = \frac{16\rho_e}{\pi^2 d^4} i^2 \tag{8.45}$$

with:

$$\rho_e = z\rho_m + (1 - z)\rho_a \tag{8.46}$$

with ρ_a, ρ_m and ρ_e representing the resistivity of the austenite, the martensite, and the biphase mixture respectively.

Furthermore, the term $\mathrm{div}\vec{q}$ corresponds to the conductive and convective heat exchanges at the surface of the wire. For a thin wire (i.e., whose length-to-diameter ratio is greater than 50), in order to be able to account for the radial temperature gradients, to begin with, $\mathrm{div}\vec{q}$ represents the density of heat lost by convection on the lateral surface of the wire, so that:

$$\mathrm{div}\vec{q} = \frac{hS_{amf}}{V_{amf}} = \frac{4h}{d}(T - T_a) \tag{8.47}$$

where h is the convection coefficient of the SMA and T_a the ambient temperature.

8.4.3 Extension to Reversible Phase Transformation: Austenite⟹Rphase for NiTi (Lexcellent et al. 1994)

This phase transformation between the austenite and a pre-martensite called the R phase as regards NiTi, is mechanically reversible, i.e., without hysteresis. As explained in Chap. 2, the existence of this transformation is dependent on the composition and or the thermomechanical treatment of the considered alloy.

Although the amplitude of the transformation deformation of the $A \Rightarrow R$ never exceeds 0.7% (a little over 10% of that corresponding to the $A \Rightarrow M$ transformation for the same alloy), because the dissipation at each cycle is very small, the fatigue resistance is excellent. The use of this phase transition is absolutely operational for NiTi wires embedded into an epoxide matrix, for instance; resulting in adaptive structure (see this chapter, Exercise 3).

It is elementary to model a simple tensile test.

Let the free energy be: (by omitting the thermal deformation ε^{th} contribution in order to simplify):

$$\rho\psi(\mathrm{E}, z, T) = u_1^0 - Ts_1^0 - z\Pi_0^f(T) + \frac{1}{2}E(\varepsilon - \gamma z)^2 + C_v\left(T - T_0 - \ln\left(\frac{T}{T_0}\right)\right) + \tag{8.48}$$

$$\psi_{it} z(1 - z) \tag{8.49}$$

avec $E_A = E_R = E$ the Young's moduli of the austenite (taken to be equal to one another), and the stress:

$$\sigma = \rho \frac{\partial \psi}{\partial \varepsilon} = E(\varepsilon - \gamma z) \tag{8.50}$$

and:

$$dD = \Pi^f dz = 0 \; z\epsilon \, [0, 1] \tag{8.51}$$

because there is no hysteresis.
This leads to:

$$\Pi^f(\sigma, z, T) = -\frac{\partial \psi}{\partial z} = \Pi_0^f(T) + \frac{\gamma \sigma}{\rho} + (1 - 2z)\psi_{it}(T) = 0 \; \forall z \tag{8.52}$$

and in particular:

$$\Pi^f(\sigma = \sigma^{ar}, \; z = 0, \; T) = -\frac{\partial \psi}{\partial z} = \Pi_0^f(T) + \frac{\gamma \sigma^{ar}}{\rho} - \psi_{it}(T) = 0 \tag{8.53}$$

$$\Pi^f(\sigma = \sigma^{ra}, \; z = 0, \; T) = -\frac{\partial \psi}{\partial z} = \Pi_0^f(T) + \frac{\gamma \sigma^{ra}}{\rho} + \psi_{it}(T) = 0 \tag{8.54}$$

We obtain the threshold stresses for the onset of the forward transformation σ^{ar} and the reverse transformation σ^{ra}:

$$\sigma^{ar} = \frac{\rho}{\gamma}(-\Pi_0^f(T) + \psi_{it}(T)) \tag{8.55}$$

$$\sigma^{ar} = \frac{\rho}{\gamma}(-\Pi_0^f(T) - \psi_{it}(T)) \tag{8.56}$$

where:

$$\sigma^{ar} - \sigma^{ra} = 2\frac{\rho}{\gamma}\psi_{it} < 0 \tag{8.57}$$

which indicates that $\psi_{it} < 0$, by contrast to the classic $A \Rightarrow M$ transformation.

The measurements of σ^{ar} *and* σ^{ra}, at a given temperature *T*, can be used to determine $\Pi_0^f(T)$ and $\psi_{it}(T)$.

It is, therefore, elementary to model the traction curve (Fig. 8.18) – that is:

- for the elastic austenitic domain, we plot a straight line from the origin $\sigma = 0$ to $\sigma = \sigma^{ar}$,

we draw a second line between σ^{ar} *and* σ^{ra},

beyond σ^{ra} we draw a line parallel to the first one and whose with slope is E.

While this modeling method of is easy, it introduces in the energy expression a term of interaction $\psi_{it} z(1 - z)$ which is negative, which is a problem in terms of the two phases mixture.

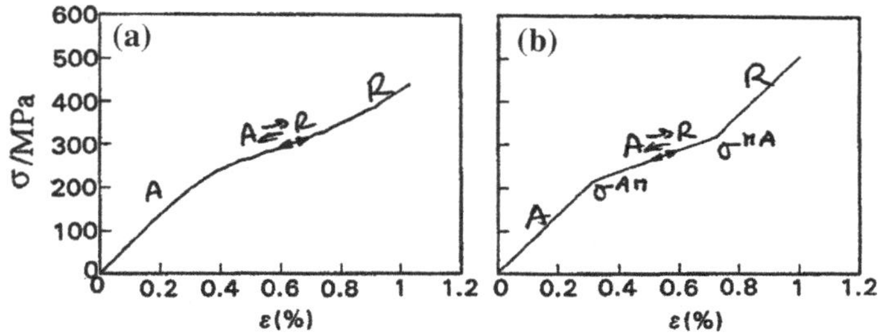

Fig. 8.18 Traction curves associated with the $A \rightleftharpoons R$ transformation: **a** experiment; **b** simulation (Lexcellent et al. 1994)

8.4.4 Multiaxial Isothermal Behavior

Shape-memory alloys used as actuators, for instance, are not simply wires subjected to a tensile stress. Other examples of structural elements made of SMAs may be beams subject to bending stress which means we need to know the material's response both under tensile and compressive stress. The use of tubes, subjected to a twisting torque, requires to be familiar with the SMAs' behavior under shearing, and so on.

More generally speaking, for an increasingly broad range of applications, it is necessary to know the SMAs' response to complex stresses.

Hitherto, models of thermomechanical behavior have been validated by way of uniaxial tests (using tension). Yet results on samples under tension and torsion (Rogueda et al. 1996; Raniecki et al. 1999; Taillard et al. 2008; Grabe and Bruhns 2009) or under triaxial loading (Gall et al. 1998) ("carrot" of SMA subjected to lateral pressure and axial compression) show that their behavior is influenced by a multiaxial stress state. Comparisons between classical models and the experimental results under nonproportional loading emphasize the need to enrich the databases in order to understand and model the mechanisms of formation and reorientation of martensite platelets (Lim and McDowell 1999; Markets and Fisher 1996; Sittner et al. 1995).

The first task was to take into account the asymmetry between tension and compression. This asymmetry was measured on Cu–Zn–Al polycrystals by Vacher and Lexcellent in 1991 (Vacher and Lexcellent 1991) and NiTi polycrystals by Orgéas and Favier in 1998 (Orgeas and Favier 1998) (Fig. 8.19).

Indeed, in the domain of pseudoelasticity ($T > A_f^0$), experimental observations show that:

(a) the threshold stress for the onset of transformation under tension σ_T^{am} has a smaller modulus than the one measured under compression σ_C^{am},

(b) the deformation for phase transformation associated with tension γ_T is greater than the one obtained under compression γ_C.

However, the energies for tension and compression are equal – meaning that:

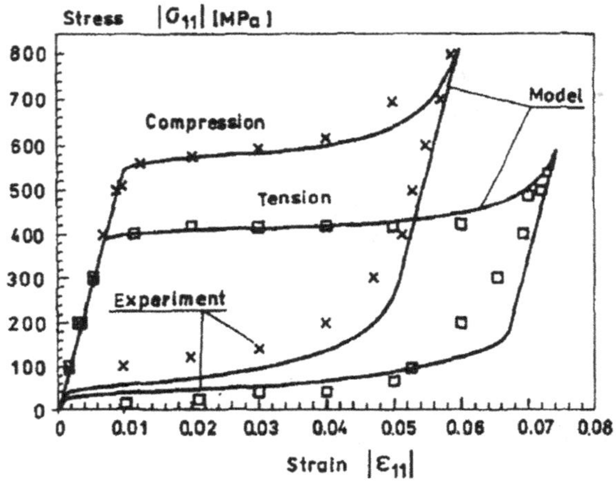

Fig. 8.19 Pseudoelastic curve for NiTi under tension and compression: experiments performed by Orgéas and Favier, simulations by Orgeas and Favier (1998); Raniecki and Lexcellent (1998)

$$\sigma_T^{am}\gamma_T = \sigma_C^{am}\gamma_C = C(T) \tag{8.58}$$

(c) as the threshold stresses σ_T^{am} and σ_C^{am} are linear functions of the temperature, it results Fig. 8.20:

$$-\frac{d\sigma_C^{am}}{dT} > \frac{d\sigma_T^{am}}{dT} \tag{8.59}$$

(d) the stress hysteresis obtained under compression is greater than that obtained under tension (Fig. 8.19);
(e) the behavior observed with pure shearing is symmetrical (see Fig. 8.21).

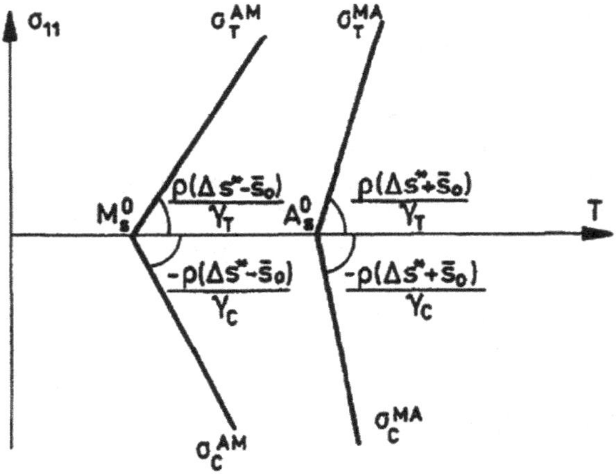

Fig. 8.20 Clapeyron diagram showing traction and compression: schematic illustration

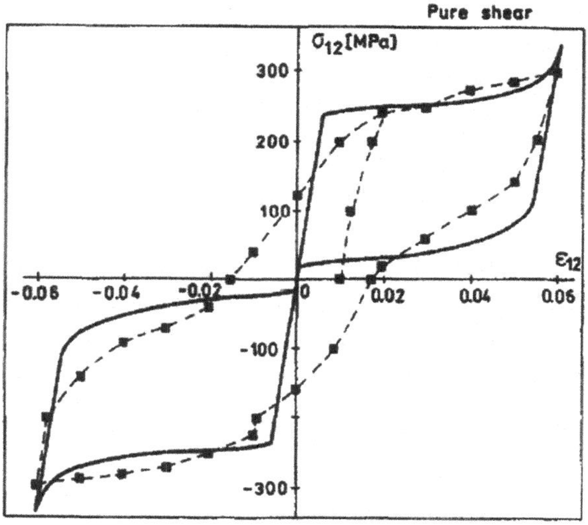

Fig. 8.21 Hysteresis loop with pure shearing: Orgéas and Favier's experiment (Orgeas and Favier 1998)

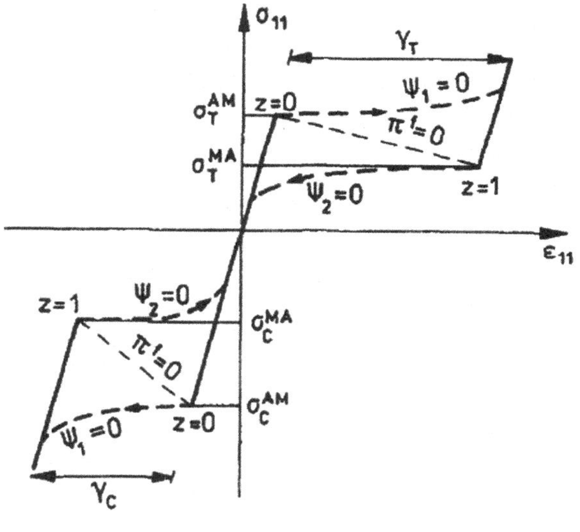

Fig. 8.22 Theoretical illustration of the pseudoelastic loop under traction/compression (Raniecki and Lexcellent 1998)

Figure 8.22 shows the theoretical modeling of the pseudoelastic loop under tension-compression.

At this stage, the key problem is to come up with an acceptable formula for $g^*(\underline{\sigma})$ which is the equation ruling the boundary of the elastic austenitic domain. The appropriate tools to do this are suggested by the plasticity.

Because the behavior of the two-phase medium is isotropic, and $\underline{\kappa}$ is independent of the hydrostatic pressure, we suggest to write $g^*(\underline{\sigma})$ under the form:

$$\rho g^*(\underline{\sigma}) = \gamma \overline{\sigma} g(y_\sigma) \tag{8.60}$$

with $g(y_\sigma = 0) = 1$.

If $\underline{s} = \mathrm{dev}(\underline{\sigma})$ is the stress deviator, then:

$$\overline{\sigma} = \left(\frac{3}{2}\underline{s} : \underline{s}\right)^{\frac{1}{2}} \tag{8.61}$$

corresponds to the equivalent Huber-von Mises stress.

y_σ is a function of $\overline{\sigma}$ and the third invariant of the stress deviator tensor $J_3' = det(\underline{\sigma})$ chosen in dimensionless form as follows:

$$y_\sigma = \frac{27\det(\underline{\sigma})}{2\overline{\sigma}^3} \tag{8.62}$$

what gives $-1 \leq y_\sigma \leq 1$.

With the introduction of the third invariant, it becomes possible to take into account the asymmetry between tension and compression (Gillet et al. 1996).

The choice of $g(y_\sigma)$ may be freely chosen, provided that it guarantees the convexity of the austenitic elastic domain.

By derivation of $\rho g^*(\underline{\sigma}) = \gamma \overline{\sigma} g(y_\sigma)$ versus $\underline{\sigma}$, one obtains:

$$\underline{\kappa} = \overline{\underline{\kappa}} + \overline{\overline{\underline{\kappa}}} \tag{8.63}$$

with:

$$\overline{\underline{\kappa}} = \gamma \sqrt{\frac{3}{2}} g(y_\sigma) \underline{\mathbf{N}} \tag{8.64}$$

$$\overline{\overline{\underline{\kappa}}} = \gamma \sqrt{6} \frac{dg(y_\sigma)}{dy_\sigma} (\sqrt{6}(\underline{\mathbf{N}}^2 - \mathbf{1}/3) - y_\sigma \underline{N}) \tag{8.65}$$

and:

$$\underline{N} = \sqrt{\frac{2}{3}} \frac{\mathbf{s}}{\overline{\sigma}} \tag{8.66}$$

We obtain the equivalent transformation strain:

$$\overline{\varepsilon}^{tr} = \sqrt{\frac{2}{3} \underline{\varepsilon}^{tr} : \underline{\varepsilon}^{\mathrm{tr}}} = z \overline{\gamma}(y_\sigma) \tag{8.67}$$

with:

$$\overline{\gamma}(y_\sigma) = \gamma \sqrt{(g(y_\sigma))^2 + 9(1 - y_\sigma^2) \left(\frac{dg(y_\sigma)}{dy_\sigma}\right)^2} \tag{8.68}$$

This formulation was verified on a micromechanical model by Aleong et al. (2002).

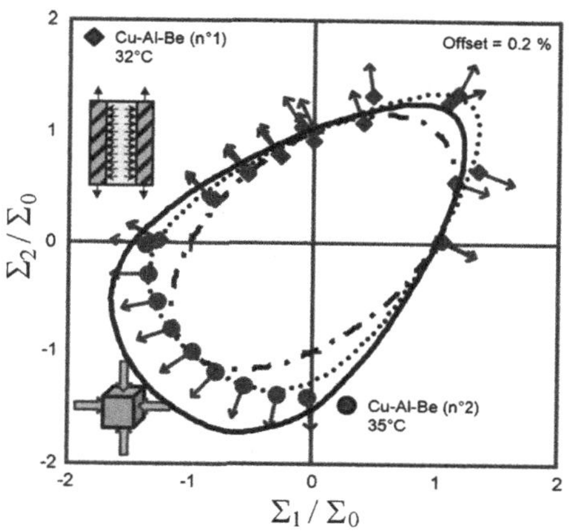

Fig. 8.23 Curve of the boundary of the elastic austenitic domain obtained by way of biaxial mechanical tests performed on Cu–Al–Be (Bouvet et al. 2002)

We have chosen two functions $g(y_\sigma)$:
– a linear one:

$$g(y_\sigma) = 1 + a y_\sigma \tag{8.69}$$

which guarantees the convexity of $g^*(\underline{\sigma})$ if and only if $0 \leq a \leq 1/8$,
– the second one under the form cos(arccos), introduced by Bouvet et al. (2002):

$$g(y_\sigma) = \cos(1/3 \arccos(1 - a(1 - y_\sigma)) \tag{8.70}$$

which guarantees the convexity of $g^*(\underline{\sigma})$ if and only if $0 \leq a \leq 1$.
The surfaces of phase transformation onset can be obtained by:

$$\rho g^*(\underline{\sigma} = \underline{\sigma}^{am}(z = 0)) = b(T - M_s^0) \tag{8.71}$$

The measurements show that the normality rule on the phase transformation initiation surface ($A \Rightarrow M$) is verified on Fig. 8.23.

8.5 Anisothermal Expansion

Leclercq and Lexcellent (1996), Lexcellent et al. (2006)

The framework is strictly the same as the R_L model, except that instead of a single internal variable z, we introduce two of them: the volume fraction of oriented martensite M_σ : (z_σ) and the volume fraction of self-accommodating martensite M_T : (z_T) as suggested by Brinson (1993).

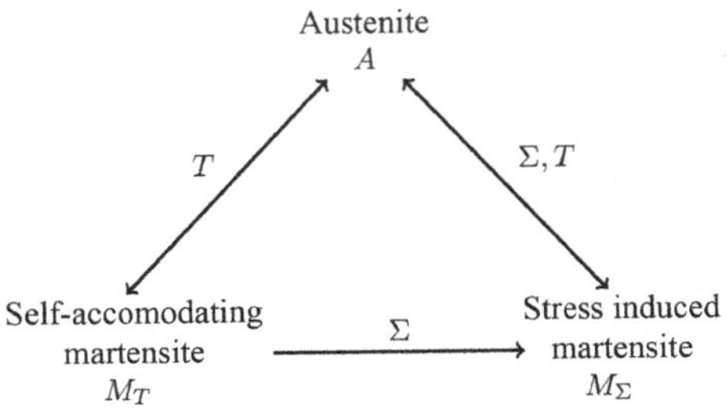

Fig. 8.24 Possible phase transformation (T: température; σ contrainte)

In the thermomechanical process, simple cooling of the austenite A gives rise to martensite M_T, and simple heating causes the reverse transformation $M_T \Rightarrow A$. An isothermal mechanical stress at $T > M_s^0$ generates oriented martensite M_σ and the mechanical unloading, the reverse transformation $M_\sigma \Longrightarrow A$. The process of reorientation of the martensite variants, induces $M_T \Rightarrow M_\sigma$ under stress. Only the direct path $M_\sigma \Rightarrow M_T$ is not possible: this transformation has to pass through the intermediary austenitic phase A. Any thermomechanical stress, particularly if it is nonproportional, will cause a martensitic transformation coupled with a reorientation of the martensite variants (Fig. 8.24).

The Helmholtz free energy of the "three-phases" system is written:

$$\psi(\underline{\varepsilon}, T, z, z_T\, z_\sigma) = (1 - z)\psi^{(1)} + z_T \psi^{(2)} + z_\sigma \psi^{(3)} + \Delta\psi \tag{8.72}$$

with $\alpha = 1$ for austenite A, $\alpha = 2$ for M_T and $\alpha = 3$ for M_σ
and also:

$$z = z_T + z_\sigma \tag{8.73}$$

where z represents the overall volume fraction of martensite (and $(1 - z)$ that of austenite). z is separated into z_T – the volume fraction of self-accommodating martensite M_T – and z_σ – the volume fraction of oriented martensite M_σ.

A "reasonable" expression of $\Delta\psi$, which accounts for both the interaction between the austenite and martensite and the one between the two martensite variants M_T *et* M_σ is written as:

$$\Delta\psi = z(1 - z)\psi_{it} + z_T z_\sigma \psi_{it}^m \tag{8.74}$$

where ψ_{it} *and* ψ_{it}^m are considered to be constant.

Let us consider that:

$$\rho\psi^\alpha(\underline{\varepsilon}_\alpha, T) = u_\alpha^0 - Ts_\alpha^0 + \frac{1}{2}(\underline{\varepsilon}_\alpha - \underline{\varepsilon}_\alpha^{\mathrm{tr}} - \underline{\varepsilon}_\alpha^{\mathrm{th}})\underline{\underline{L}}(\underline{\varepsilon}_\alpha - \underline{\varepsilon}_\alpha^{\mathrm{tr}} - \underline{\varepsilon}_\alpha^{\mathrm{th}}) + C_v\left(T - T_0 - \ln\left(\frac{T}{T_0}\right)\right) \tag{8.75}$$

with:

$$\underline{\sigma}_\alpha = \rho \frac{\partial \psi_\alpha}{\partial \underline{\varepsilon}_\alpha} = \underline{\underline{L}} : (\underline{\varepsilon}_\alpha - \underline{\varepsilon}_\alpha^{\text{tr}} - \underline{\varepsilon}_\alpha^{\text{th}}) = \underline{\underline{L}} : (\underline{\varepsilon}_\alpha^{el}) \tag{8.76}$$

We shall extend what we wrote for the two-phase case ($\alpha = 1, 2$) to the three-phase case ($\alpha = 1, 2, 3$).

- 1st hypothesis:

$$\underline{\varepsilon}_\alpha^{\text{th}} = \underline{\varepsilon}_1^{\text{th}} = \underline{\varepsilon}_2^{\text{th}} = \underline{\varepsilon}_3^{\text{th}} = \underline{\varepsilon}^{\text{th}} = \alpha(T - T_0)\underline{1} \tag{8.77}$$

- 2nd hypothesis:

$$\underline{\varepsilon}_1^{\text{tr}} = \underline{\epsilon}_2^{\text{tr}} = 0,\ \underline{\varepsilon}_3^{\text{tr}} = \underline{\kappa} \tag{8.78}$$

- 3rd hypothesis:french

$$\underline{\varepsilon} = (1 - z)\underline{\varepsilon}_1 + z_T \underline{\varepsilon}_2 + z_\sigma \underline{\varepsilon}_3 \tag{8.79}$$

- 4th hypothesis: the stress state is independent of the phase state:

$$\underline{\sigma}_\alpha = \underline{\sigma}_1 = \underline{\sigma}_2 = \underline{\sigma}_3 = \underline{\sigma} \Longrightarrow \underline{\varepsilon}_\alpha^{el} = \underline{\varepsilon}_1^{el} = \underline{\varepsilon}_2^{el} = \underline{\varepsilon}_3^{el} = \underline{\varepsilon}^{el} \tag{8.80}$$

$$\underline{\varepsilon} = \underline{\varepsilon}^{\text{el}} + \underline{\kappa} z_\sigma + \alpha(T - T_0)\underline{1} \tag{8.81}$$

- and a 5th natural hypothesis because the two martensite variants have the same crystallographic phase, i.e., $u_0^2 = u_0^3$ and $s_0^2 = s_0^3$, which gives us the expression of the free energy:

$$\rho\psi(\underline{\varepsilon}, z, T) = u_1^0 - T s_1^0 - z\Pi_0^f(T) + \tfrac{1}{2}(\underline{\varepsilon} - \underline{\kappa} z_\sigma - \alpha(T - T_0)\underline{1})\underline{\underline{L}}(\underline{\varepsilon} - \underline{\kappa} z_\sigma - \alpha(T - T_0)\underline{1}) + C_v\left(T - T_0 - \ln\left(\frac{T}{T_0}\right)\right) + z(1 - z)\psi_{it} + z_T z_\sigma \psi_{it}^m \tag{8.82}$$

with:

$$\underline{\sigma} = \rho\frac{\partial \psi}{\partial \underline{\varepsilon}} = \mathbf{L} : (\underline{\varepsilon} - \underline{\kappa} z_\sigma - \alpha(T - T_0)\underline{1});\ s = -\frac{\partial \psi}{\partial T} \tag{8.83}$$

and the thermodynamic forces associated with the internal variables:

$$\Pi_\sigma^f(\underline{\sigma}, z, z_T, T) = -\frac{\partial \psi}{\partial z_\sigma} = \Pi_0^f(T) + \frac{\underline{\kappa} : \underline{\sigma}}{\rho} - (1 - 2z)\psi_{it}(T) - z_T \psi_{it}^m$$

$$\Pi_T^f(z, z_\sigma, T) = -\frac{\partial \psi}{\partial z_T} = \Pi_0^f(T) - (1 - 2z)\psi_{it}(T) - z_\sigma \psi_{it}^m$$

In reference to Chap. 2, the dissipation of mechanical origin is written:

$$dD = \Pi_\sigma^f dz_\sigma + \Pi_T^f dz_T \geq 0 \tag{8.84}$$

In the particular case of reorientation, e.g. $M_T \Rightarrow M_\sigma$, we have:

$$dz = 0 \Longrightarrow dz_T = -dz_\sigma \tag{8.85}$$

and the mechanical dissipation term is then written:

$$\Pi_{T\sigma}^f dz_\sigma \geq 0 \tag{8.86}$$

with:

$$\Pi_{T\sigma}^f = \Pi_\sigma^f - \Pi_T^f = \frac{\underline{\kappa} : \underline{\sigma}}{\rho} - (z_T - z_\Sigma)\psi_{it}^m \tag{8.87}$$

For any thermomechanical process in the field $(\underline{\sigma}, \underline{\varepsilon}, \mathrm{T})$, the Clausius–Duhem inequality must strictly be verified.

As before, we consider that for phase transformation processes, the onset criterion will consist in complying that the associated thermodynamic force reaches the value of 0. For instance, $A \Longleftrightarrow M_\sigma$ corresponds to $\Pi_\sigma^f = 0$ or $A \Longleftrightarrow M_T$ corresponds to $\Pi_T^f = 0$.

8.5.1 Kinetics of Phase Transformation or Reorientation

As we have just seen, the explanation of the free energy enables us to determine the state equations and the thermodynamic forces associated with each internal variable. However, this information is not sufficient to describe a dissipative process. Consequently, there are no thermodynamic relations stemming from the free energy which can explicit the kinetics of the volume fractions z_σ et z_T.

For any dissipative system, the missing equations can be obtained by introducing an additional thermodynamic function, namely a pseudo-potential of dissipation (Moreau 1970; Halphen and N'Guyen 1974) or criterion functions like in elastoplasticity

Leclercq and Lexcellent (1996). We assume the existence of five criterion functions Φ_σ^F Φ_σ^R, Φ_T^F, Φ_T^R and $\Phi_{T\sigma}$ associated with the forward transformation: $F : A \Rightarrow M_\sigma, A \Rightarrow M_T$ and with the reverse $R : M_\sigma \Rightarrow A, M_T \Rightarrow A$ and with the reorientation of martensite platelets $M_T \Rightarrow M_\sigma$.

In a formalism identical to that for plasticity, these functions Φ are constant during the phase transformation associated with them; and are given by:

$$\Phi_\sigma^F = \Pi_\sigma^f - k_\sigma^F = Y_\sigma^F \tag{8.88}$$

$$\Phi_\sigma^R = -\Pi_\sigma^f + k_\sigma^R = Y_\sigma^R \tag{8.89}$$

$$\Phi_T^F = \Pi_T^f - k_T^F = Y_T^F \tag{8.90}$$

$$\Phi_T^R = -\Pi_T^f + k_T^R = Y_T^R \tag{8.91}$$

$$\Phi_{T\sigma} = \Pi_{T\sigma}^f - k_{T\sigma} = Y_{T\sigma} \tag{8.92}$$

Y_α^F, Y_α^R *et* $Y_{T\Sigma}$ $(\alpha = \sigma \ or \ T)$ are non-negative constants, and k_α^F, k_α^R *et* $k_{T\sigma}$ (Leclercq and Lexcellent 1996) are functions which have the property of being null at the onset of phase transformation (be it forward or reverse) or during reorientation.

The consistency equations $d\Phi_\alpha^F{=}0, d\Phi_\alpha^R = 0, d\Phi_{T\sigma}{=}0$ can be used to obtain the expressions of dz_σ and dz_T. The functions k_α^F k_α^R and $k_{T\sigma}$ are chosen such that the classical kinetics used by metallurgists (Koistinen and Marburger 1959) are verified:

$$dz_\sigma = (1 - z_\Sigma)\left[\frac{\gamma a_\sigma^F}{\rho \Delta s} d\sigma_{ef} - a_\sigma^F b_\sigma^F (\exp(-b^F(T - M_S^0))dT\right] \quad for \ A \Rightarrow M_\sigma \tag{8.93}$$

$$dz_T = (1 - z_T) a_T^F dz_T \quad for \ A \Rightarrow M_T \tag{8.94}$$

$$dz_\sigma = (1 - z_\sigma)\frac{\gamma a_{T\sigma}}{\rho \Delta s} d\sigma_{ef} \quad for \ M_T \Rightarrow M_\sigma \tag{8.95}$$

with $\sigma_{ef} = \overline{\sigma} g(y_\sigma)$ which is the effective stress.

Regarding the reverse phase transformation: $M_\sigma \Rightarrow A$, $M_T \Rightarrow A$, a unique criterion function can be built in the plane of the stress deviator for any proportional or radial loading (Lexcellent et al. 2006):

$$\Phi_\sigma^R = -\sigma_{ef} + \sigma_0^{ma}(z_\Sigma, z_T, T) \tag{8.96}$$

σ_0^{ma} being the threshold stress associated with the reverse transformation $M_\sigma \Rightarrow A$.

The use of the maximum dissipation principle with the aim of obtaining the complementary laws (e.g., the kinetics) requires a convex elastic domain in order to ensure its unicity. However, in the case of the reverse transformation, this elastic domain is not convex, which suggests the construction of a "non-associated" constitutive "framework" using a function $\chi(\sigma, z, z_\Sigma, T)$ which verifies:

$$\begin{aligned} \chi(\sigma, z, z_\Sigma, T) < 0 \ when \ \Phi_\sigma^R(\sigma, z, z_\Sigma, T) < 0 \\ \chi(\sigma, z, z_\Sigma, T) = 0 \ when \ \Phi_\sigma^R(\boldsymbol{\Sigma}, z, z_\Sigma, T) = 0 \end{aligned} \tag{8.97}$$

Such a function may be chosen, thus:

$$\chi(\underline{\sigma}, z, z_\Sigma, T) = -\underline{\sigma} : \frac{\underline{\varepsilon}^{tr}}{\varepsilon_{\text{ef}}^{\text{tr}}} + \sigma_0^{ma}(z_\sigma, z_T, T) \tag{8.98}$$

with $\varepsilon^{tr}_{ef} = \gamma z_\sigma$ and therefore:

$$\frac{d\underline{\varepsilon}^{\mathrm{tr}}}{dt} = \gamma \frac{dz_\sigma}{dt} \frac{\underline{\varepsilon}^{tr}}{\varepsilon^{tr}_{ef}} \ i.e.\ \underline{\kappa} = \rho \frac{\partial}{\partial \underline{\sigma}}(-\frac{\gamma}{\rho}\chi)\ et\ \varepsilon^{tr} = \underline{\kappa} z_\sigma \tag{8.99}$$

As before, dz_σ derives from the consistency condition $d\Phi^R_\sigma = 0$:

$$dz_\sigma = z_\sigma \left[\frac{\gamma a^R_\sigma}{\rho \Delta s} d\sigma_{ef} - a^R_\sigma b^R_\sigma (\exp(-b^R(T - M^0_S))dT \right] \quad pour\ M_\sigma \Rightarrow A \tag{8.100}$$

The condition $d\Phi^R_T = 0$ allows us to obtain dz_T:

$$dz_T = -a^R_T z_T dT \tag{8.101}$$

The parameters are identified from tensile curves (or compression curves), the measurements of the anisothermal curves, in the stress-free state, and also the Clapeyron diagram (Σ_{ef}, T) (Fig. 8.25). This diagram is built on the basis of experimental observations: under low stresses, it is difficult to separate the stress-induced martensite M_σ from purely thermally induced martensite M_T. In order to care with this problem in the space (σ_{ef}, T), a domain of self-accommodating martensite is defined at $T < M^f_0$ in addition to the two domains of austenite and oriented martensite.
For the practical identification of the parameters:
– the measurements of M^s_0 and A^s_0 lead to two equations:

$$\Pi^f_T(\underline{\sigma} = \underline{0}, \mathrm{T} = \mathrm{M}^s_0, z_T = 0,\ z_\sigma = 0) = (\Delta u - u_0) - (\Delta s - s_0) M^s_0 = 0\ for\ A \Rightarrow M_T \tag{8.102}$$

$$\Pi^f_T(\underline{\sigma} = \underline{0}, \mathrm{T} = \mathrm{A}^s_0, z_T = 1,\ z_\sigma = 0) = (\Delta u + u_0) - (\Delta s + s_0) A^s_0 = 0\ for\ M_T \Rightarrow A \tag{8.103}$$

– regarding the pseudoelastic behavior, the boundary values of initiation of the forward and reverse transformations can be obtained from the relationship: $\underline{\sigma} : \underline{\kappa} = \gamma \sigma_{ef}$:

$$\Pi^f_\sigma(\sigma_{ef}, \mathrm{T}, z_T = 0,\ z_\sigma = 0) = \Pi^f_\sigma(\sigma_0, \mathrm{T} = \mathrm{M}^s_0, z_T = 0,\ z_\sigma = 0)\ for\ A \Rightarrow M_\sigma \tag{8.104}$$

Thus:

$$\sigma_{ef} - \sigma_0 - \rho \frac{(\Delta s - s_0)}{\gamma}(T - M^s_0) = 0 \tag{8.105}$$

or by derivation:

$$\frac{d\sigma_{ef}}{dT} = \rho \frac{(\Delta s - s_0)}{\gamma} = C_M \tag{8.106}$$

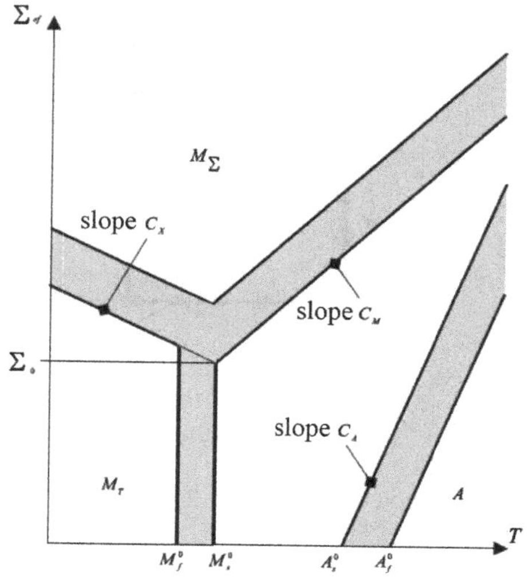

Fig. 8.25 Clausius–Clapeyron diagram in the effective stress/temperature plane

C_M can be obtained from the measurement of the transformation onset stresses for different isothermal tensile curves. $\sigma^{am_\sigma}(T)$ (Fig. 8.25).

Similarly, for the reverse transformation $M_\sigma \Rightarrow A$ we have:

$$\Pi^f_\sigma(\sigma_{ef}, \mathrm{T}, z_T = 0,\ z_\sigma = 1) = \Pi^f_\sigma(0, \mathrm{T} = \mathrm{A}^s_0, z_T = 0,\ z_\sigma = 1) \tag{8.107}$$

Thus,

$$\sigma_{ef} - \rho\frac{(\Delta s + s_0)}{\gamma}(T - A^s_0) = 0 \tag{8.108}$$

or by derivation:

$$\frac{d\sigma_{ef}}{dT} = \rho\frac{(\Delta s + s_0)}{\gamma} = C_A \tag{8.109}$$

C_A is also determined experimentally (Fig. 8.25).

With regard to the process of reorientation, the stress for reorientation initiation is governed by the following conditions:

$$\Pi^f_{\sigma T}(\sigma_{ef}, \mathrm{T}, z_T = 1,\ z_\sigma = 0) = \Pi^f_{\sigma T}(\sigma_0, \mathrm{T} = \mathrm{M}^s_0, z_T = 1,\ z_\sigma = 0) \tag{8.110}$$

or by derivation:

$$\sigma_{ef} - \Sigma_0 - \rho\frac{s^m_0}{\gamma}(T - M^s_0) = 0 \tag{8.111}$$

thus:

$$\frac{d\sigma_{ef}}{dT} = -\rho\frac{s^m_0}{\gamma} = -C_X \tag{8.112}$$

In addition, at every point of the reorientation onset, we have $\Pi^f_{\sigma T} = 0$, so that:

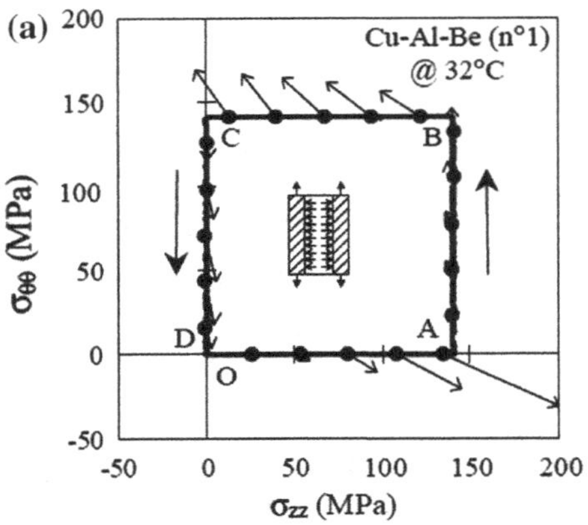

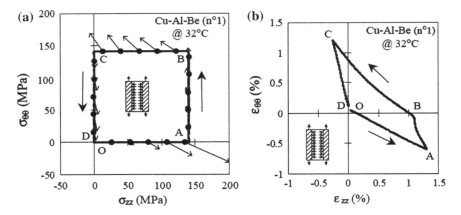

Fig. 8.26 « Square » trajectory imposed under traction/internal pressure (**a**); response in terms of strain (**b**) (Bouvet et al. 2002)

$$\Pi^{f}_{\sigma T}(\sigma_0, \mathrm{T} = \mathrm{M}^{\mathrm{s}}_0, z_T = 1,\ z_\sigma = 0\,) = \sigma_0 - \frac{\rho}{\gamma}(u^m_0 - M^s_0 s^m_0) = 0 \qquad (8.113)$$

All these equations can be used to estimate the values of the six parameters of the model: $\triangle u_0$, $\triangle s_0$, u_0, s_0, u^m_0, s^m_0, with γ being measured on one of the experimental curves.

8.5.1.1 Application of the R_L Model

So far, no developed model is capable of truly modeling nonproportional loads. However, in Bouvet's PhD, square and quarter-circle trajectories were carried out and made available for all (Figs. 8.26 and 8.27).

8.5.2 Criticism of the R_L approach

This approach, while it is very useful, does not really account for the microstructure or the changes in that microstructure during the thermomechanical loading. It assumes that the phase transformation is total, which is clearly not the case for polycrystals, because depending on their orientation in relation to the mechanical stress, some grains are not, or are scarcely, transformed!

Roughly speaking, there are phenomenological models inspired by plasticity models such as the R_L approach or the model developed by Aurrichio and Petrini (2004),Aurrichio et al. (1997).

Other models such as those of Panico and Brinson (2007), Sadjadpour and Bhattacharya (2007) and Chemisky et al. (2011) introduce internal variables in order to

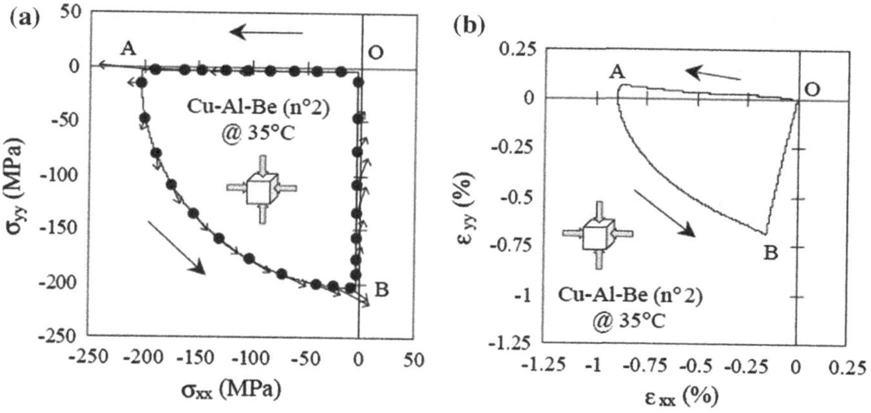

Fig. 8.27 « Quarter-circle » trajectory imposed under bi-compression (**a**); response in terms of strain (**b**) (Bouvet et al. 2002)

incorporate certain microscopic data. To this end, Chemisky et al. (2011) give an exhaustive review of these models including information about the microstructure.

Finally, there are other approaches such as the elasto-hysteresis models. These hereditary models "with discrete memory" put forward by Guelin et al. (1976) will also be examined. They are different in essence; as, it is not strains (elastic, transformation, etc.) which are added, but stresses (elastic, anelastic).

8.6 Design of SMAs Elements

8.6.1 Introduction

A simplified "strength of materials"-type method is used to perform calculations for bars or beams subjected to bending stress, torsion, etc.

For instance, a simple spring works mainly under torsion. Therefore, we need to know the mechanical response of an SMA element subjected to external mechanical loadings, causing stress and strain gradients, as the case of bending or torsion.

In order to perform calculations for slender structures made of SMAs, such as beams, bars or plates, there is a simplified method equivalent to the strength-of-materials method for elastic bodies (Gillet et al. 1996) or finite element method (FEM) calculations (Boubakar et al. 1998; Vieille et al. 2001).

8.6.2 *"Strength of Materials"-Type Calculations for Beams Subject to Bending or Torsion (Rejzner and Lexcellent 1999)*

8.6.2.1 Beam with a Rectangular Cross-section, Subjected to Pure Bending: Theoretical Study

A schematic curve of pseudoelastic behavior of a Ni–Ti alloy under isothermal tension can be described as follows (Fig. 8.28).

This pseudoelastic behavior without hardening can be qualified as "ideal".

Let us consider a rectangular beam with a cross-section area S: height $2h$ (**y** axis), width b (**z** axis), length $L = 2l$ (**x** axis) (Fig. 8.29).

The only mechanical stress consists of a surface forces density $\vec{T}\,(\text{l, y, z})\,\vec{x} = \sigma(l, y, z)\,\vec{\text{x}}$ whose resulting torsor$-M\mathbf{z}\; in\; x = l$ and another density of forces $\vec{T}\,(-l, y, z)(-\vec{x}) = \sigma(-l, y, z)(-\vec{x})$ whose resulting torsor$+ M\mathbf{z}\; in\; x = -l$

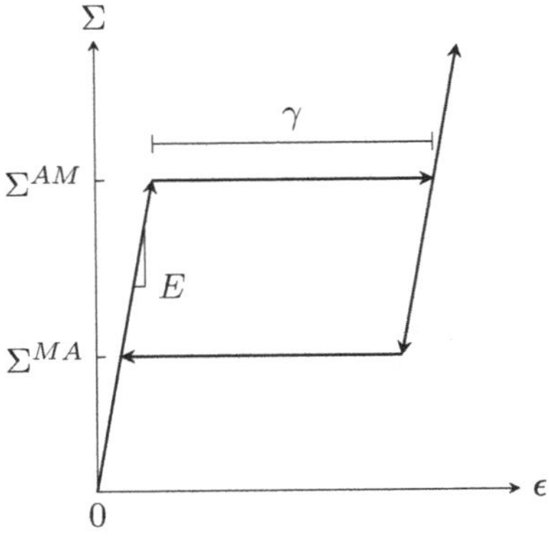

Fig. 8.28 Ideal pseudoelastic behavior of a NiTi-type shape-memory alloy

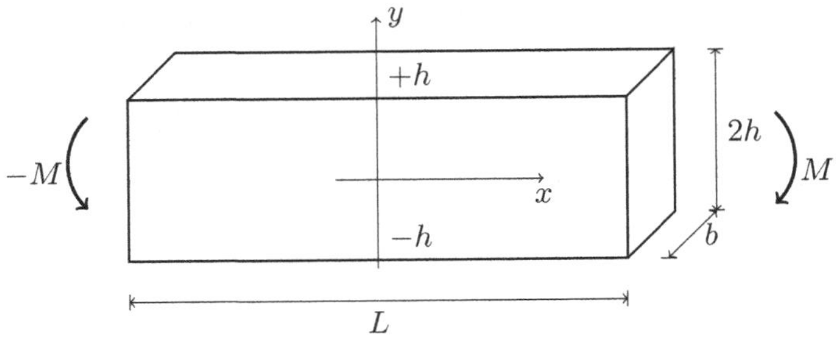

Fig. 8.29 Rectangular beam subject to pure bending

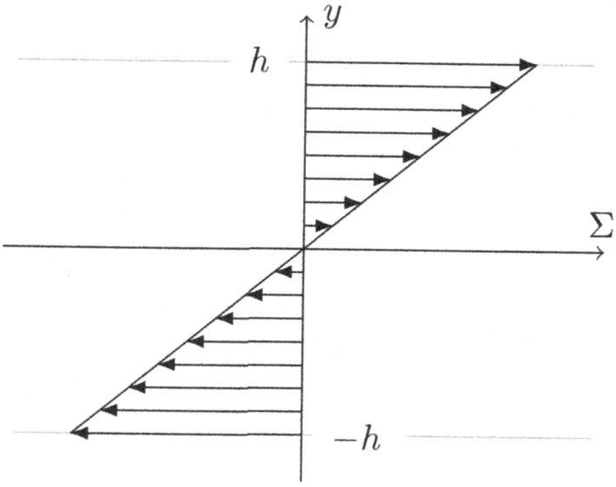

Fig. 8.30 Stress gradient associated with elastic loading

Thus,

$$M = \int\int_{x=cte} y\sigma_{xx} dydz \quad (8.114)$$

Elastic resolution (Fig. 8.30):

$$\varepsilon_{xx} = \frac{y}{R} = Ky \quad (8.115)$$

$$\sigma_{xx} = E\varepsilon_{xx} = EKy \quad (8.116)$$

where R is the curvature radius of the neutral fiber (in fact of the neutral plane $y = 0$) K ($= 1/R$), the curvature, and E the Young's modulus (Fig. 8.30).

Thus for $0 \leq \varepsilon = \varepsilon_{xx} \leq \varepsilon^{am}$, $0 \leq \sigma = \sigma_{xx} \leq \sigma^{am}$; the momentum M varies in a linear fashion with the curvature K (see Fig. 8.31):

$$M = \frac{2}{3} bh^3 EK \quad (8.117)$$

with a phase transformation onset value on both sides of the beam ($y = \pm h$)

$$M^{am}(z = 0) = \frac{2}{3} bh^2 \sigma^{am} = \frac{2}{3} bh^2 E\varepsilon^{am} \quad (8.118)$$

Pseudoelastic resolution of the biphase beam "A + M" (see Fig. 8.32).
For $\sigma = \sigma^{am}$, $\varepsilon^{am} \leq \varepsilon \leq \varepsilon^{am} + \gamma$
If $2h_p$ represents the thickness of the austenitic core, then:

$$\sigma = \left| \begin{array}{l} \frac{\sigma^{am} y}{h_p} \; pour \; 0 \leq y \leq h_p \\ \sigma^{am} \; pour \; h_p \leq y \leq h \end{array} \right. \quad (8.119)$$

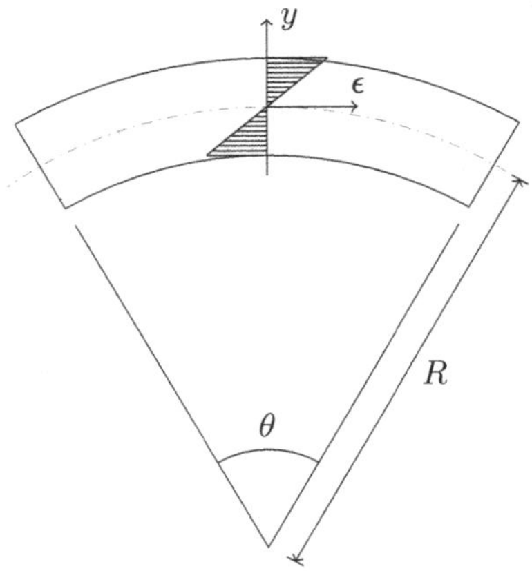

Fig. 8.31 Visualization of the elastic curvature of the rectangular beam

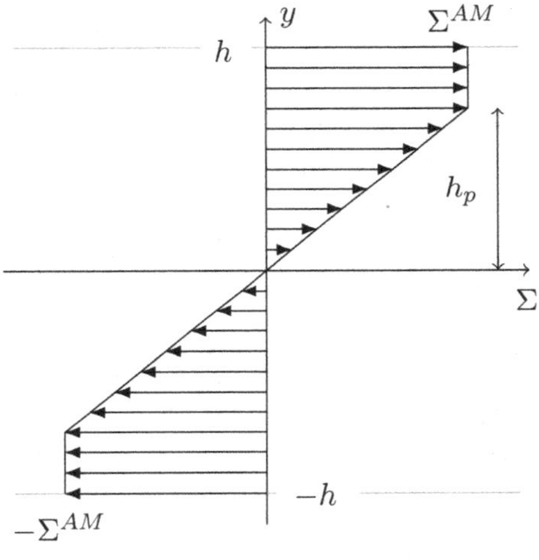

Fig. 8.32 Stress gradiant associated with pseudoelastic loading

This gives

$$M = 2b\left(\int_0^{h_p} \frac{\sigma^{am} y^2}{h_p} dy + \int_{h_p}^{h} \sigma^{am} y dy\right) \tag{8.120}$$

$$M = bh^2\sigma^{am}\left(1 - \frac{1}{3}\left(\frac{h_p}{h}\right)^2\right) = \frac{3}{2}M^{am}(z = 0)\left(1 - \frac{1}{3}\left(\frac{h_p}{h}\right)^2\right) \tag{8.121}$$

$$M = M_L\left(1 - \frac{1}{3}\left(\frac{h_p}{h}\right)^2\right) \tag{8.122}$$

h_p can be calculated as a function of K:

$$h_p = \frac{\sigma^{am}}{E} \frac{1}{K} \tag{8.123}$$

Thus, there is a nonlinear relation between M and K (see Fig. 8.33):

$$M = \frac{3}{2} M^{am}(z=0) \left(1 - \frac{1}{3h^2} \left(\frac{\sigma^{am}}{E} \right)^2 \frac{1}{K^2} \right) \tag{8.124}$$

with the limit momentum M_L defined by:

$$M_L = \frac{3}{2} M^{am}(z=0) \tag{8.125}$$

The momentum at the end of transformation $M^{am}(z=1)$ can be evaluated when $\varepsilon = \varepsilon^{am} + \gamma$ for $y = \pm h$:

$$M^{am}(z=1) = \frac{3}{2} M^{am}(z=0) \left(1 - \frac{1}{3} \left(\frac{\varepsilon^{am}}{\varepsilon^{am} + \gamma} \right)^2 \right) \tag{8.126}$$

The Fig. 8.34 gives the evolution of $^{h-h_p}/_h$ as function of the ratio $^{M}/_{M_l}$.

The stress release is more complex to define. While the loading stress-strain curve is unique, it is not true for unloading. Indeed, the response curve is different for two samples containing different volume fractions of martensite.

8.6.2.2 Solving Pure Torsion Problem: Relation Between the Twisting Torque C and the Unitary Angle of Torsion α

From the curve of ideal pseudoelastic traction, the threshold shear transformation stresses τ^{am} (τ^{ma}) and the associated shear transformation strain are extrapolated on the basis of the equivalent von Mises stress and strain:

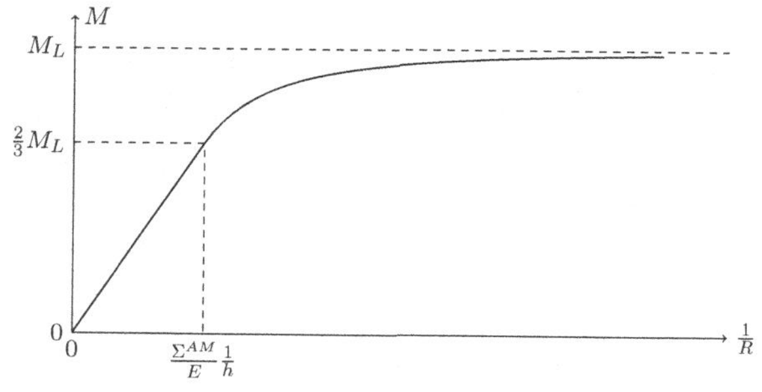

Fig. 8.33 Relation between the bending momentum M and the radius of curvature $K = 1/R$

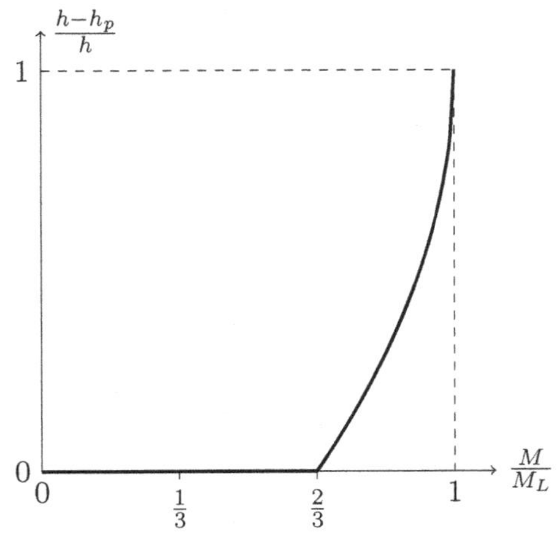

Fig. 8.34 evolution of $h-h_p/h$as function of the ratio M/M_l

$$\tau^{am} = \frac{\sigma^{am}}{\sqrt{3}}, \ \tau^{ma} = \frac{\sigma^{ma}}{\sqrt{3}}, \ \gamma_c = \sqrt{3}\gamma \tag{8.127}$$

The procedure is identical to the one used in bending, and will not be discussed in detail.

Let us consider a circular beam whose radius R and length l subject to two torques $\pm C\mathbf{z}$ on its end surfaces ($z = \pm l/2$).

In the context of a kinematic approach, the shear strain is written:

$$\varepsilon_{Z\theta} = \frac{\alpha r}{2} \tag{8.128}$$

where α is the unitary angle of torsion.

The twisting torque C is conventionally defined by

$$C = \int\int_{z=cte} r\sigma_{z\theta} r dr d\theta \tag{8.129}$$

Under loading:

- elastic torsion of the austenitic beam:

$0 \leq \varepsilon_{z\theta} \leq \varepsilon_c^{am}, 0 \leq \sigma_{z\theta} \leqq \tau^{am}$ $(\sigma_c^{am} = \frac{\tau^{am}}{2G})$

The torque C varies in a linear fashion with α:

$$C = \frac{\pi R^4}{2} G\alpha \tag{8.130}$$

where G is the torsion shear modulus or Coulomb's modulus with a value of transformation onset torque $C^{am}(z = 0)$ in $r = R$, so that:

$$C^{am}(z = 0) = \frac{\pi R^3}{2}\tau^{am} \tag{8.131}$$

- pseudoelastic torsion of the biphased A + M beam:

$\sigma_{z\theta} = \tau^{am}\ \varepsilon_c^{am} \leq \varepsilon_{z\theta} \leq \varepsilon_c^{am} + \gamma_c$
If $2r_p$ is the thickness of the austenitic core of the cylinder, then

$$\sigma_{z\theta} = \left| \begin{array}{l} \frac{\tau^{am} r}{r_p}\ pour\ 0 \leq r \leq r_p \\ \tau^{am}\ pour\ r_p \leq r \leq R \end{array} \right. \tag{8.132}$$

$$C = 2\pi \left(\int_0^{r_p} \frac{\tau^{am}}{r_p} r^3 dr + \int_{r_p}^{R} \tau^{am} r^2 dr \right) \tag{8.133}$$

$$C = \frac{2}{3}\pi R^3 \tau^{am} \left(1 - \frac{1}{4}\left(\frac{r_p}{R}\right)^3\right) = \frac{4}{3}C^{am}(z = 0)\left(1 - \frac{1}{4}\left(\frac{r_p}{R}\right)^3\right) \tag{8.134}$$

r_p can be calculated as a function of α:

$$r_p = \frac{\tau^{am}}{G\alpha} \tag{8.135}$$

Thus, there is a nonlinear relation between C and α:

$$C = C^{am}(z = 0)\left(1 - \frac{1}{4R^3}\left(\frac{\tau^{am}}{G}\right)^3 \frac{1}{\alpha^3}\right) \tag{8.136}$$

$\varepsilon_{z\theta} = \varepsilon_c^{am} + \gamma_c\ for\ r = R$:
The momentum at the end of phase transformation $C^{am}(z = 1)$ can be evaluated when $\varepsilon_{z\theta} = \varepsilon_c^{am} + \gamma_c\ for\ r = R$:

$$C^{am}(z = 1) = \frac{4}{3}C^{am}(z = 0)\left(1 - \frac{1}{4}\left(\frac{\varepsilon_c^{am}}{\varepsilon_c^{am} + \gamma_c}\right)^3\right). \tag{8.137}$$

8.7 Case Studies

8.7.1 *Study of the Flexion of a Prismatic Bar Subjected to a Punctual Force*

Statement:

Let us consider a prismatic bar of Ni–Ti (considering the constitutive law chosen) with a rectangular section (height $2h$ along $\mathbf{y}$; width b along $\mathbf{z}$) subjected to a punctual force $-P\mathbf{y}$ at the center of inertia of the end section $x = 1$. The section $x = 0$ is anchored (see Fig. 8.35).

The material behavior can be considered as « perfectly pseudoelastic » (Fig. 8.36).
(1) As P increases: give the coordinates of the points where the phase transformation begins. Calculate the corresponding value of P which will be denoted P^{am}: P^{am} will be calculated as a function of σ^{am}, b *and* h.
(2) Let us consider $P \geq P^{am}$, draw the profile of stress at the anchored point ($x = 0$). Let $2h_p(0)$ be the residual thickness of austenite in $x = 0$. Calculate P as a function of P^{am}, b, h and of the $h_p(0)/h$ ratio.
(3) For a section with fixed x ($0 \leq x \leq l$), calculate $h_p(x)$. Deduce the geometric shape of the austenitic domain.

For a fixed $P \geq P^{am}$, plot h_p as a function of x.

NOTE. In this problem, we ignore the effect of the shearing force associated with P, and therefore the associated shear stresses. If we take into account these factors, how might the problem be viewed as a new one?

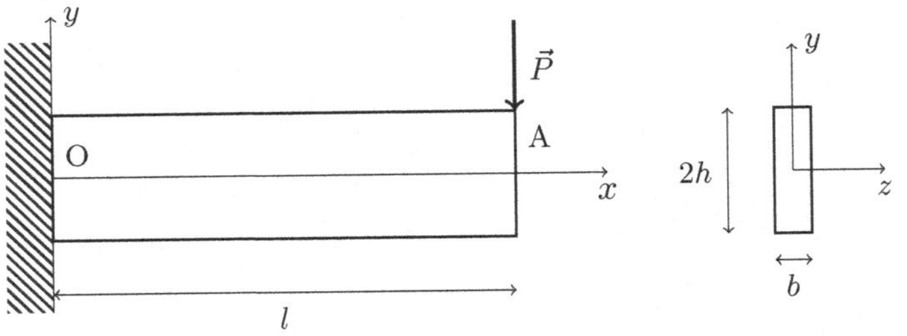

Fig. 8.35 Bending of a bar under the influence of a punctual force

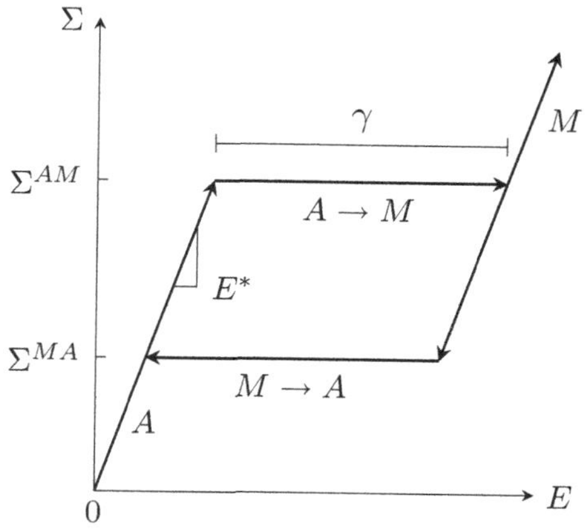

Fig. 8.36 Traction curve «without hardening» but with hysteresis

8.7.1.1 Brief Solution

"Elastic" bending stress:

$\sigma_{xx} = \frac{P(l-x)y}{I}$ where I is the quadratic bending momentum $I = \frac{2bh^3}{3}$

$\sigma^{am} = \mid \sigma_{xx}(\text{x=0, y=}\pm h) \mid = \frac{P^{am} l h}{I}$

$\Longrightarrow P^{am} = \frac{2}{3} \frac{bh^2}{l} \sigma^{am}$

$(2) P = \frac{3}{2} P^{am} \left(1 - \frac{1}{3}\left(\frac{h_p(x=0)}{h}\right)^2\right)$

$h_p(x) = \left(3h^2 - \frac{2h^2 P}{P^{am}}(l-x)\right)^{\frac{1}{2}}$

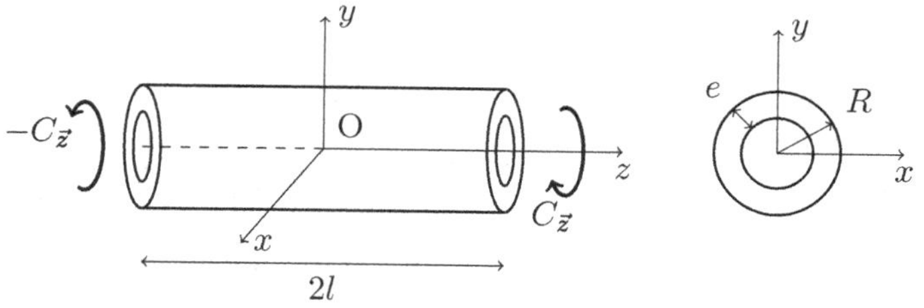

Fig. 8.37 Slender tube subjected to twisting torques

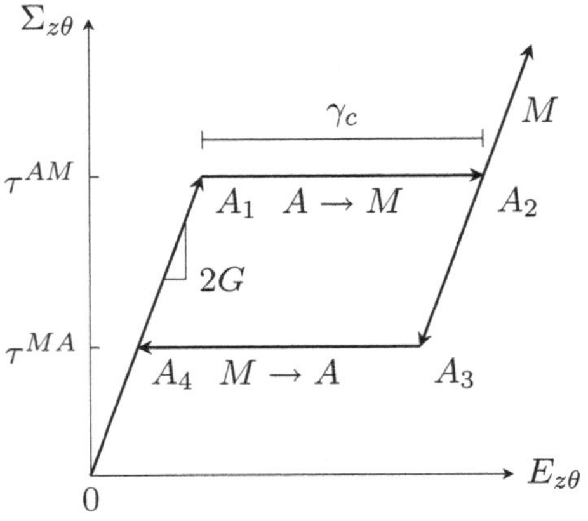

Fig. 8.38 Curve for torsion « without hardening » but with hysteresis

8.7.2 Slender Tube Subjected to Twisting Torques

8.7.2.1 Statement

Let us consider a slender tube (whose thickness e is very small in comparison to the average radius R) of length $2l$ (Fig. 8.37). At temperature T and in the stress-free state, the material is in the austenitic state ($T > A_0^f$).

The beam is subjected to a twisting torque $C\mathbf{z}$ *at* $z = l$; ($-C\mathbf{z}$ at $z = -l$).

The threshold stress for the direct transformation is τ^{am} *for* $A \Rightarrow M$ and τ^{ma} *for* $M \Rightarrow A$.

Let us also consider the behavior to be perfect pseudoelastic (i.e., without hardening) with a Coulomb's modulus G and the shear strain associated with the complete transformation γ_c (see Fig. 8.38).

(1) Classic elastic solution: with the slender tube hypothesis (e << R), calculate the quadratic momentum under torsion I_0 and the torsion stress $\sigma_{Z\theta}$ (assumed to be the same throughout the section of the tube). Also, find the strain associated with the torsion $\varepsilon_{z\theta}$. Deduce the relationship between the torque C and the unitary angle of torsion α.

(2) Calculate the value of the torque $C^{am}(z = 0)$ corresponding to the start of transformation in the tube and the corresponding angle α^{am}(z = 0) and α^{am}(z = 1).

(3) When $C \geq C^{am}(z = 1)$, what is the value of α? Plot the evolution of C with the change in α.

(4) Give a mechanical and physical explanation of what happens if:

(a) α is kept constant while the tube is heated;

(b) the torque C is eliminated without changing the temperature.

8.7.2.2 Brief Solution

(1) $I_0 = 2\pi R^3 e$, $\sigma_{z\theta} = \frac{C}{2\pi R^2 e}$ $\varepsilon_{z\theta} = \frac{\sigma_{z\theta}}{2G} = \frac{\alpha R}{2}$
$C = G I_0 \alpha$
(2) $C^{am}(z=0) = \frac{I_0 \tau^{am}}{R} = 2\pi R^2 e \tau^{am}$
$\alpha^{am}(z=0) = \frac{\tau^{am}}{GR}$, $\alpha^{am}(z=1) = \frac{1}{GR}(\tau^{am} + 2G\gamma_c)$.

8.7.3 *Study of a "Parallel" Hybrid Structure*

Statement:

Let us consider a Ni–Ti wire, with a pre-deformation γ in the martensitic state at temperature T_0.

This "straight" wire is embedded into an epoxy resin: resulting in a hybrid structure (Fig. 8.39).

Wire: medium A: Young's modulus E_A; volume fraction f_A.

Epoxy resin: medium B, whose behavior is considered to be elastic: Young's modulus E_B; volume fraction f_B.

Initial conditions: $t = 0$:

- medium A: $T = T_0$, $\varepsilon_0 = \gamma$, $z = 1$;
- medium B: considered to be "pre-deformed": $T = T_0$; $\varepsilon_0 = \gamma$.

(1) If we consider the resin/wire interface to be ideal, justify very briefly the modeling of the hybrid material as two "parallel" bars A and B.
(2) Write the equilibrium and compatibility under strain of the combination of the two bars.

(2.1) Write the expression of the global strain σ as a function $f_A, \sigma_A \cdot f_B\, \sigma_B$.

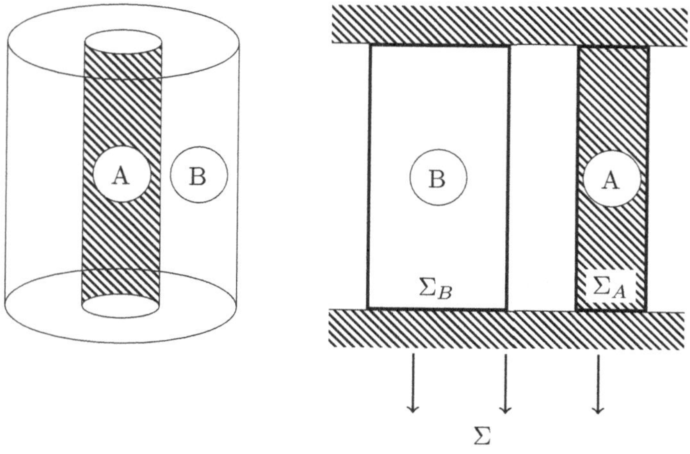

Fig. 8.39 Representation of a « parallel » hybrid structure

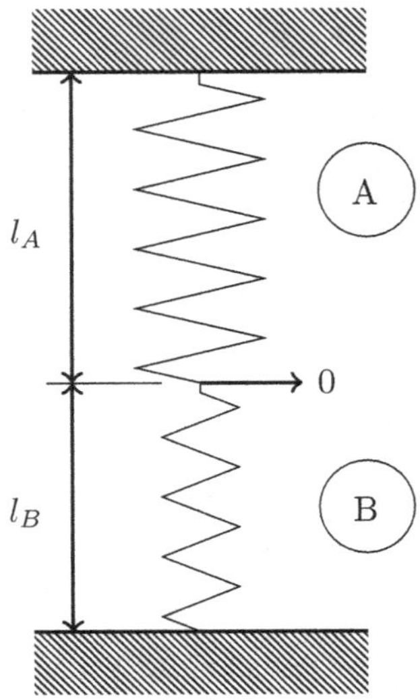

Fig. 8.40 Representation of a « series » hybrid structure

(2.2) Write $d\sigma_A$, $d\sigma_B$ and $d\varepsilon$ (the increment of total strain) as function of E_A, E_B, f_A, f_B et γ, $d\sigma$, dz.
(3) At $t = 0^+$, the Ni–Ti wire is heated by the Joule effect. Briefly describe the structure's physical response.

Give $\varepsilon(t)$ as a function of the parameters γ, E_A, E_B, f_A, f_B and z(t).

When $t \to +\infty$, give the yield value of $\varepsilon(t)$.

8.7.3.1 Brief Solution

(1) $d\varepsilon_A = d\varepsilon_B = d\varepsilon$
(2) $\sigma = f_A\sigma_A + f_B\sigma_B$

Incremental constitutive laws:

$d\sigma_A = E_A(d\varepsilon_A - \gamma dz)$

$d\sigma_B = E_B d\sigma_B$

(3) $d\varepsilon = \frac{f_A E_A}{f_A E_A + f_B E_B}\gamma dz$

$\int_0^t d\varepsilon = \varepsilon(t) - \gamma = \frac{f_A E_A}{f_A E_A + f_B E_B}\gamma(z(t) - 1)$

$\varepsilon(t = +\infty) = \frac{f_b E_b}{f_A E_A + f_B E_B}\gamma$

8.7.4 *Study of a "Series" Hybrid Structure*

8.7.4.1 Statement

Let us consider a hybrid structure consisting of two springs connected in a series (the structure is anchored at both ends) (Fig. 8.40).

- a NiTi spring, A, pre-deformed in the martensitic state ($z = 1$), with a value γ, length l_A and section S;
- a spring B with Young's modulus E_B and length l_B and section S.

Initial conditions at $t = 0$: $T = T_0$, A and B in the stress-free state.
(1) The Ni–Ti wire is heated. Briefly describe the structure's physical response.
(2) Evaluate $d\sigma_A$ as function of $d\sigma_B$
(3) Take f_A, f_B to be the volumetric proportions of A and B in relation to the total volume of both springs; evaluate $d\varepsilon$ as a function of f_A, f_B *et* $d\varepsilon_A$, $d\varepsilon_B$.
(4) Write, in the incremental form:

- the constitutive law for spring A;
- the constitutive law for spring B.

(5) Deduce then $d\varepsilon$ as function of $d\sigma$ et dz. What hypothesis can be made about $d\varepsilon$? Deduce $\sigma(t)$.
When $t \to \infty$, give the yield value of $\sigma(t)$.
(6) What might be used such a setup for?

8.7.4.2 Brief Solution

(1) $d\sigma_A = d\sigma_B = d\sigma$
(2) $d\varepsilon = f_A d\varepsilon_A + f_B d\varepsilon_B$
avec $d\varepsilon_A = \frac{d\sigma_A}{E_A} + \gamma dz$

$$dE_B = \frac{d\sigma_B}{E_B} \Longrightarrow d\varepsilon = \left(\frac{f_A}{E_A} + \frac{f_B}{E_B}\right) d\sigma + \gamma f_A dz$$

$$d\varepsilon = 0 \Longrightarrow d\sigma = -\frac{\gamma f_A dz}{\left(\frac{f_A}{E_A} + \frac{f_B}{E_B}\right)}$$

$$\int_0^t d\sigma = \sigma(t) - 0 = -\frac{\gamma f_A}{\left(\frac{f_A}{E_A} + \frac{f_B}{E_B}\right)} (z(t) - 1)$$

$$\sigma(t = +\infty) = \frac{\gamma f_A}{\left(\frac{f_A}{E_A} + \frac{f_B}{E_B}\right)}$$

Chapter 9
Behavior of Magnetic Shape Memory Alloys

Abstract The main advantage of magnetic shape memory alloys (MSMAs) over conventional shape memory alloys (SMAs) lies in the possibility for their actuation not only by stress or temperature but also by a magnetic field. This chapter is mainly devoted to the modeling of the thermo-magneto-mechanical behavior of single crystals under compressive loading (considering their fragility) and also under a magnetic field.

9.1 Introduction

The main advantage of magnetic shape-memory alloys (MSMAs) over conventional shape-memory alloys (SMAs), lies in the possibility for their actuation not only by stress or temperature, but also by a magnetic field.

MSMAs have a maximum deformation of around 6–10%, just like conventional SMAs, but the response times they exhibit are similar to those of magnetostrictive materials (around one millisecond). So far, research and applications are largely focused on two main materials. Both were developed in 1995–1996 in the United States, by a team at MIT with Ni-Mn-Ga (ULLAKKO et al. 1996) and a team at the University of Minnesota with FePd (James and Wuttig 1998). These materials exhibit the MSMA properties, the most widely used today is, without contest, Ni-Mn-Ga.

The article by Liang et al. (2006) details the three possibilities for performing magnetic actuation using magnetic shape-memory alloys. The first of these options consists of obtaining a phase transformation induced by a magnetic field; however, it requires very powerful magnetic fields and is inappropriate for actuation. The second possibility is the martensitic rearrangement induced by a magnetic field. Finally, the third possibility, referred to as the hybrid approach, consists in using a magnetic field gradient to create a sufficient force to lead to a phase transformation.

This chapter will be mainly devoted to the modeling of thermo-magneto-mechanical behavior of single crystals under compressive loading (considering their fragility) and also under a magnetic field.

C. Lexcellent, *Linear and Non-linear Mechanical Behavior of Solid Materials*, DOI 10.1007/978-3-319-55609-3_9

9.2 Cristallographie du Ni-Mn-Ga

As explained by Mullner et al. (2003), the mechanical properties depend on the microstructure of the material. Using X-ray diffraction measurements, coupled with transmission electron microscopy, it is possible: (i) to identify the austenitic and martensitic lattices, and thereby discover which crystallographic transformation is taking place, (ii) to measure the lattice parameters. This was performed on a sample of Ni-Mn-Ga by Ge et al. (2006).

We shall draw upon certain lessons learnt in Chap. 8, devoted to martensitic transformation.

The cubic austenite of side a_0 can be transformed in three quadratic martensite variants (sides *a et c*) (Fig. 9.1).

Figure 9.2 [ROU04] represents two adjoining martensite variants.

Figure 9.3 presents the general behavior of a MSMA. At high temperature, the alloy is in an austenitic state. After cooling, the sample contains the three martensite variants in equal proportions: no macroscopic deformation occurs as a result. The application of a compressive stress promotes the formation of the variant whose short axis is in the direction of that stress (M2 in Fig. 9.3) and thereby alters the geometric shape of the sample. The application of a magnetic field favors the variant whose axis of easy magnetization (identical to the short axis) is in the direction of that field (M1 in Fig. 9.3) and also alters the shape of the sample. The balance between these two effects can be used to create actuation.

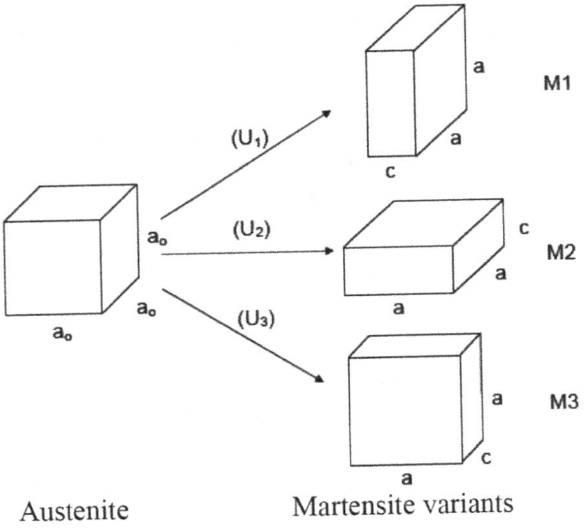

Fig. 9.1 The three martensite variants for a transformation of a cubic lattice to quadratic ones

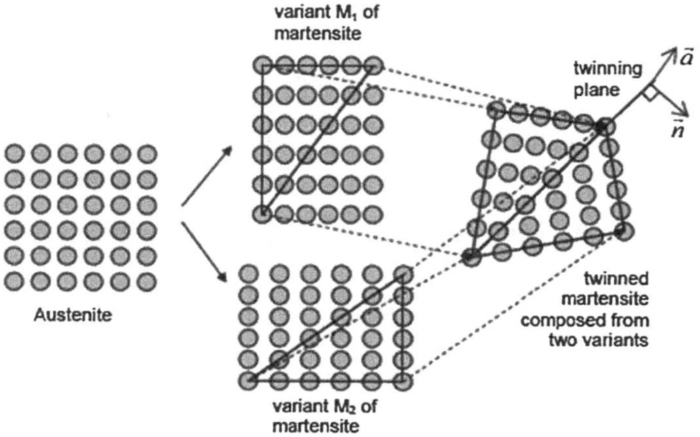

Fig. 9.2 A schematic two-dimension situation: one cubic (*left*) and two martensite variants M_1 and M_2, side by side

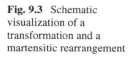

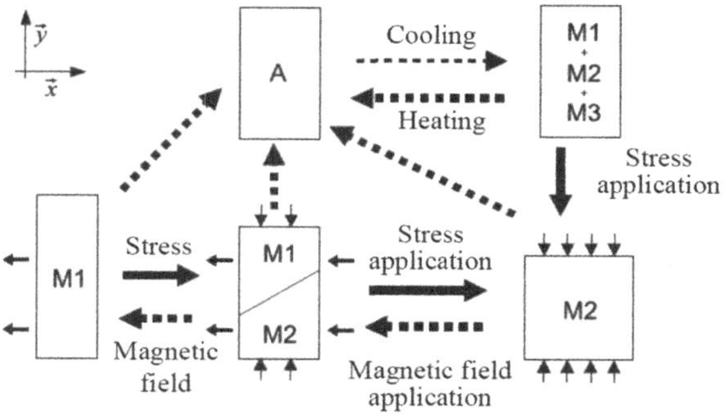

Fig. 9.3 Schematic visualization of a transformation and a martensitic rearrangement

9.2.1 Calculations of Microstructures

It is the mathematic theory of the martensitic transformation called CTM, by Ball and James that allows us to carry out these calculations. A few explanations can be found in the Chap. 3 of the book by C. Lexcellent «Shape-memory Alloys Handbook» (Wiley editor 2013).

Let's consider $\underline{F_k}$ the transformation gradient tensor of austenite A in the variant k of martensite $\underline{M}$:

$$d\overrightarrow{x}\,(M) = \underline{F_k}.d\overrightarrow{x^0}\,(A) \tag{9.1}$$

and the Green–Lagrange tensor is defined by:

$$\underline{E_k^{tr}} = \frac{1}{2}\left(\underline{{}^t\mathrm{F_k}.F_k} - \underline{1}\right) = \frac{1}{2}\left(\underline{U_k^2} - \underline{1}\right) \tag{9.2}$$

with three variants for the transformation cubic⟹quadratic:

$$\underline{U_1} = \begin{bmatrix} \beta & 0 & 0 \\ 0 & \alpha & 0 \\ 0 & 0 & \alpha \end{bmatrix}, \ \underline{U_2} = \begin{bmatrix} \alpha & 0 & 0 \\ 0 & \beta & 0 \\ 0 & 0 & \alpha \end{bmatrix} \ \underline{U_3} = \begin{bmatrix} \alpha & 0 & 0 \\ 0 & \alpha & 0 \\ 0 & 0 & \beta \end{bmatrix} \tag{9.3}$$

and $\alpha = a/a_0$, $\beta = c/a_0$. (See Fig. 9.1).

For the reorientation of the variant M_k *in the variant* M_l, the strain tensor is:

$$\underline{E_{kl}^{re}} = \frac{1}{2}\left(\underline{U_l^2} - \underline{U_k^2}\right) \tag{9.4}$$

One has to note that the interface between austenite and martensite can only exist as «twinned martensite» facing austenite (See Fig. 9.4).

Indeed, the CTM gives the solution to the «twinning equation»:

$$\underline{QU_i} - \underline{U_J} = \vec{a} \otimes \vec{n}. \tag{9.5}$$

Let us apply this to the cubic → quadratic transformation. We shall begin with variants 1 and 2. The calculations are as follows.

Let us consider a rotation matrix of 180° around the $\vec{e}$ axis denoted $\underline{R}$ and defined by:

$$\vec{e} = \frac{1}{\sqrt{2}}\begin{pmatrix} 1 \\ 1 \\ 0 \end{pmatrix} \tag{9.6}$$

It is easy to verify that $\underline{R.U_1.R} = \underline{U_2}$ and we obtain:

$$1.\ \vec{a} = \frac{\sqrt{2}(\beta^2 - \alpha^2)}{\beta^2 + \alpha^2}\begin{pmatrix} -\beta \\ \alpha \\ 0 \end{pmatrix}, \ \vec{n} = \frac{1}{\sqrt{2}}\begin{pmatrix} 1 \\ 1 \\ 0 \end{pmatrix} \tag{9.7}$$

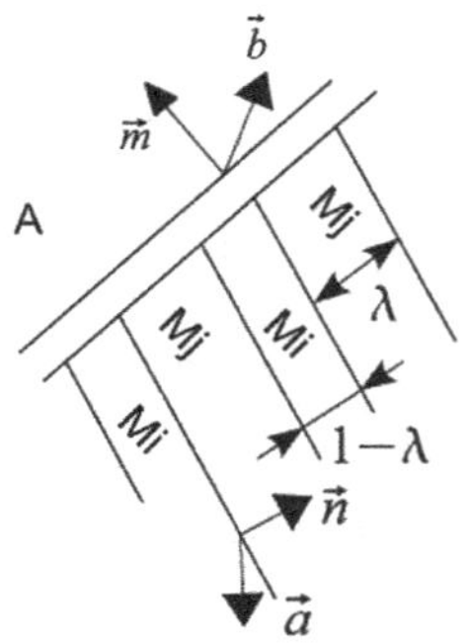

Fig. 9.4 «Twinned» martensite forming an interface with austenite

$$\mathbf{2.}\ \vec{a} = \frac{\sqrt{2}(\beta^2 - \alpha^2)}{\beta^2 + \alpha^2}\begin{pmatrix} -\beta \\ -\alpha \\ 0 \end{pmatrix},\ \vec{n} = \frac{1}{\sqrt{2}}\begin{pmatrix} 1 \\ -1 \\ 0 \end{pmatrix} \tag{9.8}$$

Note that these calculations are illustrated in Fig. 9.2.

If necessary, $\underline{Q}$ can be obtained using Eq. (9.5) (twinning equation).

The compatibility equation between austenite A and the couple of martensite variants M_i/M_j of proportion $(1-\lambda)$ and λ, respectively, is:

$$\underline{Q}'.\left(\lambda \underline{Q}.\underline{U_j} - (1-\lambda)\,\underline{U_i}\right) = \underline{1} + \vec{b} \otimes \vec{m} \tag{9.9}$$

We shall assume that the twinning equation for variants i and j has a solution. One again, Ball and James (1987, 1992) give the procedure in order to obtain the solution to the equation for the austenite/martensite interface.

(1) Calculate:

$$\delta = \vec{a}.\underline{U_i}.\left(\underline{U_i^2} - \underline{1}\right)^{-1}\vec{n} \tag{9.10}$$

and

$$\eta = tr\left(\underline{U_i^2}\right) - det\left(\underline{U_i^2}\right) - 2 + \frac{\|a\|^2}{2\delta}. \tag{9.11}$$

The equation for the austenite/martensite interface has a solution if and only if:

$$\delta \leq -2\ and\ \eta \geq 0 \tag{9.12}$$

(2) To find the solutions, calculate:

$$\lambda = \frac{1}{2}(1 - \sqrt{1 + \frac{2}{\delta}}) \tag{9.13}$$

For Ni_2-Mn-Ga $\alpha = {}^{a}/_{a_0} = 1.0188$ et $\beta = {}^{c}/_{a_0} = 0.9589$ one obtains $\lambda = 0.3083$ and:

▲for $M_1 \Longrightarrow M_2$

$$\underline{E}_{12}^{re} = \frac{1}{2}\left(\underline{U_2^2} - \underline{U_1^2}\right) = diag(0.0593, -0.593, 0) \tag{9.14}$$

▲for $A \Longrightarrow (M_1, M_2)$

$$\underline{E^{tr}} = \frac{1}{2}\left(\underline{U_{tr}^2} - \underline{1}\right). \tag{9.15}$$

with

$$\underline{U_{tr}} = \lambda \underline{U_2} + (1-\lambda)\,\underline{U_1}$$

$\underline{E^{tr}} = diag(-0.0224, 0.0004, 0.0190)$.

9.3 Model of the Magneto-Thermo-Mechanical Behavior of a Monocrystal of Magnetic Shape-Memory Alloy

As usual, we shall conduct our discussion within the context of thermodynamics of irreversible processes (TIP) for models with internal variables; the additional ingredient being magnetism.

9.3.1 *Expression of the Gibbs Free Energy Associated with Magneto-Thermo-Mechanical Loading*

We shall divide the Gibbs free energy G into four expressions: a chemical energy G_{chem}, a thermal energy G_{therm}, a mechanical energy G_{mech} and finally a magnetic energy G_{mag}. This free energy can therefore be expressed under the following form:

$$G(\underline{\sigma}, T, \overrightarrow{h}, z_0, z_1 \ldots z_{n,}, \alpha, \theta, \alpha_A) = G_{chem}(T, z_0) + G_{therm}(T) + G_{meca}(\underline{\sigma}, z_0, z_1 \ldots z_{n,}) + G_{mag}(T, \overrightarrow{h}, z_0, z_1 \ldots z_{n,}, \alpha, \theta, \alpha_A) \tag{9.16}$$

where the state variables are as follows:

- $\underline{\sigma}$ the stress tensor for the stresses applied to the sample;
- $\overrightarrow{h} = \mathrm{H}\overrightarrow{x}$: the magnetic field applied;
- T: the temperature.

The internal variables chosen are:

- z_0: the volume fraction of austenite;
- z_k: the volume fraction of the variant M_k of martensite ($k = 1 : : : 3$). We shall consider that the sample allows three martensite variants, which, by the crystallographic calculations demonstrated above, gives us;

with

$$\sum_{k=0}^{k=3} z_k = 1 \tag{9.17}$$

- α: the proportion of the Weiss domain within a REV of a martensite variant (see Fig. 9.5);
- α_A: the proportion of the Weiss domain within the austenite;
- θ: the angle of rotation of the magnetization vector $\overrightarrow{m}$ under the influence of a magnetic field $\overrightarrow{h}$ within a REV.

It should be noted, here, that no magneto-mechanical or thermo-mechanical energy term is present; the connection between the energies is obtained by an appropriate choice of internal variables.

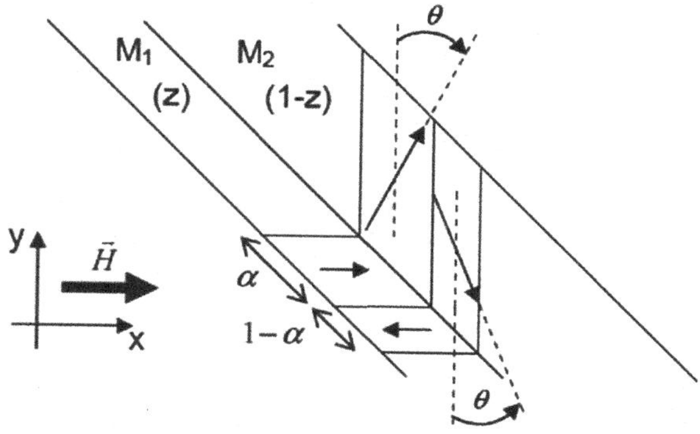

Fig. 9.5 Representative Elementary Volume with two variants M_1 and M_2 ($z_1 = z$, $z_2 = 1 - z$)

9.3.2 *Choice of the Representative Elementary Volume*

Clearly, in order to introduce internal variables, we need to define them. A choice of a representative elementary volume (REV) was made by Hirsinger and Lexcellent (2002) for a sample containing two variants M_1 and M_2 (see Fig. 9.5).

z represents the volume fraction of M_1 martensite and $(1 - z)$ to the volume fraction of M_2 martensite. Within each of the variants, two Weiss domains are represented, with respective fractions α and $1-\alpha$. θ corresponds to the angle between the direction of weak magnetization and the real direction of magnetization within M_2.

The Fig. 9.6 illustrates the evolution of the REV under the influence of a magnetic field $\vec{h} = H\vec{x}$. For $\vec{h} = \vec{0}$, the total magnetization of the sample is null, because the magnetization of one Weiss domain is compensated by the other domain ($\alpha = 0.5$) (Fig. 9.6a). When a relatively weak magnetic field is applied, the size of the Weiss domain whose axis of magnetization lies in the direction of the magnetic field increases, the consequence of which is that the domain size for the other variant also increases, due to the continuity of the magnetic flux. This growth continues until the other Weiss domain disappears completely ($\alpha = 1$) (Fig. 9.6b). When the field becomes stronger, two situations may occur. The first, illustrated in Fig. 9.6c, occurs when the sample is subjected to weak or no stress; a reorientation takes place, increasing the proportion of magnetization along the axis of the magnetic field. The second situation occurs when the sample is mechanically blocked, meaning that the alteration of the respective proportions of the two martensite variants is not possible. We then see a rotation of the magnetization so that it is orientated along the axis of the magnetic field applied (Creton 2004; Hirsinger et al. 2004).

However, a problem exists relativity to the definition of this REV in terms of the continuity of the magnetic flux when there is a rotation of the magnetization. Indeed, when the angle θ expands, the magnetic flux at the interface between two variants is reversed in M_2, but remains the same in M_1.

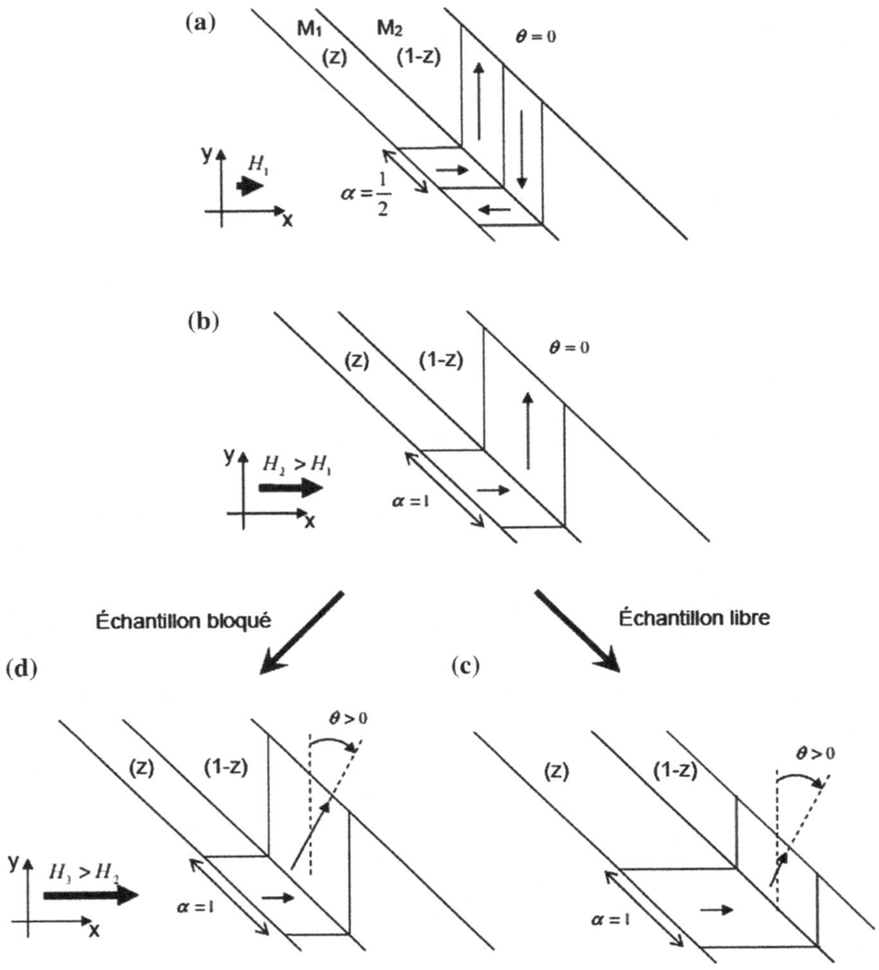

Fig. 9.6 Evolution of the representative elementary volume under the influence of a magnetic field

This detail shows that the REV chosen constitutes a mesoscopic approximation of more complex micromagnetic phenomena.

9.3.3 Expression of Chemical Energy

This energy is associated with the latent heat corresponding to a $A \Longrightarrow M$ phase transformation. As all the martensite variants form a single phase in the crystallographic sense of the word, they have the same chemical energy. With reference to z_0, the volume fraction of austenite, we can write the chemical contribution to the free energy:

$$G_{chem}(T, z_0) = \left(u_0^A - Ts_0^A\right) z_0 + \left(u_0^M - Ts_0^M\right)(1 - z_0) = u_0^M - Ts_0^M + z_0 \Pi_0^f(T) \tag{9.18}$$

with $\Pi_0^f(T) = \Delta u - T\Delta s$ and $\Delta u = u_0^A - u_0^M$; $\Delta s = s_0^A - s_0^M$.

This formulation is the same as for conventional SMAs.

9.3.4 Expression of Thermal Energy

If we make the hypothesis that the specific heats are the same, regardless of the state of phase of the material, by definition, we have:

$$C_p = -T\frac{d^2 G_{therm}}{dT^2}. \tag{9.19}$$

The expression of this energy obtained after integration is written:

$$G_{therm} = C_p\left[(T - T_0) - T\ln\left(\frac{T}{T_0}\right)\right]. \tag{9.20}$$

9.3.5 Expression of Mechanical Energy

For what remains in this chapter, we shall limit ourselves to the case where $\vec{h} = H\vec{x}$ and a uniaxial compression in direction $\vec{y}$ (Fig. 9.7), meaning that :

$$\underline{\sigma} = diag\,(0, \sigma, 0). \tag{9.21}$$

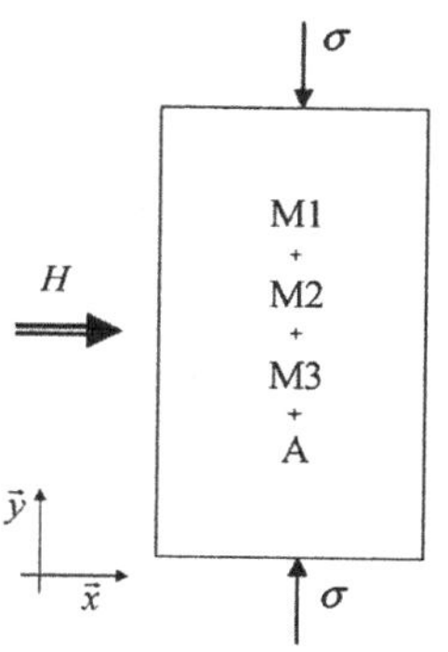

Fig. 9.7 Application of a magneto-mechanical stress to a Ni-Mn-Ga monocrystal (cubic-to-quadratic)

In this simple case, the equation of ρG_{meca} is reduced to:

$$\begin{aligned}\rho G_{meca}(\sigma,\ z_0\ ,z_1\ ,z_2\ z_3\ ) = &-\tfrac{\sigma}{2}\left[(z_1+z_3)\,(\alpha^2-1)+z_2(\beta^2-1)\right]\\ &-\tfrac{1}{2}\tfrac{\sigma^2}{E}+Az_0(1-z_0)+K(z_1z_2+z_2z_3+z_3z_1)\end{aligned} \tag{9.22}$$

where E is the Young's modulus, and considering the interactions between the martensite variants which all have the same weight:

$$K_{ij} = K\ \forall i\ ,\ j = 1\ ,2\ 3 \tag{9.23}$$

In addition, a restriction needs to be considered:

$$\sum_{k=0}^{k=3} z_k = 1. \tag{9.24}$$

This means that among the four volume fractions, only three are independent.

9.3.6 *Expression of Magnetic Energy*

9.3.6.1 Magnetization of Martensite: Axis of Easy Magnetization

$$m_1(H) = m_s(2\alpha(H) - 1) \tag{9.25}$$

where m_s is the saturation magnetization and $\alpha \in [0, 1]$ represents the proportion of the Weiss domain (see Fig. 9.5).

Thus, α is chosen as a linear function of H:

$$(2\alpha(H) - 1) = \frac{\chi_a H}{m_s} \tag{9.26}$$

with $m_1(H) = \chi_a H$.

9.3.6.2 Magnetization of Martensite: Axis of Difficult Magnetization

Magnetization along the axis of difficult magnetization corresponds to a rotation of magnetization within the considered variant. On the basis of the choice of REV made above (see Fig. 9.5):

$$m_2(H) = m_3(H) = m_s \sin(\theta\ (H)) \tag{9.27}$$

where $\theta \in \left[-\frac{\pi}{2}, \frac{\pi}{2}\right]$ represents the angle of rotation of the magnetization. We shall choose the function $sin(\theta\,(H))$ as being linear under H in the form:

$$\sin(\theta\,(H)) = \frac{\chi_t H}{m_s} \qquad (9.28)$$

Therefore meaning, m_2=m$_3$=χ_tH.

9.3.6.3 Magnetization of Austenite

With a service temperature lower than the Curie temperature of the material, the behavior is considered to be similar to the one of variant M_1:

$$m_0(H) = m_s(2\alpha_A(H) - 1) \qquad (9.29)$$

where $\alpha_A \in [0, 1]$ represents the proportion of the Weiss domain in the austenite. α_A is chosen as a linear function of H:

$$(2\alpha_A(H) - 1) = \frac{\chi_A H}{m_s}. \qquad (9.30)$$

9.3.6.4 Mixture Law

Then a mixture law gives the global magnetization of the material:

$$m(H) = \sum_{k=0}^{k=3} z_k M_k \qquad (9.31)$$

$$m(H) = m_s\,(z_0(2\alpha_A(H) - 1) + z_1(2\alpha(H)\text{-}1) + (z_2 + z_3)\sin(\theta\,(H)\,))\,. \qquad (9.32)$$

The curves given by Likhachev et al. (Fig. 9.8) (Likhachev et al. 2004) show that when $z = 1, m = M$ is linear in $H_0 = H$ with slope χ_t, and for $z = 0$ m linear in H with slope χ_a.

In this two-variants model ($z_3 = 0$), $z_1 =$ z; $z_2 = 1 - z$ (Gauthier et al. 2007) we write:

$$m = m_x = \chi_a H z + \chi_t H\,(1 - z)\,. \qquad (9.33)$$

Using Eq. (9.32) and by integration of the different parts of mdH, as done in Gauthier et al. (2007) and in Gauthier's PhD (2007), the expression of G_{mag} is established:

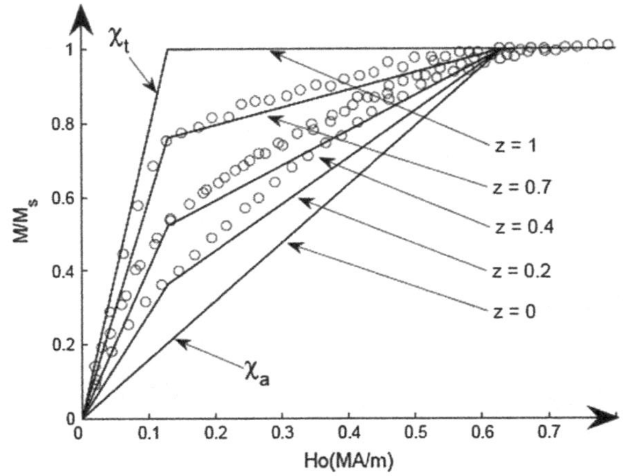

Fig. 9.8 Magnetization curves for different fractions z of variant 1 (model with two variants: 1 and 2) model: lines; experiments (o) (Experiments performed by Likhachev)

$$\rho G_{mag}(H, z_0, z_1, z_2, z_3, \alpha, \theta, \alpha_A) =$$
$$-\mu_0 m_s \left[z_1((2\alpha - 1)\, H - \tfrac{m_s}{2\chi_a} (2\alpha - 1)^2) + (z_2 + z_3) \left(\sin(\theta)\, H - \tfrac{m_s}{2\chi_t} sin^2(\theta) \right) \right]$$
$$-\mu_0 m_s \left[z_0((2\alpha_A - 1)\, H - \tfrac{m_s}{2\chi_A} (2\alpha_A - 1)^2) \right]. \tag{9.34}$$

Observations of the experimental curves show that parameter m_s is not constant, but rather depends on the temperature (Heczko and Ullako 2001). For ferromagnetic materials, the Weiss theory gives the dependency of m_s on T by an implicit equation, made explicit by Zuo et al. (1998):

$$\frac{m_s(T)}{m_s^0} = \tanh\left(\frac{m_s(T)}{m_s^0} \frac{T_c}{T} \right) \tag{9.35}$$

where T_c is the Curie temperature and m_s^0 the magnetization at 0°K. In order to simplify the calculation, the parameters m_s^0 and T_c are considered as identical for austenite and martensite, although they are slightly different in reality.

9.3.7 General Expression of the Free Energy

For a monocrystal containing three martensite variants (for a cubic $\Longrightarrow$ quadratic transformation) and an austenitic phase, under the influence of magneto-thermo-mechanical stresses, the expression of the Gibbs free energy is written by superposition of the four contributions: chemical, thermal, mechanical and magnetical.

Thereafter, the thermodynamics forces are given by:

$$\underline{\varepsilon} = -\rho \frac{\partial G}{\partial \underline{\sigma}}$$

$$\mu_0 m = -\rho \frac{\partial G}{\partial H}$$

$$s = -\frac{\partial G}{\partial T}.$$

A magneto-thermal effect is present in the expression of the entropy, due to the temperature dependence of m_s.

The thermodynamics forces associated with the variables α, α_A and θ are taken to be equal to zero, e.g.,

$$\rho \frac{\partial G}{\partial \alpha} = 0\,,\ \rho \frac{\partial G}{\partial \alpha_A} = 0\,,\ \rho \frac{\partial G}{\partial \theta} = 0 \tag{9.36}$$

The choice of the free energy expression confirms that purely magnetic behavior is considered to be reversible

One also can write the thermodynamic forces associated to austenite and martensite variant fractions:

$$\pi_i^f = -\rho \frac{\partial G}{\partial z_i}.$$

The behavior is irreversible, and the Clausius–Duhem inequality can be written:

$$dD = -\rho dG - \mu_0 m dH - \varepsilon d\sigma - s dT \geq 0 \tag{9.37}$$

where dD constitutes the dissipation increment. With respect to Chap. 2, its expression reduces to:

$$dD = \sum_{i=0}^{3} \pi_i^f dz_i \geq 0 \; avec \; \sum_{i=0}^{3} dz_i = 1. \tag{9.38}$$

It still remains to write the phase transformation kinetics or reorientation one.

9.4 Applications

9.4.1 Confrontation Modeling-Experience

Figure 9.9 illustrates the experimental results obtained by Straka et al. 2006.

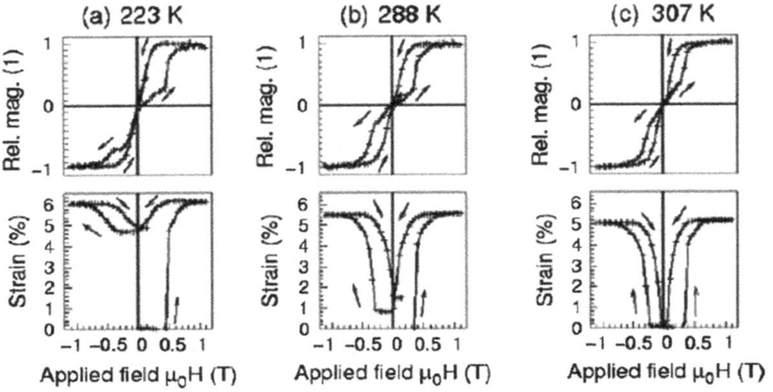

Fig. 9.9 Evolution of deformation magnetization under magnetic field action at constant external stress $\sigma = -1$ MPa. Experiments of Straka et al. (2006)

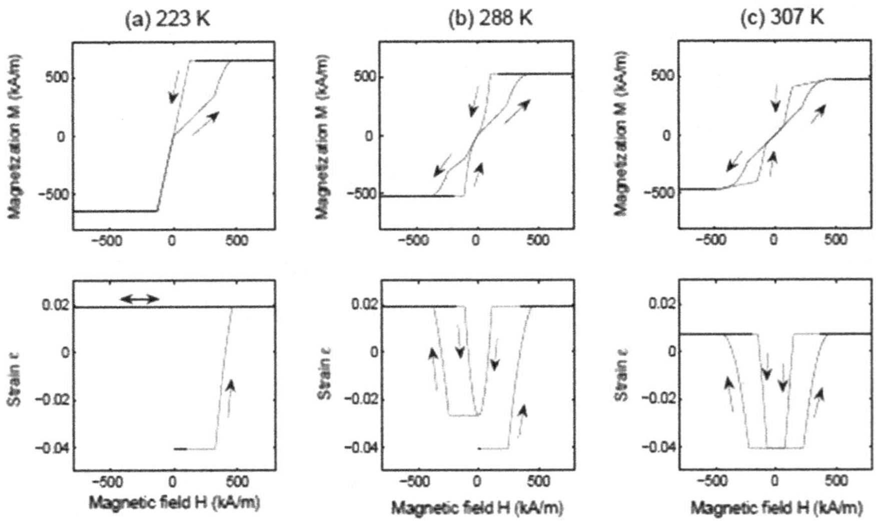

Fig. 9.10 Evolution of deformation magnetization under magnetic field action at constant external stress $\sigma = -1$ MPa. Experiments of Straka et al. (2006). Modeling by Gauthier et al. (2011)

Straka et al. (2006). Figure 9.10 shows the modeling of experimental curves presented in Fig. 9.9. The curves are obtained with the applied magnetic field with a "minus" and "plus" to show the maximum and minimum deformation for the first cycle and repeated cycles (useful for actuation).

At low temperatures (T = 223 K), an increase in H generates a deformation that is, in part, maintained when the magnetic field is canceled. At higher temperature (T = 288 K), a repeated operation is possible and the amplitude of deformation slightly increases with temperature (T = 307 K). One can note some discrepancies

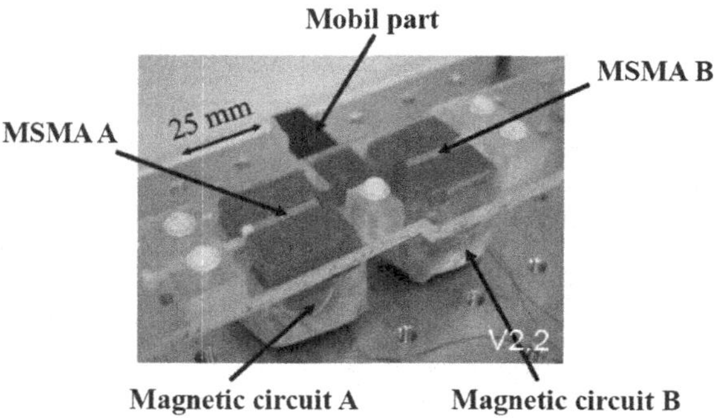

Fig. 9.11 Picture of the «Push-Pull» actuator

between experiments and simulations but agreement is fair. First, the magnetic field associated with the start of the shift, decreases with temperature due to the increase of the critical force. Second, the maximum achievable deformation also decreases because of the decrease of the saturation magnetization.

9.4.2 Application: Creation of a «Push/Pull» Actuator

Figure 9.11 presents version V2.2 of this actuator. A horizontal stack consists of a sample of an MSMA (MSMA *A*), a moving part made of plastic and another sample of MSMA (MSMA *B*). This stack is held in position using a non-ferromagnetic material. The total length can be adjusted by means of a screw. Two V2-type magnetic circuits are used to create two magnetic fields H_a and H_b, applied to samples MSMA *A* and MSMA *B*, respectively.

The functioning principle of such an actuator is presented in Fig. 9.12. A magnetic field is applied to MSMA *A*. A martensitic rearrangement occurs, and a deformation

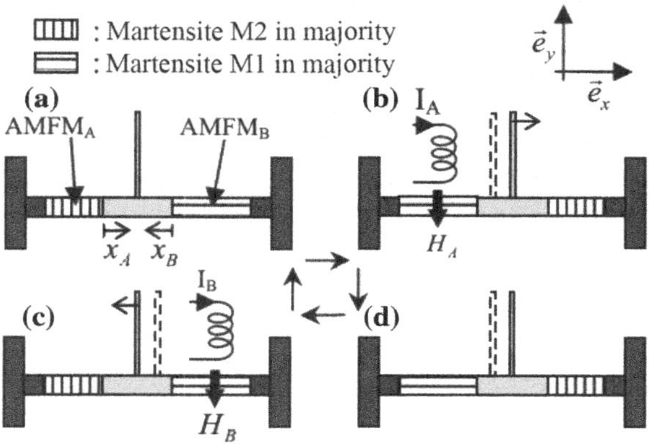

Fig. 9.12 Principle of operation of the «Push-Pull» actuator

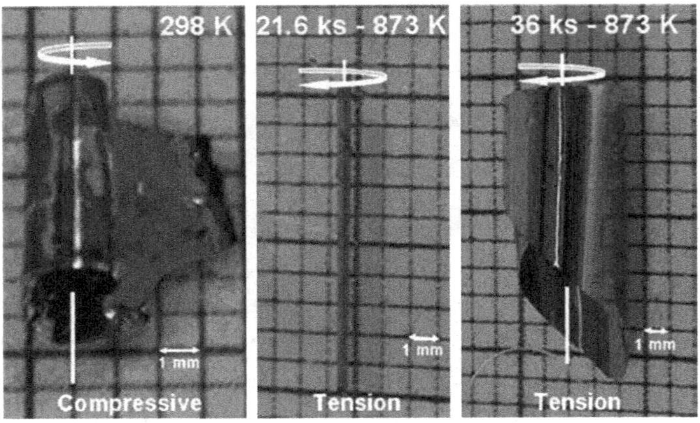

Fig. 9.13 Picture of three films applied at 298 K at reheated for 21.6 and 36 ks at 873 K, respectively (the *arrows* indicate the directions of rolling)

appears, causing the moving part to shift to the right. Similarly, if we wish to move it to the left, we apply a field to MSMA *B*. By modulating the two fields, it is possible to achieve different stable positions.

9.5 Conclusion

The aim of this chapter was to put forward a simulation of a monocrystal of a magnetic shape-memory alloy taking account of the thermo-magneto-mechanical pairing. The attempt is successful, although the formulation is already complicated (see the expression of the free energy).

For polycrystals, homogenization techniques can be considered, as are calculations using the finite element method.

Although we have managed to design and develop an actuator, the main obstacle for an industrial use of massive MSMAs is their fragility.

One possible solution consists in elaborating of thin MSMA films obtained by RF sputtering. These films represent potential materials for micro- and nanosystems. However, their properties are highly dependent on the structures and the internal stresses induced by their manufacture (Bernard et al. 2009) (Fig. 9.13).

Chapter 10
Mechanical Elements of Fracture and Damage

Abstract The purpose of the fracture mechanics is to study and predict the initiation and propagation of macroscopic cracks in solids. The damage manifests itself by micro-cracks which have surface discontinuities and cavities such as volume is continuities. Apparently, between damage and fracture, it is a question of scale! The « linear fracture mechanic » assumption, that is to say brittle fracture, is considered. For the resolution of damage, a so-called internal variable « mechanical damage variable » is introduced.

10.1 Introduction

The purpose of the fracture mechanics is to study and predict the initiation and propagation of macroscopic cracks in solids. Indeed, discontinuities in the material affect the homogeneity of the medium, hence the state of stress, strain, and displacement (Cailletaud et al. 2011).

The damage manifests itself by micro-cracks which have surface discontinuities and cavities such as volume discontinuities (Lemaitre et al. 2009).

Apparently, between damage and fracture, it is a question of scale!

In general, the damage is before the fracture. The separation of a body into two separate parts occurs after a starting stage including the development of micro-cavities, microcracks ... under the action of thermal, chemical, and mechanical stress. Thus, the propagation of macroscopic cracks may lead to the collapse of the structure, or on the contrary, cracks may stop. A current example is offered by the cracked tanks of the Flamanville EPR or Eurodif. The resolution of this technical problem uses strong metallurgical knowledge (optimal carbon content of the steel used) in fracture mechanic and damage.

The failure mode can be brittle. The fracture then occurs without plastic deformation. Ductile fracture is associated with significant plastic deformation or viscoplastic ones. The energy required to produce fracture, characterized by resilience (ability of materials to withstand shocks) is much more significant in ductile rupture. This resilience depends on temperature, transition temperature ductile $\Longleftrightarrow$ brittle. The

C. Lexcellent, *Linear and Non-linear Mechanical Behavior of Solid Materials*, DOI 10.1007/978-3-319-55609-3_10

failure mode also depends on the stress state, particularly the stress triaxiality (ratio between the first and the second invariant).

Depending on the thermomechanical loading and the material considered, if the medium has a plastic or viscoplastic behavior, it is the "nonlinear fracture mechanics," or "local approach" where a description of the stress state and crack tip deformation as precise as possible necessary.

In the absence of plasticity or if it is very confined, the medium can be considered as elastic everywhere and the "linear fracture mechanics" will be used.

Cailletaud et al. (2011) point out two dates that mark the development of fracture mechanics.

In 1920, Griffith shows that the rupture of an elastic-fragile medium can be characterized by a global variable G which will be later called the"energy release rate."

In 1956, from the study of the singularities of the stress field, Irwin introduces the concept of "stress intensity factor".

10.2 Stress Intensity Factor

In this section, we consider the "linear fracture mechanics" assumption, that is to say brittle fracture.

Strictly speaking, mechanical fracture problems are always three-dimensional. However, their resolution is difficult and we will study two simpler specific cases, namely the plane strain and antiplane situation.

10.2.1 Plane Elasticity

Chapter 3 provides the solution with the Airy function $A\,(r, \theta)$.

Let's consider the dynamic volumic force density $\overrightarrow{f^\star} = \overrightarrow{Cte}$, then the solution in polar coordinates is:

$$\left\{\begin{aligned} \sigma_{rr} &= \frac{1}{r}\frac{\partial A}{\partial r} + \frac{1}{r^2}\frac{\partial^2 A}{\partial \theta^2} \\ \sigma_{\theta\theta} &= \frac{\partial^2 A}{\partial r^2} \\ \sigma_{r\theta} &= -\frac{\partial}{\partial r}\left(\frac{1}{r}\frac{\partial A}{\partial \theta}\right) \end{aligned}\right. \tag{10.1}$$

with: $\Delta\Delta A = 0$.

10.2.2 Antiplane Elasticity

This case is known as "antiplane" if:

$$\vec{u} \begin{cases} u_1 = 0 \\ u_2 = 0 \\ u_3 = u_3(x_1, x_2) \end{cases} \Longrightarrow$$

$$\underline{\varepsilon} = \begin{pmatrix} 0 & 0 & \varepsilon_{13} = \frac{1}{2}u_{3,1} \\ 0 & 0 & \varepsilon_{23} = \frac{1}{2}u_{3,2} \\ \varepsilon_{31} & \varepsilon_{32} & 0 \end{pmatrix} \Longrightarrow$$

$$\underline{\sigma} = \begin{pmatrix} 0 & 0 & \sigma_{13} = \mu u_{3,1} \\ 0 & 0 & \sigma_{23} = \mu u_{3,2} \\ \sigma_{31} & \sigma_{32} & 0 \end{pmatrix}.$$

In the absence of volume forces, the equilibrium equations reduce to:

$$\sigma_{31,1} + \sigma_{32,2} = 0 \iff \Delta u_3(x_1, x_2) = 0.$$

Thus u_3 must be a harmonic function, we can write (Leblond 2003):

$$\mu u_3 = Re\ f(z)$$

where $f(z)$ is an analytic function of $z = x_1 + ix_2$.

By posing $f = h + ik$ (with h et k reals), $\sigma_{31} = h_{,1} = k_{,2}$ and $\sigma_{32} = h_{,2} = -k_{,1}$ (According to the conditions of Cauchy–Riemann); one can synthesize these writings into a single equation:

$$\sigma_{31} - i\sigma_{32} = f'(z).$$

10.2.3 Stress Singularity in Mode I and II (Fig. 10.1)

First point out the difference in severity between an angular defect (zero radius at the crack tip) and a regular defect (finite radius of curvature). The difference between stress concentration and stress singularity can be used to remove the stress singularity near an angular defect. In order to stop the crack growth, a simple method consists in drilling a circular hole at the crack tip (technique known as the SNCF method! (Suquet 2003)).

We choose the calculations performed by Leblond (2003) considering that they are reference results in the literature items.

Let's consider a straight crack in plane strain. In order to determine the stress field near the crack tip, we need to define the landmark (O, x_1, x_2) with O located at the crack tip and Ox_1 located along the extension of the crack, and the corresponding coordinates (Fig. 10.2).

The classic approach developed by Williams (1952) in the study consists of study the function which complies with the following equations:

– Beltrami equation $\Delta\Delta A = 0$

– Free boundaries conditions on the crack lips: $\begin{cases} \sigma_{\theta\theta}(r, \pm\pi) = A_{,rr}(r, \pm\pi) = 0 \\ \sigma_{r\theta}(r, \pm\pi) = -\left(\frac{1}{r}A_{,\theta}\right)_{,r}(r, \pm\pi) = 0 \end{cases}$

Let research a function A under the form:

$$A(r, \theta) = r^{\alpha+2}\psi(\theta)$$

It follows that the stress will be in $r^{\alpha \cdot}$. We can easily calculate:

$$\Delta A = \frac{\partial^2 A}{\partial r^2} + \frac{1}{r}\frac{\partial A}{\partial r} + \frac{1}{r^2}\frac{\partial^2 A}{\partial \theta^2} = r^{\alpha}\left((\alpha+2)^2\psi(\theta) + \psi''(\theta)\right)$$

$$\Delta\Delta A = r^{\alpha-2}\left[\alpha^2(\alpha+2)^2\psi(\theta) + \left(\alpha^2 + (\alpha+2)\right)^2\right)\psi''(\theta) + \psi''''(\theta)\right].$$

The condition $\Delta\Delta A = 0$ is thus written:

$$\alpha^2(\alpha+2)^2\psi(\theta) + \left(\alpha^2 + (\alpha+2)\right)^2\right)\psi''(\theta) + \psi''''(\theta) = 0.$$

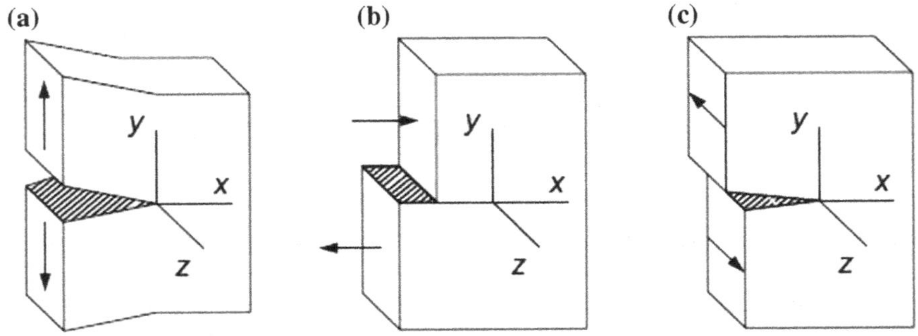

Fig. 10.1 Schematic representation of the three class of fracture modes: **a** mode I (opening), **b** mode II (in plane shear), and **c** mode III (tearing)

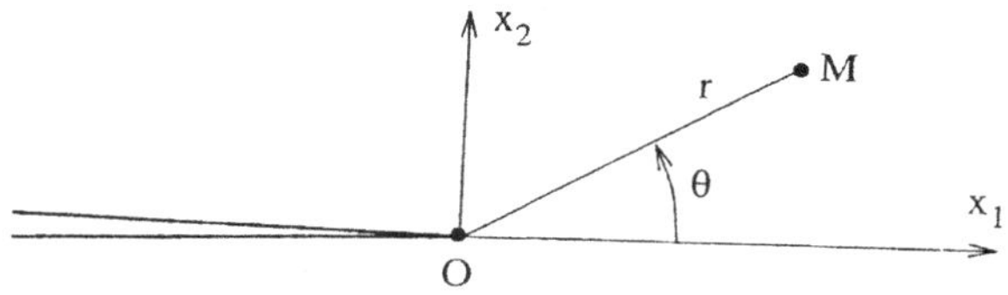

Fig. 10.2 Crack in plane situation (Leblond 2003)

If we look for the solution in the exponential form $e^{ik\theta}$, we obtain the equation:

$$k^4 - \left[\alpha^2 + (\alpha+2)^2\right]k^2 + \alpha^2(\alpha+2)^2 = 0 \Longleftrightarrow \left(k^2 - \alpha^2\right)\left(k^2 - (\alpha+2)^2\right) = 0.$$

The roots are $k = \pm\alpha,\ k = \pm(\alpha+2)$ and the solutions ψ under the exponential form $\exp(i\alpha\theta)$, $\exp(-i\alpha\theta)$, $\exp(i(\alpha+2)\theta)$, $\exp(-i(\alpha+2)\theta)$ such as:

$$\psi(\theta) = \mathrm{A}'\sin(\alpha\theta) + \mathrm{B}\cos(\alpha\theta) + \mathrm{C}\sin[(\alpha+2)\theta] + \mathrm{D}\cos[(\alpha+2)\theta]$$

where $A',\ B,\ C,\ D$ are four real constants.

The boundary conditions are written as:

$$A_{,rr}(r, \pm\pi) = 0 \Longleftrightarrow (\alpha+1)(\alpha+2)r^{\alpha}\psi(\pm\pi) = 0 \Longleftrightarrow \psi(\pm\pi) = 0$$

which implicitly assumes that $\alpha \neq -1$ and $\alpha \neq -2$, we shall see later that these cases are excluded:

$$\left(\frac{1}{r}A_{,\theta}\right)_{,r}(r, \pm\pi) = 0 \Longleftrightarrow (\alpha+1)r^{\alpha}\psi'(\pm\pi) = 0 \Longleftrightarrow \psi'(\pm\pi) = 0$$

with $\alpha \neq -1$.

So they give:

$$\left\{\begin{array}{c} \psi(\pm\pi) = \mathrm{A}'\sin(\pm\alpha\pi) + \mathrm{B}\cos(\pm\alpha\pi) \\ +\mathrm{C}\sin[\pm(\alpha+2)\pi] + \mathrm{D}\cos[\pm(\alpha+2)\pi] = 0 \\ \psi'(\pm\pi) = \mathrm{A}'\alpha\cos(\pm\alpha\pi) - \mathrm{B}\alpha\sin(\pm\alpha\pi) \\ +C(\alpha+2)\cos(\pm(\alpha+2)\pi) - \mathrm{D}(\alpha+2)\sin(\pm(\alpha+2)\pi) = 0 \end{array}\right\}$$

or

$$\left\{\begin{array}{c} \left(\mathrm{A}' + \mathrm{C}\right)\sin(\pm\alpha\pi) + (\mathrm{B} + \mathrm{D})\cos(\pm\alpha\pi) = 0 \\ \left(\mathrm{A}'\alpha + \mathrm{C}(\alpha+2)\right)\cos(\pm\alpha\pi) - (\mathrm{B}\alpha + \mathrm{D}(\alpha+2))\sin(\pm\alpha\pi) = 0 \end{array}\right.$$

$$\Longleftrightarrow \begin{cases} \left(A' + C\right)\sin(\pm\alpha\pi) = 0 \\ (B + D)\cos(\pm\alpha\pi) = 0 \\ \left(A'\alpha + C(\alpha + 2)\right)\cos(\pm\alpha\pi) = 0 \\ (B\alpha + D(\alpha + 2))\sin(\pm\alpha\pi) = 0 \end{cases}$$

if α is not integer or half-integer, one has $\sin(\pm\alpha\pi) \neq 0$ and $\cos(\pm\alpha\pi) \neq 0$, e.g., $A' = B = C = D = 0$, trivial solution without interest.

If α is integer $\alpha = n$ one has $sin\,(n\pi) = 0$, so the equations reduce to:

$$\begin{cases} B + D = 0 \\ A'n + C(n + 2) = 0 \end{cases} \Longleftrightarrow \begin{cases} C = -A'\frac{n}{n+2} \text{ avec } n \neq -2 \\ D = -B \end{cases}.$$

This solution is not trivial.

If α is half-integer: $\alpha = n + \frac{1}{2}$ one has $cos\left[\left(n + \frac{1}{2}\right)\pi\right] = 0$, so the equations reduce to:

$$\begin{cases} A' + C = 0 \\ B\left(n + \frac{1}{2}\right) + D\left(n + \frac{5}{2}\right) = 0 \end{cases} \Longleftrightarrow \begin{cases} C = -A' \\ D = -B\frac{n+1/2}{n+5/2} \end{cases}$$

Again, this solution is not trivial.

Let us now consider the reasonable assumption that "the elastic energy of the body, resulting from the applied load is FINISHED." This means that the integral $\int \frac{1}{2}\underline{\sigma} : \underline{\varepsilon^{el}}\, dv$ should be convergent in $r = 0$. Or, $\underline{\sigma}$ is proportional to r^{α}, as well as $\underline{\varepsilon^{el}}$ and the integration element, in polar coordinates, in r. The quantity to integrate is hence proportional to $r^{2\alpha+1}$. Thus, the dominant term in the development of stresses corresponds to $\alpha = -1/2$. It is singular (infinite) at the crack tip.

It appears a contradiction with the very foundations of the theory of classical elasticity that involve infinitesimal deformation. Despite this drawback, the simplistic theory developed turns out to be useful, as the predictions are fair (by Leblond 2003).

We can find the asymptotic expression of the stresses (that is to say the dominant term development), e.g., $\alpha = -1/2$ and by using formulas giving σ_{rr}, $\sigma_{\theta\theta}$, $\sigma_{r\theta}$ as function of A. By choosing $K_I = B\sqrt{2\pi}$ and $K_{II} = A'\sqrt{2\pi}$ Leblond obtains:

$$\begin{cases} \sigma_{rr} = \frac{K_I}{4\sqrt{2\pi r}}\left(5\cos\frac{\theta}{2} - \cos\frac{3\theta}{2}\right) + \frac{K_{II}}{4\sqrt{2\pi r}}\left(-5\sin\frac{\theta}{2} + 3\sin\frac{3\theta}{2}\right) \\ \sigma_{\theta\theta} = \frac{K_I}{4\sqrt{2\pi r}}\left(3\cos\frac{\theta}{2} + \cos\frac{3\theta}{2}\right) + \frac{K_{II}}{4\sqrt{2\pi r}}\left(-3\sin\frac{\theta}{2} - 3\sin\frac{3\theta}{2}\right) \\ \sigma_{r\theta} = \frac{K_I}{4\sqrt{2\pi r}}\left(\sin\frac{\theta}{2} + \sin\frac{3\theta}{2}\right) + \frac{K_{II}}{4\sqrt{2\pi r}}\left(\cos\frac{\theta}{2} + 3\cos\frac{3\theta}{2}\right). \end{cases} \quad (10.2)$$

The coefficients K_I *and* K_{II} are called « stress intensity factors » of modes I and II, respectively.

We chose this analytical method from the Airy function $A(r, \theta)$ because it seems easier than the complex potentials given in literature Westergaard (1939), Sun and Jin (2012).

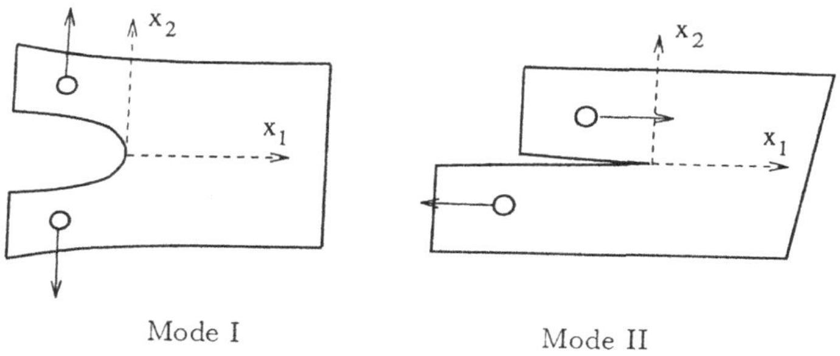

Fig. 10.3 Cracks situation for mode I and mode II (Leblond 2003)

This important result shows that regardless of the geometry of the body studied, regardless of the applied load, the asymptotic expression stress in plane strain conditions depends only on K_I *et* K_{II}.

From stresses expressions, we can calculate the deformations and the displacements. The results in polar coordinates are the following:

$$\begin{cases} u_r = \frac{K_I}{4\mu}\sqrt{\frac{r}{2\pi}}\left[(5-8\nu)\cos\frac{\theta}{2} - \cos\frac{3\theta}{2}\right] + \frac{K_{II}}{4\mu}\sqrt{\frac{r}{2\pi}}\left[(-5+8\nu)\sin\frac{\theta}{2} + 3\sin\frac{3\theta}{2}\right] \\ u_\theta = \frac{K_I}{4\mu}\sqrt{\frac{r}{2\pi}}\left[(-7+8\nu)\sin\frac{\theta}{2} + \sin\frac{3\theta}{2}\right] + \frac{K_{II}}{4\mu}\sqrt{\frac{r}{2\pi}}\left[(-7+8\nu)\cos\frac{\theta}{2} + 3\cos\frac{3\theta}{2}\right]. \end{cases} \tag{10.3}$$

It results immediately the following properties on the crack lips ($\theta = \pm\pi$)
– for mode I:

$$u_1(r, \pm\pi) = 0,\ u_2(r, \pm\pi) = \pm\frac{2(1-\nu)K_I}{\mu}\sqrt{\frac{r}{2\pi}}$$

$$\| u_1 \|(r) = 0,\ \| u_2 \|(r) = \frac{4(1-\nu)K_I}{\mu}\sqrt{\frac{r}{2\pi}}$$

– for mode II:

$$u_2(r, \pm\pi) = 0,\ u_1(r, \pm\pi) = \pm\frac{2(1-\nu)K_{II}}{\mu}\sqrt{\frac{r}{2\pi}}$$

$$\| u_2 \|(r) = 0,\ \| u_1 \|(r) = \frac{4(1-\nu)K_{II}}{\mu}\sqrt{\frac{r}{2\pi}}$$

One has: $[[u_i]](r) = u_i(r, \pi) - u_i(r, -\pi)$. So for mode I, u_1 is continuous through the crack and u_2 has a discontinuity proportional to K_I.

For mode II, u_2 is continuous and u_1 has a discontinuity proportional to K_{II}.

Only the mode I causes an "opening" of the crack; Mode II causes an implane shear (see Fig. 10.3).

10.2.4 Stress Singularity in Mode III

Let's consider now the « antiplane » case. In order to obtain stress proportional to r^{α}, it is necessary that the analytical function $f(z)$ is proportional to $r^{\alpha+1}$, and therefore of the form $Cz^{\alpha+1}$ where $C = A' + iB$ is a complex constant. Then, we have:

$$f'(z) = (\alpha + 1)\, Cz^{\alpha} = (\alpha + 1)\left(\mathrm{A}' + \mathrm{iB}\right) r^{\alpha} \exp(i\alpha\theta)$$

hence

$$\begin{aligned}\sigma_{31} &= Ref'(z) = (\alpha + 1)\left(\mathrm{A}'\cos(\alpha\theta) - \mathrm{B}\sin(\alpha\theta)\right) r^{\alpha}\\ \sigma_{32} &= -Imf'(z) = -(\alpha + 1)\left(\mathrm{A}'\sin(\alpha\theta) - \mathrm{B}\cos(\alpha\theta)\right) r^{\alpha}\end{aligned}$$

The boundary condition on the crack lips is written (assuming $\alpha \neq -1$, if not the elastic energy is infinite):

$$\sigma_{32}(r, \pm\pi) = 0 \Longleftrightarrow \mathrm{A}'\sin(\pm\alpha\pi) + \mathrm{B}\cos(\pm\alpha\pi) = 0 \Longleftrightarrow \begin{cases}\mathrm{A}'\sin(\alpha\pi) = 0\\ \mathrm{B}\cos(\alpha\pi) = 0\end{cases}$$

The most singular term development of stress components (given the requirement that the elastic energy is finite) is $\alpha = -1/2$. Let's pose $B = -2K_{III}/\sqrt{2\pi}$, one obtains:

$$\begin{aligned}\sigma_{31} &= -\frac{K_{III}}{\sqrt{2\pi R}}\sin\left(\frac{\theta}{2}\right)\\ \sigma_{32} &= \frac{K_{III}}{\sqrt{2\pi R}}\cos\left(\frac{\theta}{2}\right)\end{aligned} \quad (10.4)$$

K_{III} is called the stress intensity factor in mode III.

One obtains the displacement u_3: $u_3 = \frac{1}{\mu} Re\, f(z) = \frac{1}{\mu} Re\left(-2iK_{III}\sqrt{\frac{r}{2\pi}}e^{i\theta/2}\right)$, e.g.,

$$u_3 = \frac{2K_{III}}{\mu}\sqrt{\frac{r}{2\pi}}\sin\left(\frac{\theta}{2}\right) \quad (10.5)$$

This implies:

$$[[u_3]](r) = u_3(r, \pi) - u_3(r, -\pi) = \frac{4K_{III}}{\mu}\sqrt{\frac{r}{2\pi}} \quad (10.6)$$

As the II mode, the mode III does not cause crack opening, but causes a « tearing mode » (See Fig. 10.4).

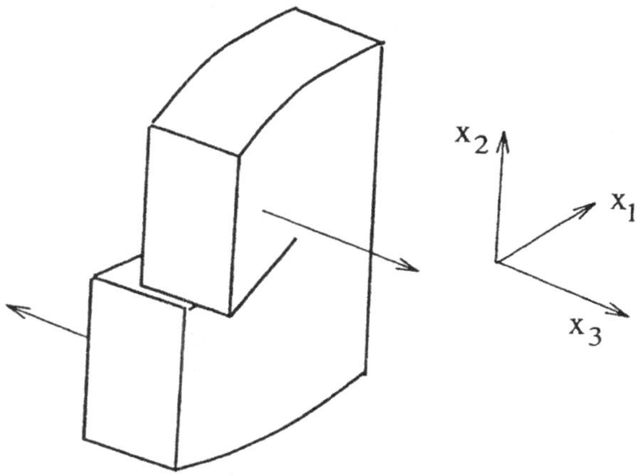

Fig. 10.4 Crack in mode III situation « tearing mode » (Leblond 2003)

10.2.5 Theory of Irwin for Mode I

We only consider crack propagation in mode I ($K_{II} = K_{III} = 0$) case in which propagation is a straight line without interruption of the tangent.

The idea of Irwin (1958) is to bring the criterion of propagation on the intensity factor K_I characterizing the "intensity" of the singularity. He postulated the existence of a quantity called "tenacity" and denoted K_{Ic} as a characteristic of the material (as well as the coefficients of elasticity and yield elasticity stress σ_y).

$$\begin{array}{c} K_I < K_{Ic} \Longrightarrow \textit{no propagation} \\ K_I = K_{Ic} \Longrightarrow \textit{possible propagation} \end{array} \tag{10.7}$$

This Irwin hypothesis is hard to justify except by its effectiveness and the concept of toughness is now widely used in structural design.

The following provides in the table orders of magnitude of the tenacity for a few common materials (Suquet 2003):

$$\left\{ \begin{array}{cc} \textit{Aluminum alloy} & K_{Ic} \simeq 30\,\mathrm{MPa}\sqrt{\mathrm{m}} \\ \textit{Alloy of titanium} & K_{Ic} \simeq 100\,\mathrm{MPa}\sqrt{\mathrm{m}} \\ \textit{Quenched steel} & K_{Ic} \simeq 120\,\mathrm{MPa}\sqrt{\mathrm{m}} \\ \textit{Polymer} & K_{Ic} \simeq 3\,\mathrm{MPa}\sqrt{\mathrm{m}} \\ \textit{Wood} & K_{Ic} \simeq 2\,\mathrm{MPa}\sqrt{\mathrm{m}} \\ \textit{Concrete} & K_{Ic} \simeq 1\,\mathrm{MPa}\sqrt{\mathrm{m}}. \end{array} \right\}. \tag{10.8}$$

10.2.6 Example of Application: Selection of a Pressure Vessel Steel

Suquet (2003).

Statement:

A pressurized tank consists of a tube whose radius is $R = 2m$, height is H and has thin wall with a thickness e (Fig. 10.5). One assumes that $H \gg R$.

The tank contains a fluid whose pressure p. It must withstand a maximum internal pressure $p_{max} = 50\,\text{MPa}$. The designer must choose between three grades of steel with different ultimate stress σ_u values (The highest stress bearable by the steel within the Huber–Von Mises criterion) and tenacity K_{Ic}

$$\left\{\begin{array}{lll} grade\ A & \sigma_u = 1250\,MPa & K_{Ic} = 90\,\text{MPa}\sqrt{\text{m}} \\ grade\ B & \sigma_u = 900\,MPa & K_{Ic} = 120\,\text{MPa}\sqrt{\text{m}} \\ grade\ C & \sigma_u = 650\,MPa & K_{Ic} = 190\,\text{MPa}\sqrt{\text{m}}. \end{array}\right\}$$

One wonders about the choice of steel grade according to the defects to be feared in this type of structure.

The most dangerous defects (because hardly detectable) are opening cracks on the inner wall of the tank. RX does not detect cracks of less than 0.5 cm in diameter.

(1) Let estimate the stress σ_{rr}, $\sigma_{\theta\theta}$, σ_{zz}. Note that the shear stresses are zero by essence.

It can be estimated through the "coppersmith reasoning" (see Fig. 10.6) by the balance of the upper half of the tank:

$$pX\pi R^2 = \sigma_{zz} X 2\pi Re \Longrightarrow \sigma_{zz} = \frac{pR}{2e}$$

and by the balance of the right half of the tank:

$$pX2R = \sigma_{\theta\theta} X2e \Longrightarrow \sigma_{\theta\theta} = \frac{pR}{e}.$$

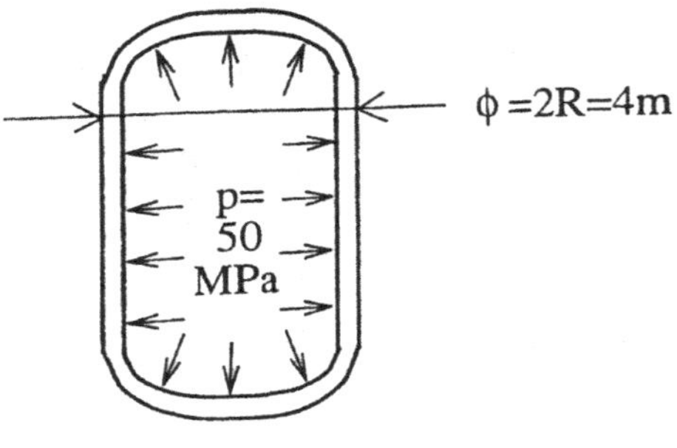

Fig. 10.5 Tank under pressure (Leblond 2003)

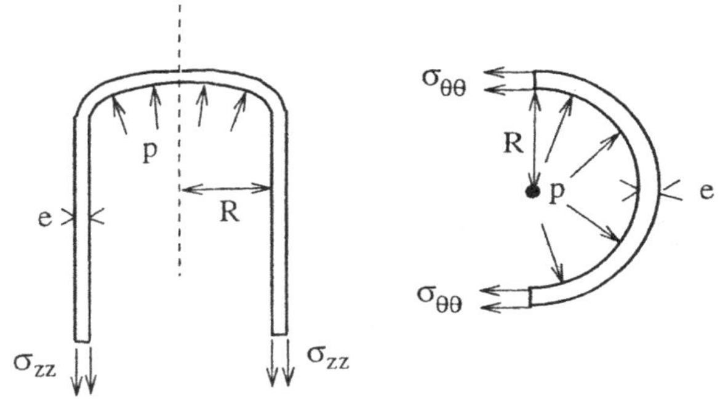

Fig. 10.6 Tank equilibrium (Leblond 2003)

The stress σ_{rr} is not uniform; it varies from $-p\ to\ 0$. As $R/e \gg 1$, it will be neglected in front to σ_{zz}, $\sigma_{\theta\theta}$. One has:

$$\underline{\sigma} \simeq \frac{pR}{e}\begin{pmatrix} 0 & 0 & 0 \\ 0 & 1 & 0 \\ 0 & 0 & 1/2 \end{pmatrix}$$

$$\Longrightarrow \text{dev}\underline{\sigma} \simeq \frac{pR}{e}\begin{pmatrix} -1/2 & 0 & 0 \\ 0 & 1/2 & 0 \\ 0 & 0 & 0 \end{pmatrix}$$

$$\overline{\sigma} = \frac{\sqrt{3}}{2}\frac{pR}{e}.$$

The condition $\overline{\sigma} \le \sigma_u$ gives the minimum value of e to avoid failure by plasticity: $e_{min} = \frac{\sqrt{3}}{2}\frac{pR}{\sigma_U}$. They obtain for the three grades of steel: 6.9, 9.6, 13.3 cm (Leblond 2003), (Suquet 2003).

Let's consider the second failure mechanism, the highest stress is $\sigma_{\theta\theta}$, and the most dangerous cracks are those in the plane perpendicular to $\overrightarrow{e_\theta}$. We show that the configuration giving the highest K_{Ic} is when the crack propagates inside:

$$K_i = 1.12\sigma_{\theta\theta}\sqrt{\pi a}\ with\ a\ crack\ length.$$

Its use implies that $a \ll e$; so it is possible to assess the stress $\sigma_{\theta\theta}$ in the absence of cracks.

The condition $K_I \le K_{Ic}$ for the largest possible crack $\{a = 0.5\ cm\}$ imposes the minimum value of e vis-a-vis the second mechanism:

$$K_{Ic} = 1.12\frac{pR}{e'_{min}}\sqrt{\pi a} \Longrightarrow e'_{min} = 1.12\frac{pR}{K_{Ic}}\sqrt{\pi a}.$$

For steels A, B, C: we finally obtain 15.6, 11.7, 7.4 cm.

Thus, the minimum thickness (considering the two failure mechanisms) calculated for steels A, B, C is 15.6, 11.7, 13.3 cm respectively.

Grade B gives the best compromise.

10.3 Energetic Theory of Fracture

10.3.1 *Original Reasoning of Griffith (1920)*

Griffith assumed the existence of "a positive connection energy" per unit area γ. This means that the energy of the material is lower when the crack lips are in contact than when they are separated.

Let's denote l the length of crack created.

Assuming quasi-static propagation under constant loading of a crack, the energy balance between 0 *and dt* is written:

$$\delta W_{ext} = dW_{ext} + 2\gamma dl.$$

The power of external forces « under constant loading » $\dot{\phi}$ is:

$$\phi = \int_{\partial \Omega_T} \overrightarrow{T^d}.\overrightarrow{u}\, dS.$$

Example $\partial \Omega_T$ constitutes the boundary part where the stress vector is imposed $\overrightarrow{T^d} = \underline{\sigma}\,\overrightarrow{n}$ (with $\overrightarrow{n}$ unit normal to the contour) and $\overrightarrow{u}$ displacement vector:

$$\dot{\phi} = \int_{\partial \Omega} \overrightarrow{T}.\dot{\overrightarrow{u}}\, dS = \int_{\partial \Omega_T} \overrightarrow{T^d}.\dot{\overrightarrow{u}}\, dS = \frac{d}{dt}\int_{\partial \Omega_T} \overrightarrow{T^d}.\overrightarrow{u}\, dS$$

as $\dot{\overrightarrow{u}} = \dot{\overrightarrow{u^d}} = \overrightarrow{0}$ on the Ω_u part, where the displacement are imposed $\dot{\overrightarrow{T^d}} = \overrightarrow{0}$ on $\partial \Omega_T$.

It comes therefore, by denoting W instead W_{el} the elastic energy:

$$\dot{\phi} = \dot{W} + 2\gamma \dot{l}.$$

One concludes that:

$$-\dot{W} + \dot{\phi} = 2\gamma \dot{l} \Longrightarrow G\dot{l} = 2\gamma \dot{l}$$

where

$$G = -\frac{d}{dt}(W - \phi) = -\frac{dP}{dl} \tag{10.9}$$

G is called "the strain energy release rate" and represents the opposite of the derivative with respect to the crack length (at constant load) of "the total potential energy" $P = W - \phi$.

The condition of propagation is written:

$$G = 2\gamma. \tag{10.10}$$

10.3.2 Improved Griffith Model

The thermodynamics of irreversible processes is used in this model.

The growth of the crack causes a power dissipation G_c (> 0) by crack length unit.

Equations resulting from the two principles of thermodynamics:

$$\begin{cases} (G - G_c)\,\dot{l} = 0 \\ G_c \dot{l} \geq 0 \end{cases}.$$

The criterion is:

$$G = G_c \tag{10.11}$$

with G_c constant characteristic of the material as well as the Young's modulus and yield strength.

10.3.3 Various Expressions of G: Equivalence of Theories of Griffith and Irwin

The objective is twofold:
(1) provide explicit expressions of G for a practical use of the criterion $G = G_c$,
(2) establish the equivalence of Irwin theory K_{Ic} and the Griffith of G_c.

The Leblond (2003) calculations give:

$$G = \frac{1}{2} \int_{\partial\Omega} \left(\vec{T} . \frac{\partial \vec{u}}{\partial l} - \frac{\partial \vec{T}}{\partial l} . \vec{u} \right) dS \tag{10.12}$$

Interesting application:

Let's consider Q the generalized force and q the associated kinematical variable, hence:

$$G = \frac{1}{2} \left(Q \frac{\partial q}{\partial l} - \frac{\partial Q}{\partial l} q \right).$$

Let's denote $R\,(l)$ the stiffness and $C\,(l)$ the compliance which depend only on the crack length of the and not on the loading conditions such as:

$$Q = R\,(l)\,q\ ;\ q = C\,(l)\,Q.$$

We have then:

$$G = -\frac{1}{2}q^2\frac{\partial}{\partial l}\left(\frac{Q}{q}\right).$$

Thus:

$$G = -\frac{1}{2}q^2\frac{dR}{dl}\ stiffness\ formulae \tag{10.13}$$

And:

$$G = \frac{1}{2}Q^2\frac{\partial}{\partial l}\left(\frac{q}{Q}\right)$$

Thus:

$$G = \frac{1}{2}Q^2\frac{dC}{dl}\ compliance\ formulae \tag{10.14}$$

Let remark as done by Leblond (2003) that G depends only on mechanical fields (characterized by the quantities $Q\ and\ q$) that through their values when the crack length is l.

Two consequences:

(1) To measure G: only the value of the load at time t is required (therefore giving the values of $Q\ and\ q$).
(2) G does not depend on the nature of the loading.

Remark:

If Q is maintained constant; then $\partial Q/\partial l = 0$; hence:

$$G = \frac{1}{2}Q\frac{\partial q}{\partial l} = \frac{1}{2}Q^2\frac{dC}{dl}$$

If q is maintened constant; then $\partial q/\partial l = 0$; hence:

$$G = -\frac{1}{2}\frac{\partial Q}{\partial l}q = -\frac{1}{2}\left(-\frac{q}{C^2}\frac{dC}{dl}\right)q = \frac{1}{2}Q^2\frac{dC}{dl}$$

We find the same result which was not obvious « a priori ».

The formulas of stiffness and compliance have a very simple graphical interpretation on the force-displacement curve (Fig. 10.7) from which results in a method for measuring G ($= G_c$ when there is propagation).

Leblond (2003), Suquet (2003) established an important formula called "formula Irving (1958)" (Irwin 1958) which connects the rate of energy G to the stress intensity

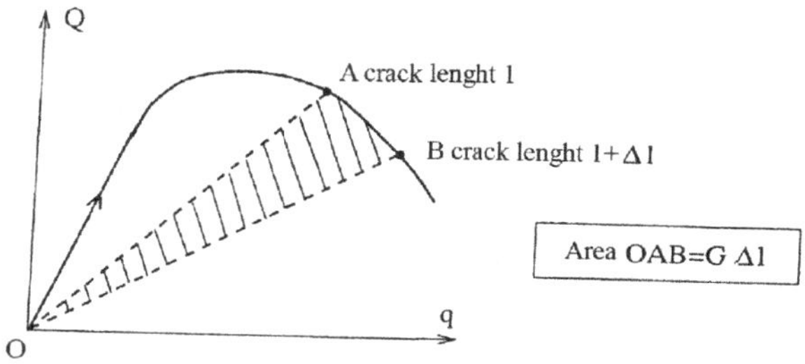

Fig. 10.7 Interpretation of G on the curve «strength-displacement » (Leblond 2003)

factors $K_I,\ K_{II},\ K_{III}$:

$$G = \frac{1-\nu^2}{E}\left(K_I^2 + K_{II}^2\right) + \frac{K_{III}^2}{2\mu} = \frac{1-\nu^2}{E}\left(K_I^2 + K_{I.}^2\right) + \frac{1+\nu}{E}K_{III}^2. \quad (10.15)$$

10.3.4 Rice Integral

Rice (1968) gave another expression of G as a contour integral. This one is very useful in practice for materials whose behavior is ductile.

Let's consider a body Ω containing a straight crack length l. It is known that G only depends on the local stress field near the crack tip. G is unchanged if we replace Ω by a smaller body Ω' limited by an outline Γ surrounding the crack tip (Fig. 10.8) and preserving the value of displacements in Γ. Indeed, the mechanical fields inside Γ are not changed.

The calculations give:

$$G = J \equiv \int_\Gamma \left(wn_1 - \sigma_{ij}u_{i,1}n_j\right) ds. \quad (10.16)$$

The integral is called "Rice integral" and denoted J.

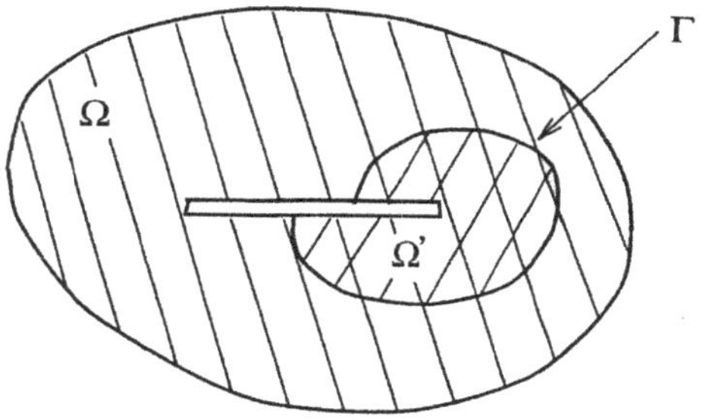

Fig. 10.8 Notations for the Rice integral (Leblond 2003)

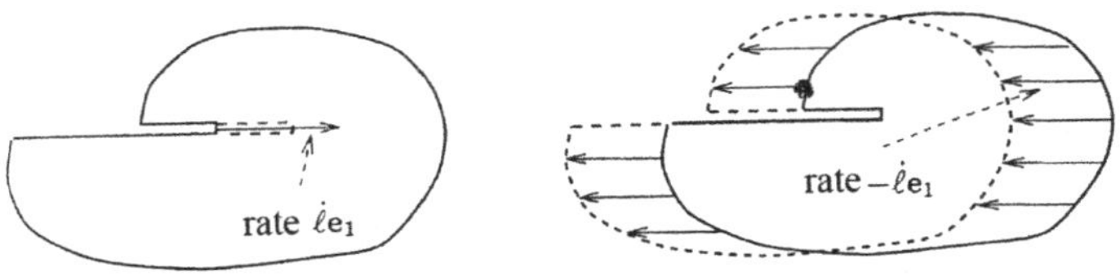

Fig. 10.9 Fixed reference system and reference system related to the crack tip (Leblond 2003)

Let's recall that w represents the elastic energy density. The horizontal velocity of the crack is $\vec{U} = -\dot{l}\vec{e_1}$ on Γ in the mobile reference system (see Fig. 10.9). Furthermore $\vec{U}.\vec{n} = -\dot{l}\vec{e_1}.\vec{n} = -\dot{l}n_1$

Let note that $\vec{e_1}$ is the unit vector collinear to the crack.

Obtaining the 10.16 formula is mainly due to Bui (1978).

He set two properties:

(1) Expression of G;

(2) invariance of J vis-à-vis the contour Γ.

10.3.5 *Analytical Application of Rice Integral*

Statement:

Let's consider a plate in plane strain states, height 2h infinite in direction x_1 containing a semi-infinite crack (Fig. 10.10).

On the boundaries $x_2 = \pm h$, the following displacements are $u_1 = 0,\ u_2 = \pm\delta$. Resolution: Let calculate G by using Rice integral on the contour Γ indicated on Fig. 10.10. On the horizontal parts $n_1 = 0$ and $u_{i,1} = 0$ hence the integral is nul. On the left vertical part $\underline{\sigma}\vec{x_1} = \vec{0}$ hence the integral is nul. On the right vertical part, $\underline{\varepsilon}$ *is* almost uniform; with $\varepsilon_{22} = \delta/h$ the others $\varepsilon_{ij} = 0$; moreover $u_{j,1} = 0$. The integration reduces to:

$$w = \left(\frac{\lambda}{2} + \mu\right)\frac{\delta^2}{h^2} = E\frac{1-\nu}{2(1+\nu)(1-2\nu)}\frac{\delta^2}{h^2}$$

$$\Longrightarrow J = G = \frac{E(1-\nu)\delta^2}{(1+\nu)(1-2\nu)h} \Longrightarrow K_I = \sqrt{\frac{EG}{1-\nu^2}} = E\frac{\delta}{(1+\nu)\sqrt{(1-2\nu)h}}.$$

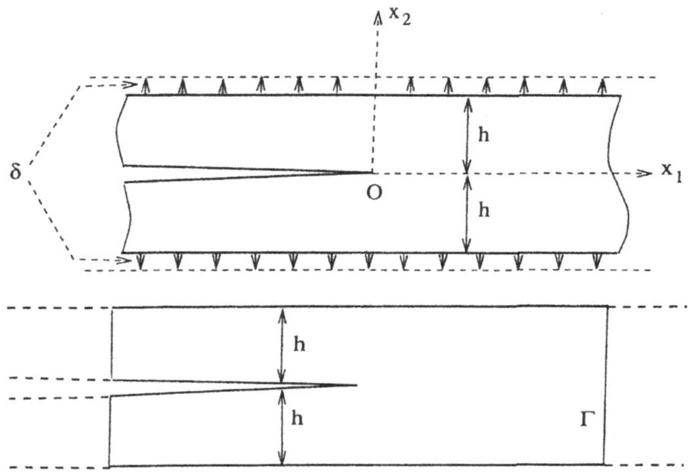

Fig. 10.10 Rice integral on a cracked plate (Leblond 2003)

10.3.6 *Some Exercises Among the 120 of the Book by Guy Pluvinage*

Pluvinage (1995).

10.3.6.1 Calculation of the Ultimate Load Versus Strain Hardening Coefficient

Statement:

A cylinder is loaded in tension and its plastic behavior is described by a power law:

$$\sigma = K\left(\varepsilon^P\right)^n \; with \;\; n = 0.16,\; K = 1300$$

Calculate the maximum load P_{max} that can withstand the sample.

Resolution steps:

– Establish the instability condition associated with the onset of necking,
– Express the volume invariance accompanying the plastic deformation,
– Calculate the strain at necking as function of the strain hardening coefficient.

Statement:

– Instability condition
If A represents the sample cross section ($A = A_0$ for $\varepsilon^p = 0$):

$$P = \sigma A \Longrightarrow dP = \sigma dA + Ad\sigma = 0 \Longrightarrow \frac{d\sigma}{\sigma} = -\frac{dA}{A}$$

– Volume invariance during the plastic deformation (*l useful lenght of the sample; V useful volume*):

$$dV = ldA + Adl = 0 \Longrightarrow d\varepsilon^p = \frac{dl}{l} = -\frac{dA}{A} = \frac{d\sigma}{\sigma}$$

with: $\sigma = K\left(\varepsilon^P\right)^n$ *et* $d\sigma = Kn\left(\varepsilon^p\right)^{n-1} d\varepsilon^p$

One obtains:

$$\varepsilon^p = n$$

The « Considére » relation gives us the critical plastic deformation:

$$\varepsilon^p_{cr} = n \tag{10.17}$$

– Calculation of the maximum load:

$$P_{max} = \sigma_{cr} A_{cr} = Kn^n \frac{A_{cr}}{A_0} A_0$$

$$\varepsilon^p_{cr} = n = \ln\frac{l_{cr}}{l_0} = -\ln\frac{A_{cr}}{A_0} \Longrightarrow \exp(-n) = \frac{A_{cr}}{A_0}$$

$$P_{max}/A_0 = Kn^n e^{-n} \tag{10.18}$$

Numerical application: $P_{max}/A_0 = 826\,\text{MPa}$.

10.3.6.2 Determination of the Plastic Zone in Mode III

Statement:

We consider a perfect elastic–plastic material with a yield tensile shear τ_y and a crack whose tip corresponds to the origin of the coordinate axes.

Under these conditions, in mode III, the stress field in the vicinity of the crack tip is:

$$\begin{aligned}\sigma_{31} &= -\frac{K_{III}}{\sqrt{2\pi r}}\sin\left(\frac{\theta}{2}\right)\\ \sigma_{32} &= \frac{K_{III}}{\sqrt{2\pi r}}\cos\left(\frac{\theta}{2}\right)\end{aligned} \quad with\ \sigma_{11} = \sigma_{22} = \sigma_{33} = \sigma_{12} = 0$$

(1) What is the shape of the plastic zone when the onset of plasticity terms obey the Tresca criterion ?
(2) Calculate the size of the plastic zone around the crack.

Numerical data:

$\tau_y = 210\,\text{MPa}, \ K_{III} = 21\,\text{MPa}\sqrt{\text{m}}$

Statement:

Geometric shape of the plastic zone.
Calculation of the eigenstress:

$$\left\{\begin{matrix}\sigma_1\\ \sigma_2\end{matrix}\right. = \pm\sqrt{\sigma_{31}^2+\sigma_{32}^2} = \pm\frac{K_{III}}{\sqrt{2\pi r}}$$

With the Mohr circle, one finds:

$$\tau_{max} = \frac{1}{2}(\sigma_1-\sigma_2) = \frac{K_{III}}{\sqrt{2\pi r}}$$

The elastic yield is attained when $\tau_{max} = \tau_y$, that is to say, the plastic zone is a circle of center "the crack tip" and radius r_y (Fig. 10.11) such that:

$$r_y = \frac{K_{III}^2}{2\pi\tau_y^2}$$

Note: Prediction using the Huber–Von Mises criterion:

$$\overline{\sigma} = \sqrt{3\left(\sigma_{13}^2+\sigma_{23}^2\right)} = \sqrt{3}\left(\frac{K_{III}}{\sqrt{2\pi r}}\right).$$

the elastic yield is defined by $\overline{\sigma} = \sigma_y$, the radius of the plastic zone is $r = r_y$ such that:

$$r_y = 3\frac{K_{III}^2}{2\pi\sigma_y^2}$$

If one takes $\sigma_y = \sqrt{3}\tau_y$ what is usual, we find the same prediction.

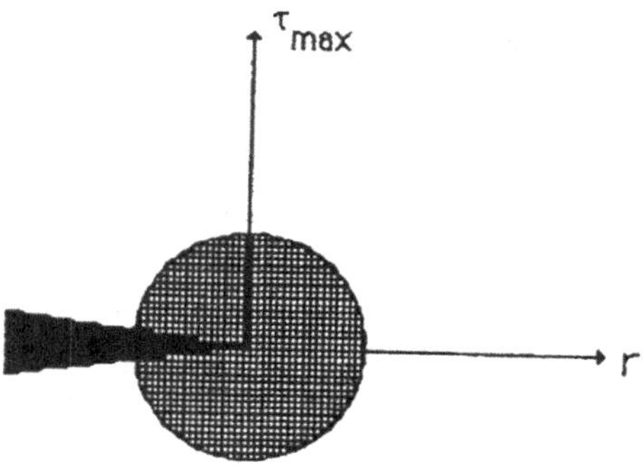

Fig. 10.11 Plastic zone around the crack tip (Pluvinage 1995)

10.4 Damage

10.4.1 Introduction

As we have already mentioned, the damage phenomenon represents surface discontinuities for microcracks and also volumic discontinuities for cavities. This is a very different deformation process.

Damage has a very strong character of irreversibility. Indeed, conventional thermal treatments make only partially disappear the defects created. The macroscopic fracture has been studied for a long time with many failure criteria (Coulomb, Rankine, Tresca, Huber–Von Mises, Mohr, Caquot...). These are the same criteria as detailed in Chap. 4, with the notable difference that the elastic limit is replaced by the tensile strength.

The first modeling of the progressive deterioration of the material which precedes the macroscopic rupture was structured by Kachanov in 1958. This approach offers a continuous variable of damage D when creep rupture of metals takes place under uniaxial stress. An extension to the ductile fracture or fatigue was performed. A generalization to 3D isotropic cases, as part of the thermodynamics of irreversible processes, will be described. The case of anisotropic damage remains an open problem.

10.4.2 Scope and Employment

The final stage of the damage is the failure of the volume element, that is to say the existence of a macroscopic crack size REV (Representative Volume Elementary); beyond, the field of fracture mechanics.

Size of the REV:

- 0.1 to 1 mm: in metals and polymers
- 1 to 10 cm: in concrete.

The theory of damage therefore describes the evolution of the phenomena between the virgin state and the onset of macroscopic cracks.

This is due to several mechanisms:

- Ductile plastic damage (at low temperatures)
- The viscoplastic brittle damage (or creep) (at middle and high temperatures)
- Fatigue damage (or microplasticity)
- The micro brittle damage (concrete).

Other damaging factors: corrosion, oxidation, radiation.

10.4.3 Phenomenological Aspects

It is necessary to imagine a so-called internal variable "mechanical damage variable" representative of the state of deterioration in the material. Several types of damage actions can be considered:
– Measurements at microstructural scale (density micro-cracks or cavities)
$\Longrightarrow$ integration on the macroscopic volume element using homogenization techniques.
– Overall physical measurements (density, resistivity). There is a need for a comprehensive model to correlate the characteristics of strength—another type of damage assessment is related to the residual life and the overall mechanical measurements particularly change in the elastic constants are easier to interpret in term of damage variable (see following paragraphs).

10.4.4 Damage Variable

Let's consider a surface S with normal $\vec{n}$. Inside S, cracks, and cavities which constitute the damage generally represent the area S_D (Fig. 10.12).

More specifically S_D is the total area of the set of traces defects, corrected for stress concentration effects.

One can define the effective resistance area $\tilde{S}$:

$$\tilde{S} = S - S_D$$

By definition:

$$D_n = \frac{S_D}{S}.$$

D_n is the mechanical measured local damage relative to the direction $\vec{n}$.

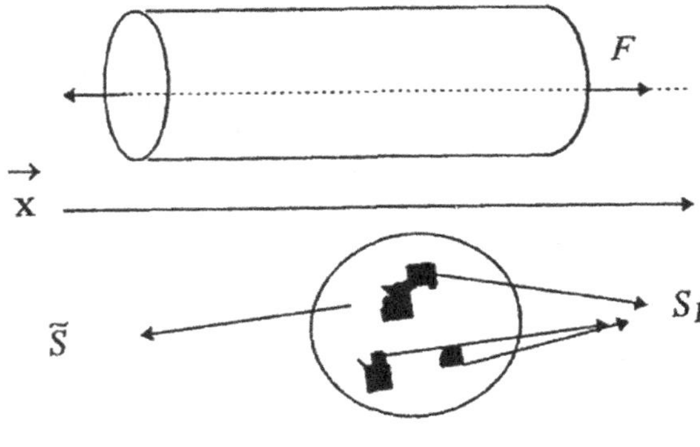

Fig. 10.12 Damaged sample loaded in tension

$D_n = 0$ that is to say $S_D = 0$: virgin state (or without damage)
$D_n = 1$: $S_D = S$ volume element is broken into two parts with a normal plane $\overrightarrow{n}$

Generally D_n depends on $\overrightarrow{n}$. If it does not depend on $\overrightarrow{n}$, it means that the damage is isotropic (which is usually an approximation).

10.4.4.1 Effective Stress

In the case of a tensile test in the direction $\overrightarrow{n}$:
– virgin material: section S; uniaxial stress $\sigma = \frac{F}{S}$
– damaged material: section $\widetilde{S}$; effective stress $\widetilde{\sigma} = \frac{F}{\widetilde{S}}$
with: $\widetilde{S} = S - S_D = S(1-D) \Longrightarrow \widetilde{\sigma} = \frac{\sigma S}{\widetilde{S}}$
Thus:

$$\widetilde{\sigma} = \frac{\sigma}{1-D}. \tag{10.19}$$

10.4.4.2 Strain Equivalence Principle (Fig. 10.13)

We assume that the behavior law that expresses the relationship between $\underline{\sigma}$ *and* $\underline{\varepsilon}$ (At least in elasticity and plasticity) remains the same by substituting to $\underline{\sigma}$ the effective stress tensor $\underline{\widetilde{\sigma}}$.

For example the isotropic elasticity of the virgin material:

$$\underline{\varepsilon} = \frac{1+\nu}{E}\underline{\sigma} - \frac{\nu}{E}\left(\mathrm{tr}\underline{\sigma}\right)\underline{1}$$

becomes for the damaged material:

$$\underline{\varepsilon} = \frac{1+\nu}{E}\underline{\widetilde{\sigma}} - \frac{\nu}{E}\left(\mathrm{tr}\underline{\widetilde{\sigma}}\right)\underline{1}$$

with an isotropic damage:

$$\underline{\varepsilon} = \frac{1}{1-D}\left(\frac{1+\nu}{E}\underline{\sigma} - \frac{\nu}{E}\left(\mathrm{tr}\underline{\sigma}\right)\underline{1}\right)$$

Example in the case of simple tension, the axial strain is written:

$$\varepsilon_e = \frac{\widetilde{\sigma}}{E} = \frac{\sigma}{(1-D)E} = \frac{\sigma}{\widetilde{E}} \Longrightarrow D = 1 - \frac{\widetilde{E}}{E}. \tag{10.20}$$

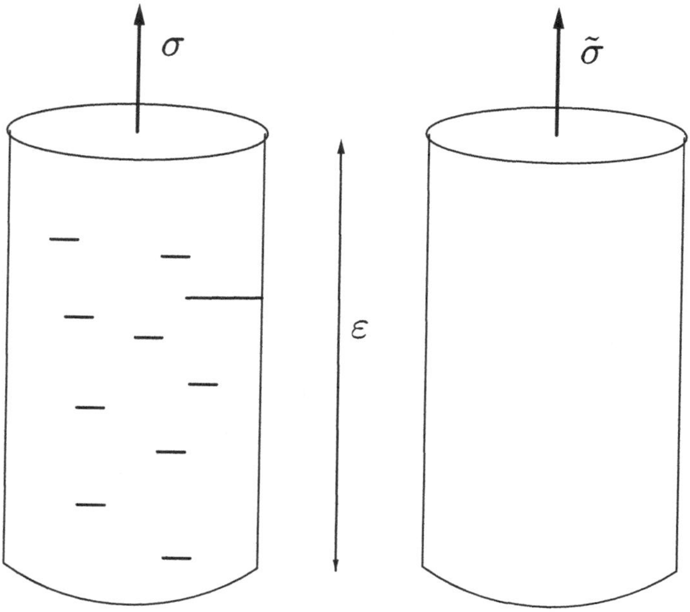

Fig. 10.13 Strain equivalence (traction test); on *left*: damaged material; on *right*: equivalent virgin material

10.4.4.3 The Principle of Energy Equivalence (Fig. 10.14)

Assuming that in the 1D case, on a test piece subjected to tensile stress that the elastic energy stored ω is the same for a virgin sample under the effective stress $\widetilde{\sigma}$ as the damaged bar subjected to the usual stress σ.

If $\widetilde{\varepsilon}$ is the deformation of the equivalent virgin material and ε the damaged material are ω is obtained by:

$$\omega = \frac{1}{2}\widetilde{\rho}\sigma\varepsilon = \frac{1}{2}\rho\widetilde{\sigma}\widetilde{\varepsilon}$$

with ρ *and* $\widetilde{\rho}$ densities of virgin and damaged materials.

By supposing that $\rho = \widetilde{\rho}$ (which is realistic because damage still occupies a low volume).

$$\Longrightarrow \sigma\varepsilon = \widetilde{\sigma}\widetilde{\varepsilon} = \frac{\sigma}{1 - D_e}\widetilde{\varepsilon} \Longrightarrow \varepsilon = \frac{\widetilde{\varepsilon}}{1 - D_e}$$

where D_e: damage variable corresponding to energy equivalence.

By taking into account that: $\varepsilon = \frac{\sigma}{E}$ *et* $\widetilde{\varepsilon} = \frac{\widetilde{\sigma}}{E}$

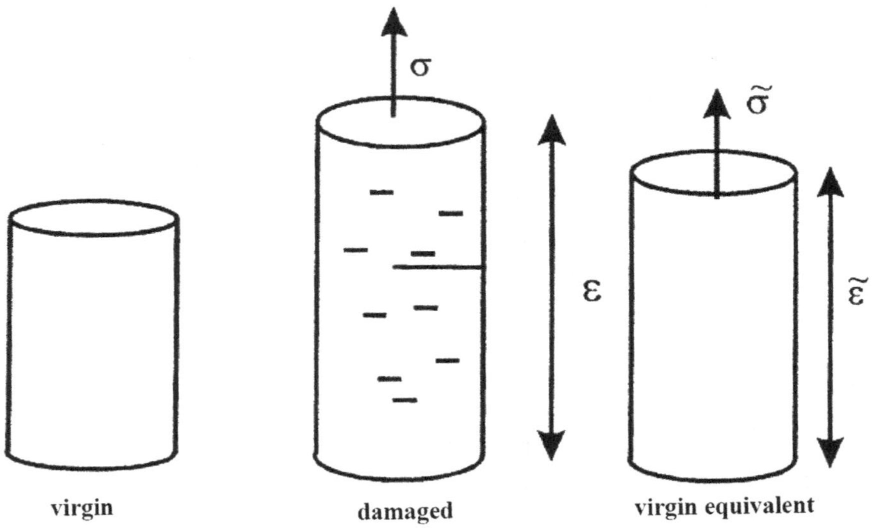

Fig. 10.14 Principle of energy equivalence

$$\Longrightarrow \varepsilon = \frac{\sigma}{\tilde{\tilde{E}}} = \frac{\sigma}{(1 - D_e)^2\, E} \Longrightarrow D_e = 1 - \sqrt{\frac{\tilde{E}}{E}}. \tag{10.21}$$

Comparison of equivalences for the elastic behavior:

The relationship necessary for the mechanics is between σ ("apparent" applied stress) and ε. The use of equivalent virgin material is only served to satisfy the rules of continuum mechanics as the damaged material can no longer be considered continuous.

In the 1D case (single tension) for strain equivalence:

$$\dot{\varepsilon} = \frac{\dot{\sigma}}{E\,(1 - D)} + \frac{\sigma\,\dot{D}}{E\,(1 - D)^2} \; or \; \dot{\varepsilon} = \frac{\dot{\tilde{\sigma}}}{E} = \left(\frac{\dot{\sigma}}{\tilde{E}}\right).$$

For energetic equivalence, it is written:

$$\dot{\varepsilon} = \frac{\dot{\sigma}}{E\,(1 - D_e)^2} + \frac{2\sigma\,\dot{D}_e}{E\,(1 - D_e)^3}.$$

Exercise:

Show that if the appropriate definitions of D *and of* D_e are used; the previous two relationships lead to the same result (which is reassuring and consistent!).

Opening problem and defects closure:

It is obvious that the compression action on a crack is in no way similar to a pulling tension on the crack. One tends to open the crack whereas the other at the tends to close it.

This simple observation is difficult to take into account in the models. It is possible, but difficult, to take into account the sign of stress in the damage evolution laws.

We will limit ourselves to "problems promoting the opening of defects."

10.4.4.4 Critical Damage for Breaking

We define the critical value of the damage D_c, as the value corresponding to the inter-atomic debonding.

Let's consider $\tilde{\sigma_u}$ the effective stress fracture by debonding and σ_u the usual ultimate stress fracture, one has:

$$\tilde{\sigma_u} = \frac{\sigma_u}{1 - D_c} \quad or \quad D_c = 1 - \frac{\sigma_u}{\tilde{\sigma_u}}.$$

Solid state physics shows that $\tilde{\sigma_u}$ is of the order of $E/25$ to $E/50$ for covalent binding materials.

The value of σ_u measured directly on the material is much lower, of the order of $E/100 \; to \; E/250$. This implies that D_c is of the order of $0.5 \; to \; 0.9$.

10.4.4.5 Extension to the Three-Dimensional Case

In the case of isotropic damage, one writes:

$$\underline{\tilde{\sigma}} = (1 - D)^{-1} : \underline{\sigma}.$$

Similarly, one postulates a relationship for the anisotropic damage:

$$\underline{\tilde{\sigma}} = \left(1 - \underline{\underline{D}}\right)^{-1} : \underline{\sigma}$$

where by rules of tensor calculus, it is necessary to choose $\underline{\underline{D}}$ as a 4 order tensor. Expression of $\underline{\underline{D}}$:

*Strain equivalence:

The 3D extension of the one-dimensional writing:

$$\varepsilon = \frac{\sigma}{\tilde{E}} = \frac{\tilde{\sigma}}{E}$$

gives $\quad \underline{\widetilde{\sigma}} = \underline{\underline{C}} : \underline{\varepsilon}$ *and* $\underline{\sigma} = \underline{\underline{C}} : \underline{\varepsilon} \Longrightarrow \underline{\varepsilon} = \underline{\underline{C}}^{-1} : \underline{\sigma} \Longrightarrow \underline{\widetilde{\sigma}} = \underline{\underline{CC}}^{-1} : \underline{\sigma} =$
$\left(1 - \underline{\underline{D}}\right)^{-1} : \underline{\sigma}$

$$\Longrightarrow \underline{\underline{D}} = \underline{\underline{1}} - \underline{\underline{\widetilde{C}\widetilde{C}}}^{-1} \tag{10.22}$$

Note: $\underline{\underline{\widetilde{C}}}$ is not necessarily symmetrical

*Energy equivalence:

In this case, an expression will be obtained which ensures the symmetry of $\underline{\underline{\widetilde{C}}}$ but do not express directly the damage tensor.

$$\underline{\underline{\widetilde{C}}} = \left(\underline{\underline{1}} - \underline{\underline{D_e}}\right)^{-T} : \underline{\underline{C}} : \left(\underline{\underline{1}} - \underline{\underline{D_e}}\right)^{-1} \tag{10.23}$$

However, this expression is not a problem for numerical evaluation of the components of damage this bound to the selected tensor.

10.4.4.6 Damage Measures

Damage is not directly accessible from measurements. Its quantitative evaluation is associated with the choice of variable to represent the phenomenon.
– Variation of the elastic modulus
Example for equivalence deformation

$$D = 1 - \frac{\widetilde{E}}{E}.$$

It is therefore required simply to measure $\widetilde{E}$ the modulus associated with the unloading during a tensile test (Fig. 10.15). But the measurement is difficult because the damage is usually very localized and requires a measurement base for 0.5 to 5 mm.

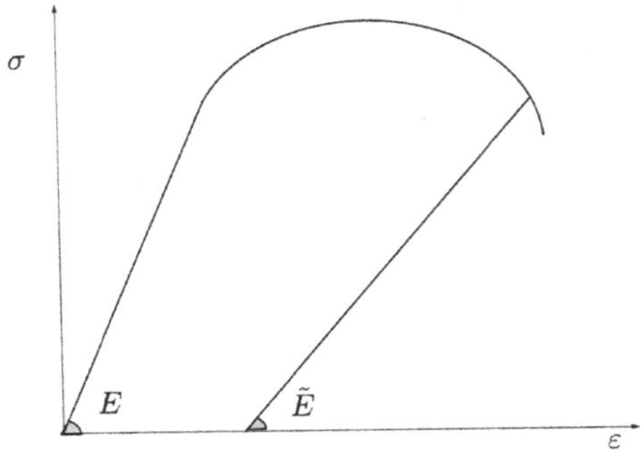

Fig. 10.15 Young modulus measurement $\widetilde{E}$ during unloading of a tensile test

Dynamic ultrasonic measurement:

A rate measurement or the time for wave propagation in a cylinder of a damaged material provides a measurement of $\widetilde{E}$:

$$\widetilde{E} = \rho \overline{v}_T^2 \frac{3\overline{v}_L^2 - 4\overline{v_T}^2}{\overline{v}_L^2 - \overline{v_T}^2}$$

where $\overline{v}_L$ is the longitudinal rate propagation and $\overline{v_T}$ the transversal one.

As part of the isotropy hypothesis of the damage (constant Poisson's ratio) and by neglecting the variation of ρ (Error <5%) the following expressions are obtained:

$$\overline{v}_L^2 = \frac{\widetilde{E}}{\rho} \frac{1-\nu}{(1+\nu)\{1-2\nu\}}, \quad v_L^2 = \frac{E}{\rho} \frac{1-\nu}{(1+\nu)\{1-2\nu\}}$$

one deduces:

$$D = 1 - \frac{\overline{v}_T^2}{v_L^2}$$

– Variation in plasticity characteristics
* Monotonous work hardening characteristics

This method is particularly interesting to characterize ductile plastic damage in addition to the measurements of changes in elasticity modulus.

This type of damage is characterized by the stress drop (observed from $\frac{d\sigma}{d\varepsilon} = 0$) due to a combination of the geometric effect (necking) and the effect of the damage (reduction in the effective section $\tilde{S} = S - S_D$).

We introduce the work hardening law of the undamaged material:

$$\varepsilon = \varepsilon^{el} + \varepsilon^p \; with \; \varepsilon^p = < \frac{\sigma - \sigma_y}{K_y} >^{M_y} .$$

When damage becomes sensitive($\varepsilon > \varepsilon^\star$) the assumption of equivalence in associated strain gives:

$$\Longrightarrow \varepsilon^p = \frac{1}{(K_y)^{M_y}} < \frac{\sigma}{1-D} - \sigma_y >^{M_y} .$$

This expression allows us to obtain an indirect measurement of D from the graph (σ, ε^p) where K_y *and* M_y are determined on the same material for $\varepsilon < \varepsilon^\star$:

$$D = 1 - \frac{\sigma}{K_Y (\varepsilon^p)^{1/M_y} + \sigma_y}$$

10.4.5 Basic Laws of Damage

• Linear ductile plastic damage law in strains formulation.

In many metallic materials subjected to increasing monotonous one-dimensional loading, there is a linear variation of damage D with strain.

A simple law that provides a rather good agreement with the experimental results is:

$$D = D_c < \frac{\varepsilon - \varepsilon_D}{\varepsilon_R - \varepsilon_D} >$$

where ε_D is the true deformation: yield damage value below which the damage is null or negligible and ε_R the true fracture strain to which the damage is equal to its critical value D_C (Three coefficients have to be identified: $\varepsilon_R,\ \varepsilon_D\ and\ D_C$).

• Kachanov creep damage law

It is a brittle viscoplastic damage law. An example of damage evolution is given in Fig. 10.16.

In 1958, for 1D loading, Kachanov proposed to model this type of evolution as:

$$\dot{D} = \left(\frac{\sigma}{A_0\,(1 - D)}\right)^r$$

where $A_0\ and\ r$ are both characteristic coefficients of the creep damage for each material.

The time to failure at constant stress creep test is expressed by the integration of

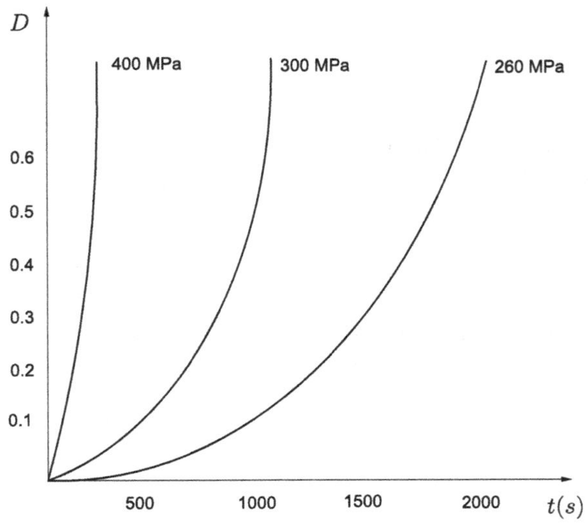

Fig. 10.16 Creep damage evolution for three stress levels

the differential equation model with the following conditions: $t = 0 \Longrightarrow D = 0$, $t = t_c \; for \; D = D_c$.

$$dD\,(1-D)^r = \left(\frac{\sigma}{A_0}\right)^r \Longrightarrow \left(-\frac{(1-D)^{r+1}}{r+1}\right)_{D=0}^{D=D_c} = \left(\frac{\sigma}{A_0}\right)^r (t)_0^{t_c}.$$

The usual values of r authorizing to neglect$(1-D)^{r+1}$ in relation to 1, D can be expressed by simple integration of the differential equation between 0 and D.

$$D = 1 - \left(1 - \frac{t}{t_c}\right)^{\frac{1}{r+1}} \quad with \; t_c = \frac{1}{r+1}\left(\frac{\sigma}{A_0}\right)^{-r}$$

• Tri-dimensional damage criteria.

For uniaxial loading, damage thresholds as deformation or strain can be introduced under which the damage is null or negligible. This concept can be generalized to 3D like the plasticity criteria. Similarly to it can be used for time-dependent damage (equipotential surfaces introduced for viscoplasticity).

For 1D situation, the damage threshold (in term of stress) defines the resistance range of the material:

$$-\sigma_D < \sigma < \sigma_D \Longrightarrow \dot{D} = 0$$

when $|\sigma|$ reaches or exceeds the threshold σ_D there is damage.

For 3D situation, a damage threshold surface is defined $f_D\left(\underline{\sigma}, D\right)$ when:

$$f_D < 0 \; \dot{D} = 0$$

$$f_D \geq 0 \; \dot{D} > 0$$

• Criterion of density release rate elastic energy.

To formulate an isotropic criterion, one can postulate that the damage mechanism is governed by the energy of total elastic deformation (distortion energy + shear deformation energy density: cavities):

$$\underline{\sigma} = \text{dev}\underline{\sigma} + \left(\frac{1}{3}\text{tr}\underline{\sigma}\right)\underline{1}$$

and

$$\underline{\varepsilon}^{\text{el}} = \text{dev}\underline{\varepsilon}^{\text{el}} + \left(\frac{1}{3}\text{tr}\underline{\varepsilon}^{\text{el}}\right)\underline{1}$$

with

$$\underline{\varepsilon}^{el} = \frac{1}{1-D}\left(\frac{1+\nu}{E}\underline{\sigma} - \frac{\nu}{E}\left(\text{tr}\underline{\sigma}\right)\underline{1}\right)$$

$$\Longrightarrow \text{dev}\underline{\varepsilon}^{el} = \frac{1}{1-D}\left(\frac{1+\nu}{E}\text{dev}\underline{\sigma}\right) \; et \; \text{tr}\underline{\varepsilon}^{el} = \frac{1}{1-D}\left(\frac{1-2\nu}{E}\text{tr}\underline{\sigma}\right)$$

Let's consider $\sigma_H = {}^{tr\underline{\sigma}}/3$ and $\varepsilon_H = {}^{tr\underline{\varepsilon}^{el}}/3$ the hydrostatic stress and strain:

$$\varepsilon_H = \frac{1-2\nu}{E}\frac{\sigma_H}{1-D}$$

assuming that the damage does not vary within the elastic range: $\frac{dD}{d\underline{\sigma}} = 0$

$$W^{el} = \int_0^{el} \underline{\sigma} : d\underline{\varepsilon}^{el}$$

The calculations give:

$$W_e = \frac{1}{2E(1-D)}\left(\frac{2}{3}(1+\nu)\overline{\sigma} + 3(1-2\nu)\sigma_H^2\right).$$

As for the stress yield point for plasticity, we define the equivalent stress damage $\sigma^\star$ such that the energy of a three-dimensional state is equal to the energy of an equivalent one-dimensional state defined by: $\sigma^\star$ $(\overline{\sigma} = \sigma^\star,\ \sigma_H = \frac{1}{3}\sigma^\star)$.

$$\sigma^\star = \overline{\sigma}\left(\frac{2}{3}(1+\nu)\overline{\sigma} + 3(1-2\nu)\left(\frac{\sigma_H}{\overline{\sigma}}\right)^2\right)^{1/2} \tag{10.24}$$

It is the ratio ${}^{\sigma_H}/_{\overline{\sigma}}$ that expresses the triaxiality stress state.

- Three invariants criterion.

The criterion of elastic energy density restitution is very simple and will be formally justified by thermodynamics. In some cases it may be set in particular defect for the creep damage where it gives the same response in tension and pure compression.

Still considering the frame of isotropy, the criterion is expressed from 3 elementary invariants:

$$\left\{\begin{array}{c} \sigma_H = \frac{1}{3}\text{tr}\left(\underline{\sigma}\right) = \frac{J_1(\underline{\sigma})}{3} \\ \overline{\sigma} = \left(\frac{3}{2}\text{dev}\underline{\sigma} : \text{dev}\underline{\sigma}\right)^{1/2} = J_2\left(\underline{\sigma}\right) \\ J_3\left(\underline{\sigma}\right) = \left(\frac{27}{2}\text{tr}\left(\text{dev}\underline{\sigma}\right)^3\right)^{1/3} \end{array}\right\}$$

Instead of $J_3\left(\underline{\sigma}\right)$ it is more useful to choose $J_0\left(\underline{\sigma}\right) = max\left(\sigma_I\right)$.
If one chooses a linear combination (α et β being phenomenological coefficients):

$$X\left(\underline{\sigma}\right) = \alpha J_0\left(\underline{\sigma}\right) + \beta J_1\left(\underline{\sigma}\right) + (1-\alpha-\beta)\, J_2\left(\underline{\sigma}\right)$$

For an uniaxial state, one obtains $X\left(\underline{\sigma}\right) = \sigma$.

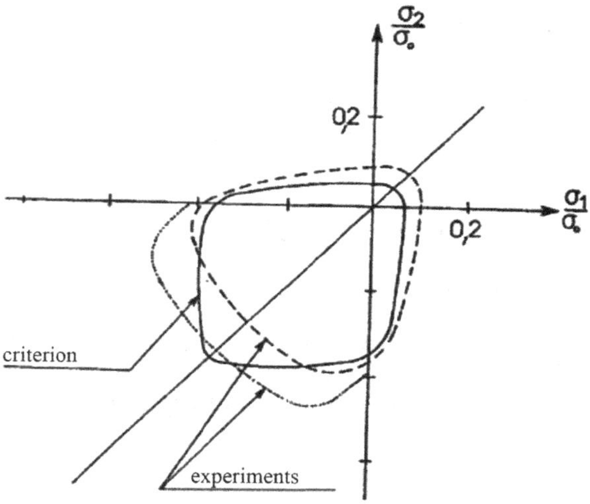

Fig. 10.17 Graphic presentation of the stress criterion (for concrete) following J. Mazars

Lemaitre et al. in their book (Lemaitre et al. 2009), illustrate the description of possible isochronous creep surfaces (instead of stress conditions that cause the same time to creep rupture).

A particularly interesting case is obtained when $\alpha = 0$:

$$\sigma_\beta^\star = \beta J_1\left(\underline{\sigma}\right) + (1-\beta)\, J_2\left(\underline{\sigma}\right)$$

or :

$$\sigma_\beta^\star = \overline{\sigma}\left((1-\beta) + \xi\beta\frac{\sigma_H}{\overline{\sigma}}\right)$$

• Asymmetrical Criterion in strain

This test was developed for concrete which has the characteristic of being substantially more resistant to damage in compression than in tension. Microscopic observations show that micro-cracks still have a normal preferential orientation to the eigen directions of extension thus comes the idea of introducing a test that depends only on $\varepsilon_{1,}$, $\varepsilon_2, \varepsilon_3$.

$$\varepsilon^\star = \left(<\varepsilon_1>^2 + <\varepsilon_2>^2 + <\varepsilon_3>^2\right)^{\frac{1}{2}}$$

See illustration of Fig. 10.17.

10.4.6 Thermodynamic Formulation of Damage

As part of the thermodynamics of irreversible processes, one adds an additional variable which is damage (See Chap. 2).

10.4.6.1 Theory of Isotropic Damage

Let's consider $\psi = \psi\left(\underline{\varepsilon}^{el}, T, D, V_k\right)$

V_k representing internal variables other than damage (such as hardening variables).

Like in elasticity, we can make the hypothesis of decoupling hardening and other effects represented by V_k and the effects associated with damage:

$$\psi = \psi_e\left(\underline{\varepsilon}^{el}, T, D\right) + \psi'\left(T, V_k\right)$$

$$\rho\psi_e = \frac{1}{2}(1 - D)\,\underline{\underline{C}} : \underline{\varepsilon}^{el} : \underline{\varepsilon}^{el}.$$

The thermal dilatation thermal is neglected. The data from this thermodynamic potential provides elasticity law of the damaged material:

$$\underline{\sigma} = \rho\frac{\partial\psi}{\partial\underline{\varepsilon}^{el}} = (1 - D)\,\underline{\underline{C}} : \underline{\varepsilon}^{el} = \underline{\underline{\tilde{C}}} : \underline{\varepsilon}^{el}$$

which is under the form:

$$\underline{\tilde{\sigma}} = \frac{\underline{\sigma}}{1 - D} = \underline{\underline{C}} : \underline{\varepsilon}^{el}.$$

The variable associated with damage is the scalar defined by:

$$Y = \rho\frac{\partial\psi^{el}}{\partial D} = -\frac{1}{2}\underline{\underline{C}} : \underline{\varepsilon}^{el} : \underline{\varepsilon}^{el} \tag{10.25}$$

Exercise:

From the expression of elastic energy density $dW_e = \underline{\sigma} : d\underline{\varepsilon}^{el}$ let show that Y is identified, with the reverse sign, to half of the variation of elastic energy induced by a damage variation (under constant stress and temperature) that is to say:

$$-Y = \frac{1}{2}\left(\frac{dW_e}{dD}\right)_{\underline{\sigma},T} \tag{10.26}$$

– Y: energy release rate

In the case of isotropic damage and when the material is elastically isotropic, there is a simple relationship between the Y variable associated with damage and the equivalent stress $\sigma^{\star}$ introduced in preceding paragraph:

$$- Y = \frac{\sigma^{\bigstar 2}}{2E\,(1-D)^2} = \frac{1}{2} E \sigma^{\bigstar 2}. \tag{10.27}$$

10.4.6.2 Dissipation Potential

The intrinsic dissipation is:

$$\underline{\sigma};\, \underline{\dot{\varepsilon}^p} - A_k \dot{V}_k - Y \dot{D} \geq 0.$$

The plastic flow phenomenon can occur without damage. The same damage can occur without appreciable macroscopic flow. One must have separately:

$$\underline{\sigma};\, \underline{\dot{\varepsilon}^p} - A_k \dot{V}_k \geq 0 \; et \; - Y \dot{D} \geq 0$$

As, $-Y = \frac{1}{2} \underline{\underline{C}} : \underline{\varepsilon^{el}} : \underline{\varepsilon^{el}} \geq 0 \; so \; \dot{D} \geq 0$

This means that our theory of the damage has not provided the material the ability to "repair"!

According to the thermodynamic of irreversible process, the evolution of the damage variable law derives from a dissipation potential whose existence is postulated:

$$\varphi \left(\underline{\dot{\varepsilon}}^P, \dot{V}_k, \dot{D}, \vec{q}\,; \underline{\varepsilon^{el}}, V_k, D, T \right)$$

by partial transformation of Legendre–Fenchel, we construct the dual potential equivalent that can express $\dot{D}$ as function of Y rather than the reverse:

$$\varphi^{\bigstar} \left(\underline{\sigma}, A_k, Y, \vec{g} = \overrightarrow{gradT}; \underline{\varepsilon^{el}}, V_k, D, T \right)$$

$$\dot{D} = -\frac{\partial \varphi^{\bigstar}}{\partial Y}. \tag{10.28}$$

10.4.6.3 Failure Criterion of the Volume Element

The foregoing criterion justifies an energy criterion defining the priming of a macro-crack:

$$-Y = \frac{1}{2} \underline{\underline{C}} : \underline{\varepsilon^{el}} : \underline{\varepsilon^{el}}.$$

The test is to postulate that the initiation of a macroscopic crack corresponds to a critical value of this energy, intrinsic to each material.

$$-Y = \mid Y \mid = Y_c \Longleftrightarrow fracture.$$

This energy must be identified to the cohesive energy of the material which can be evaluated by one-dimensional tensile test:

$$-Y = \frac{1}{2}E\left(\varepsilon^{el}\right)^2$$

By replacing ε^{el} by the damaged elasticity law:

$$\varepsilon^{el} = \frac{\sigma}{(1-D)E}$$
$$-Y = \frac{-\sigma^2}{2(1-D)^2E}$$

and for the fracture conditions:

$$\left\{\begin{array}{c} -Y = Y_c \\ \sigma = \sigma_u \\ D = D_c \end{array}\right\} \Longrightarrow Y_c = \frac{\sigma_u^2}{2E\,(1-D_c)^2}$$

or

$$D_c = 1 - \frac{\sigma_u}{(2EY_c)^{1/2}}$$

The criterion identifies the critical damage to break the sample previously calculated:

$$D_c = 1 - \frac{\sigma_u}{\widetilde{\sigma_u}}\ with\ \widetilde{\sigma_u} = (2EY_c)^{1/2}$$

The anisotropic damage or fatigue will not be treated in this book; elements can be found in the book by Lemaitre and colleagues (Lemaitre et al. 2009).

Bibliography

Aleong D, Dumont C, Arbab Chirani S, Patoor E, McDowell D (2002) Transformation surfaces of a textured pseudoelastic polycristalline Cu-Zn-Al shape memory alloy. J Int Mat Syst Struct 13:783–793

Ashby M (1972) A first report on déformation-mechanism maps. Acta Metallurgica 20(7):887–897

Aurrichio F, Petrini L (2004) A three-dimensional model describing sress-temperature induced solid phase transformations: solutions algorithm boundary value problems. Int J Numer Meth Eng 61(6):807–836

Aurrichio F, Taylor R, Lubliner J (1997) Shape-memory alloys: Macromodelling and numerical simulations of the superelastic behavior. Comput Methods Appl Mech Eng 146(3,4):281–312

Ball JM, James RD (1987) Fine phase mixtures as minimizers of energy. Arch Rational Mech Anal 100:13–52

Ball JM, James RD (1992) Proposed experimental tests of a theory of fine microstructure. Philos Trans R Soc Lond 338A:389–450

Benzaoui H, Lexcellent C, Chaillet N, Lang B, Bourjault A (1997) Experimental study and modeling of a TiNi shape memory alloy wire actuator. J Int Mat Syst Struct 8:619–629

Bernard F, Delobelle P, Rousselot C, Hirsinger L (2009) Miicrostructural mechanical and magnetic properties of shape memory aloy Ni55-Mn23-Ga22 thin films deposited by radio-frequency magnetron sputtering. Thin Solid Films 518:399–412

Berthelot J (1999) Materiaux Composites: comportement mécanique et analyse des structures. TEC.DOC 3ième édition

Besson J, Cailletaud G, Chaboche J, Forest S (2001) Mécanique non linéaire des matériaux. Hermés

Bhattacharya K (2003) Microstructure of martensite. Oxford materials

Boubakar LM, Lexcellent C (2007) SMA structures computations: plenary lecture. In: 2nd IASME international conference on continuum mechanics, Portoroz, Slovenia

Boubakar M, Vieille B, Lexcellent C et al (1998) Modélisation des alliages à mémoire de forme (Revue européenne des éléments finis volume 7 no 8

Boukamel A (2009) Physique des sons et vibrations: viscoélasticité. Ecole centrale Marseille (cours Master recherche: Mécanique physique et modélisation; Spécialité: Acoustique)

Bouvet C, Calloch S, Lexcellent C (2002) Mechanical behavior of a Cu-Al-Be shape memory alloy under nonproportional loadings. J Eng Mat Tech 124:112–124

Brinson LC (1993) One-dimensional constitutive behavior of shape memory alloys: thermomechanical derivation with non constant material functions and redefined martensite internal variable. J Int Mat Syst Struct 4:229–242

Brown J, Angel R, Ross R (2006) Elasticity of plagioclase feldspars

Buelher W, Gilfrich J, Wiley R (1963) Effect of low temperature phase changes on the mechanical properties of alloys near composition TiNi. J Appl Phys 34(5):1475–1477

C. Lexcellent, *Linear and Non-linear Mechanical Behavior of Solid Materials*, DOI 10.1007/978-3-319-55609-3

Bui H (1978) Mécanique de la rupture fragile. Masson
Burlet H, Cailletaud G, Pineau G (1988) Formation materiaux-structures: "Comportement inélastique". Centre des matériaux de l"Ecole des Mines de Paris
Cailletaud G, Tijani M, Cantournet S, Corte l, El Arem S, Forest S, Herve-Luanco E, Mazier M, Proudhon H, Ryckelink D (2011) Mécanique des Matériaux solides (Notes de cours). Mines-Paris-Tech
Casimir JB, Chevalier Y (2014) Elasticité anisotrope. sup-méca Saint Ouen
Chaboche J (2008) A review of some plasticity and viscoplasticity constitutive theories. Int J Plast 24:1642–1693
Chang L, Read T (1951) Plastic deformation and diffusionless phase changes in metals-the gold-cadmium beta phase. Trans AIME J Met 191:47–52
Chemisky Y, Duval A, Patoor E, Ben Zineb T (2011) Constitutive model for shape memory alloys including phase transformation, martensitic reorientation and twins accommodation. Mech Mater 43(7):361–376
Chrysochoos A, Louche H (2000) An infrared image processing to analyse the calorific effects accompanying strain localisation. Int J Eng Sci 38(16):1759–1788
Chrysochoos A, Wattrice B, Murra-Ciole J, El Kaim Y (2009) Fields of stored energy associated with localized necking of steel. J Mech Mat Struct 4(2):245–262
Chu C (1993) Thèse de doctorat. University de Minnesota, Minneapolis
Contesti E, Cailletaud G (1989) Description of creep-plasticity interaction with non-unified constitutive equations, application to an austenitic stainless steel. Nuclear Eng Des 116:265–280
Creton N (2004) Etude du comportement magnéto-mécanique des alliages à mémoire de forme de type heusler Ni-Mn-Ga. Mémoire de D.E.A., Ph.D. Thesis. Université de Franche Comté. Besançon (France)
Delobelle P (1988) Sur les lois de comportement viscoplastiques à variables internes. exemples de deux alliages industriels : inoxydable austénitique 17–12 sph et superalliage inco718. Revue Phys Appl 23:1–61
Delobelle P, Mermet A, Oytana C (1982) Biaxial dip tes measurements of internal stresses during high temperature plastic flow. Strenght of Metals and Alloys (ICSMA 6°), pp 675–680
Duvaut G (1984) Modélisation du comportement mécanique des matériaux. In: Seminaire d'analyse numérique de Besançon
Estrin Y (1996) Dislocation-density-related constitutive modeling. In: Krauss AS, Krauss K (eds) Unified constitutive laws of plastic deformation. Academics Press Inc., New York, pp 69–106
Feng Z (1991) Mécanique non linéaire. Université d'Evry
Fremond M (1998) L' éducation des matériaux à mémoire de forme. Revue Européenne de Eléments Finis 7(8):35–46
Fremond M (2002) Non-smooth thermomechanics. Springer, Berlin (physics and Astronomy)
Friedel J (1964)Dislocations. Pergamon Press (Gauthier-Villars)
Frost H, Ashby M (1982) Deformation-mechanism maps. Pergamon Press, Oxford
Gall K, Sehitoglu H, Maier H, Jacobus K (1998) Stress induced martensite phase transformation in polycristalline Cu-Al-Zn shape memory alloy under different stress states. Mettall Mat Trans 29A:755–763
Gauthier J (2007) Modelisation des alliages à mémoire de forme magnetiques pour la conversion d'énergie et leur commande. Université de Franche Comté, Thése de doctorat
Gauthier JY, Lexcellent C, Hubert A, Abadie J, Chaillet N (2007) Modelling rearrangement process of martensite platelets in a magnetic shape memory alloy Ni2MnGa single crystal under magnetic field and (or) stress action. J Int Mater Syst Struct 43(3–4):288–299
Gauthier JY, Lexcellent C, Hubert A, Abadie J, Chaillet N (2011) Magneto-thermo-mechanical modeling of a magnetic shape memory alloy Ni-Mn-Ga single crystal. Ann Solids Struct Mech 2(1):19–31
Ge Y, Heczko Soderberg O, Hannula S (2006) Magnetic domain evolution with applied field in a Ni-Mn-Ga magnetic shape memory alloy. Scripta Materialia 54:2155–2160
Germain P (1973) Cours de Mécanique des Milieux Continus, vol. 1. Masson

Gillet Y, Meunier M, Brailowski V, Trochu F, Patoor E, Berveiller M (1996). Comparison of thermomechanical models for shape memory alloys springs. In: Proceedings of ICOMAT 95, Journal De Physique IV
Gourgues-Lorenzo A (2008) Travaux dirigés de fluage (chapitre XIX). Ecole des Mines Paris
Grabe C, Bruhns O (2009) Path dependence and multiaxial behavior of a polycrystalline NiTi alloy within the pseudoelastic and pseudoplastic temperature regimes. Int J Plast 25(3):513–545
Guelin P, Terriez J, Wack B (1976) Identification of an elastoplastic constitutive relation. Mech Res Commun 3(4):325–330
Halphen B, N'Guyen Q (1974) Plastic and viscoplastic matrerials with generalized potentials. Mech Res Com 1(1):43–47
Halphen B, N'Guyen Q (1975) Sur les matériaux standards généralisés. J de Mécanique 14(1):39–63
Heczko Ullako K (2001) Effect of temperature on mangnetic properties of Ni-Mn-Ga magnetic shape memory (MSM) alloys. IEEE Trans Mag 37(4):2672–2674
Hill R (1948) A variational principle of maximum plastic work in classical plasticity. Q J Mech Appl Math 1:18–28
Hirsinger L, Lexcellent C (2002) Modelling detwinning of martensite platelets under magnetic and (or) stress actions in Ni-Mn-Ga alloys. J Magn Magn Mat 254–255:275–277
Hirsinger L, Creton N, Lexcellent C (2004) From crystallographic properties to macroscopic detwinning strain and magnetisation of Ni-Mn-Ga magnetic shape memory alloys. J Phys IV 115:111–120
Irwin G (1958) Fracture. Handbuch der Physik, vol VI. Springer, Berlin, pp 551–590
James R, Wuttig M (1998) Magnetostriction of martensite. Philos Mag A 77:1273–1299
Jaoul B (1965) Etude de la plasticité et application aux métaux. Dunod
Koistinen D, Marburger R (1959) A general equation prescribing the extent of the austenite-martensite transformation in pure iron-carbon and plain carbon steels. Acta Metallurgica 7:59–70
Kundu S, Bhadeshia H (2007) Crystallographic texture and intervening transformations. Scripta Materialia 57:869–872
Laverhne K, Poncelet M (2013) TD N° 8 MASTER MAGIS. ENS CACHAN
Laverhne Taillard K, Calloch S, Arbab Chirani S, Lexcellent C (2008) Multiaxial shape memory effect and superelasticity strain. Strain 45(1):77–84
Laydi M, Lexcellent C (2010) Yield criteria for shape memory materials: convexity conditions and surfaces transport. Math Mech Solids 15(2):165–208
Leblond J (2003) Mécanique de la rupture fragile et ductile. Hermes-Lavoisier
Leclercq S, Lexcellent C (1996) A general macroscopic description of the thermomechanical behaviour of shape memory alloys. J Mech Phys Solids 44(6):953–980
Lemaitre J (1971) Sur la détermination des lois de comportement des matériaux élasto-viscoplastiques. Université Paris VI, Thèse de doctorat d' état
Lemaitre J, Chaboche J, Benallal A, Desmorat R (2009) Mécanique des matériaux solides (3ième édition). Dunod
Lexcellent C (1989) Tentative de correlation entre le fluage et la diffusion des solutions solides fortement concentrées (type Cu-Zn (beta)). Acta Metall 37(8):1685–1696
Lexcellent C (2013a) Les alliages à mémoire de fome. Hermes-Lavoisier
Lexcellent C (2013b) Shape-memory Alloys Handbook. ISTE-Wiley, New York
Lexcellent C, Tobushi H, Ziolkowski A, Tanaka K (1994) Thermodynamical model of reversible r-phase transformation in NiTi shape memory alloy. Int J Press Vessel Pip 58:51–57
Lexcellent C, Boubakar ML, Bouvet C, Calloch S (2006) About modelling the shape memory alloy behaviour based on the phase transformation surface identification under proportional loading and anisothermal conditions. Int J Solids Struct 43(3–4):613–626
Liang C, Rogers C (1990) One-dimensional thermomechanical consttitutive relations for shape memory materials. J Int Mat Syst Struct 1:207–234
Liang Y, Kato H, Taya M (2006) Model calculation of 3D-phase transformation diagram of ferromagnetic shape memory alloys. Mech Mat 38:564–570
Licht C (1998) x(1-x) j'aime assez. In: Actes MECAMAT - AUSSOIS

Likhachev A, Sozinov A, Ullakko K (2004) Different modeling concepts of magnetic shape memory and their comparison with some experimental results obtained in Ni-Mn-Ga. Mat Sci Eng A 378:513–518

Lim J, McDowell D (1999) Mechanical behavior of an Ni-Ti shape memory alloy under axial-torsional proportional and non proportional loading. J Eng Mat Tech 121:9–18

Mandel J (1974) Introduction à la mécanique des milieux continus deformables. Academic Polonaise des Sciences

Markets F, Fisher FD (1996) Modelling the mechanical behavior of shape memory alloys under variant coalescence. Comput Mat Sci 5:210–226

Maugin GA (1992) The thermomechanics of plasticity and fracture. Cambridge University Press, Cambridge

Moreau J (1970) Sur les lois de frottement, de viscosité et de plasticité. C R Acad Sci Paris 271:608–611

Muller I (1989) On the size of the hysteresis in pseudo-elasticity. Contin Mecha Thermodyn 1:125–142

Muller I, Xu H (1991) On pseudo-elastic hysteresis. Acta Metallurgica Mat 39:263–271

Mullner P, Chernenko V, Kostorz G (2003) A microscopic approach to the magneticeld-induced deformation of martensite (magnetoplasticity). J Magn Magn Mat 267:325–334

Nguyen QS (1982) Problémes de plasticité et de rupture. Publications Mathématiques d'Orsay (Université de Paris-Sud)

Orgeas L, Favier D (1998) Stress-induced martensitic transformation of a NiTi alloy in isothermal shear, tension and compression. Acta Mater 46(15):5579–5591

Panico M, Brinson L (2007) A three dimensional phenomenological model for martensite reorientation in shape memory alloys. J Mech Phys Solids 55(11):2491–2511

Patoor E, Berveiller M (1990) Les alliages à mémoire de forme (Technologie de pointen°45). Hermés-Lavoisier

Patoor E, Eberhadt A, Berveiller M (1987) Potentiel pseudoélastique et plasticité de transformation martensitique dans les mono et polycristaux. Acta Metallurgica 35(11):2779–2789

Pattofatto S (2009) TD- Visco-élasticité 1D Master Magis 2009/2010 ENS Cachan. ENS CACHAN

Pluvinage G (1995) 120 exercices de mécanique élastoplastique de la rupture. Cépadues

Prager W (1949) Recent developments in mathematical theory of plasticity. J Appl Phys 20(3):235–241

Rabotnov Y (1969) Creep problems in structural members. North-Holland, Amsterdam

Raniecki B, Lexcellent C (1998) Thermodynamics of isotropic pseudo-elasticity in shape memory alloys. Eur J Mech A Solids 17(2):185–205

Raniecki B, Lexcellent C, Tanaka K (1992) Thermodynamic model of pseudoelastic behaviour of shape memory alloys. Arch Mech 44(3):261–288

Raniecki B, Miyazaki S, Tanaka K, Dietrich L, Lexcellent C (1999) Deformation behavior of Ti-Ni SMA undergoing r-phase transformation in tension-torsion (compression) tests. Arch Mech 51:745–784

Rejzner J, Lexcellent C (1999) Calculs d'actionneurs en alliages à mémoire de forme sous sollicitation mécanique élémentaire (flexion-torsion). In Journée d'étude du 17 Juin 1999. Récents développements et applications des matériaux adaptatifs pour capteurs et actionneurs, pp 25–38

Rice J (1968) A path-independent integral and the approximate analysis of strain concentrationby notches and cracks. ASME J Appl Mech 35:379–386

Rogueda C, Lexcellent C, Bocher L (1996) Experimental study of pseudoelastic behavior of a Cu-Al-Zn polycristalline shape memory alloy under tension-torsion proportional and non-proportional loading tests. Arch Mech 48:1025–1045

Sadjadpour A, Bhattacharya K (2007) A micromechanics-inspired constitutive model for shape memory alloys. Smart Mater Struct 16:1751

Sidoroff F (1980) Mécanique des solides: tome 1: mécanique des milieux continus. Ecole Centrale Lyon

Sittner P, Hara Y, Tokuda M (1995) Experimental study on the thermoelastic martensitic transformation in shape memory alloy polycristal induced by combined external forces. Metall Mater Trans 26A:2923–2935

Straka L, Heczko O, Hannula S (2006) Temperature dependence of reversible field-induced strain in Ni-Mn-Ga single crystal. Scripta Materialia 54:1497–1500

Sun CT, Jin ZH (2012) Fracture mechanics. Academic Press/Elsevier, New York

Suquet P (2003) Rupture et Plasticité. Ecole Polytechnique

Taillard K, Arbab Chirani S, Calloch S, Lexcellent C (2008) Equivalent transformation strain and its relation with martensite volume fraction for isotropic and anisotropic shape memory alloys. Mech Mater 40(4–5):151–170

Tanaka K, Nagaki S (1982) A thermodynamical description of materials with internal variables with the process of phase transitions. Ing Archiv 51:287–299

Tanaka K, Kobayashi K, Sato Y (1986) Thermomechanics of transformation pseudoelasticity and shape memory effect in alloys. Int J Plast 2:59–72

Tobushi H, Shimeno Y, Hachisuka T, Tanaka K (1998) Influence of strain rate on superelastic properties of Ti-Ni shape memory alloy. Mech Mater 30:141–150

Ullakko K, Huang J, Kantner C, O'Handley R, Kokorin V (1996) Large magneticeld-induced strains in Ni2MnGa single crystals. Appl Phys Lett 69:1966–1968

Vacher P, Lexcellent C (1991) Study of pseudo-elastic behavior of polycristalline shape memory alloys by resistivity measurements and acoustic emission. In: Proceedings of ICM VI, Kyoto Japan

Vieille B, Bouvet C, Moyne S, Boubakar L, Lexcellent C (2001) Modélisation numérique du comportement de structures minces en alliages à mémoire de forme à grandes déformations pseudoelastiques. Actes du XV iéme Congrés Français de Mécanique

Vieille B, Boubakar L, Lexcellent C (2003) Non-associated superelasticity in rotating frame formulation. J Theor Appl Mech 41(3):675–691

Westergaard H (1939) Bearing pressures and cracks. J Appl Mech 6:49–53

Williams ML (1952) Stress singularities resulting from various boundary conditions in angular corners of plates in extension. ASME J Appl Mech 19:526–528

Zuo F, Su X, Wu J (1998) Magnetic properties of the premartensitic transition in Ni2-Mn-Ga alloys. Phys Rev B 58(17):11127–11130

Index

C. Lexcellent, *Linear and Non-linear Mechanical Behavior of Solid Materials*, DOI 10.1007/978-3-319-55609-3

Printed in the United States
By Bookmasters